Order Statistics and Inference

Estimation Methods

This is a volume in
STATISTICAL MODELING AND DECISION SCIENCE

Gerald L. Lieberman and Ingram Olkin, editors
Stanford University, Stanford, California

Order Statistics and Inference

Estimation Methods

N. Balakrishnan

Department of Mathematics and Statistics
McMaster University
Hamilton, Ontario
Canada

and

A. Clifford Cohen

Department of Statistics
University of Georgia
Athens, Georgia

ACADEMIC PRESS, INC.
Harcourt Brace Jovanovich, Publishers

Boston San Diego New York
London Sydney Tokyo Toronto

This book is printed on acid-free paper. ♾

ACADEMIC PRESS, INC.
1250 Sixth Avenue, San Diego, CA 92101

United Kingdom Edition published by
ACADEMIC PRESS LIMITED
24–28 Oval Road, London NW1 7DX

Library of Congress Cataloging in Publication Data

Balakrishnan, N., date.
Order statistics and inference: estimation methods/N. Balakrishnan, A. Clifford Cohen.
p. cm. — (Statistical modeling and decision science)
Includes bibliographical references (p. ??) and index.
ISBN 0-12-076948-4 (alk. paper)
1. Order statistics. 2. Estimation theory.
I. Cohen, A. Clifford. II. Title. III. Series
QA278.7.B35 1990 90-824
519.2—dc20 CIP

90 91 92 93 9 8 7 6 5 4 3 2 1
Printed in the United States of America

To my mother, Lakshmi, and my father, Narayanaswamy

N. Balakrishnan

To the memories of Professors Cecil C. Craig and Paul S. Dwyer

A. Clifford Cohen

Contents

Acknowledgments xi
Preface xiii
List of Tables xv
List of Figures xix

Chapter 1 Introduction **1**

1.1 Introductory Remarks 1
1.2 The Role of Order Statistics in Practical Applications 2
1.3 Scope of This Volume 4

Chapter 2 Basic Theory **7**

2.1 Introduction 7
2.2 Joint Distribution of n Order Statistics 7
2.3 Joint Distribution of Two Order Statistics 8
2.4 Distribution of a Single Order Statistic 11
2.5 Distribution of Range and Some Other Statistics 17

Chapter 3 Moments and Other Expected Values **21**

3.1 Introduction 21
3.2 Some Basic Formulas 22
3.3 Recurrence Relations and Identities 23
3.4 Results for the Uniform Distribution 30
3.5 Results for the Exponential Distribution 34

3.6 Results for the Logistic Distribution 38
3.7 Results for the Gamma Distribution 43
3.8 Results for the Weibull Distribution 47
3.9 Results for the Normal Distribution 51
3.10 Results for the Half Logistic Distribution 63
3.11 David and Johnson's Approximation 68

Chapter 4 Linear Estimation Based on Order Statistics **73**

4.1 Preliminary Remarks 73
4.2 BLUE of the Scale Parameter 74
4.3 BLUE for the One-Parameter Exponential Distribution 76
4.4 BLUEs of the Location and Scale Parameters 80
4.5 BLUEs for the Two-Parameter Exponential Distribution 87
4.6 Gupta's Simplified Linear Estimators 94
4.7 Blom's Unbiased Nearly Best Linear Estimators 100
4.8 Downton's Linear Estimators with Polynomial Coefficients ... 109
4.9 Details of Other Related Work 119

Chapter 5 Maximum Likelihood Estimation **121**

5.1 Preliminary Remarks 121
5.2 The Weibull Distribution 122
5.3 The Lognormal Distribution 126
5.4 The Inverse Gaussian Distribution 131
5.5 The Gamma Distribution 135
5.6 The Rayleigh Distribution 139
5.7 The Exponential Distribution 144
5.8 Complete, Truncated, and Censored Samples from the Normal Distribution 146

Chapter 6 Approximate Maximum Likelihood Estimation **161**

6.1 Preliminary Remarks 161
6.2 Estimation for the Rayleigh Distribution 162
6.3 Estimation for the Normal Distribution 167
6.4 Estimation for the Logistic Distribution 177
6.5 Estimation for the Extreme Value Distribution 186
6.6 Estimation for the Type I Generalized Logistic Distribution ... 197
6.7 Estimation for the Half Logistic Distribution 208

Chapter 7 Optimal Linear Estimation Based on Selected Order Statistics **215**

7.1 Introduction 215
7.2 Bennett's and Jung's Optimal Asymptotic Estimators 216
7.3 Ogawa's Optimal Estimators Based on Selected Order Statistics 233
7.4 Dixon's Simplified Linear Estimators 249
7.5 Balakrishnan's Approximate Maximum Likelihood Estimation 255
7.6 Estimation of Population Quantiles 264
7.7 Details of Other Related Work 270

Chapter 8 Cohen–Whitten Estimators: Using Order Statistics **273**

8.1 Preliminary Remarks 273
8.2 The Weibull Distribution 274
8.3 The Lognormal Distribution 278
8.4 The Inverse Gaussian Distribution 281
8.5 The Gamma Distribution 286
8.6 The Exponential Distribution 289
8.7 Illustrative Examples 290

Chapter 9 Estimation in Regression Models **299**

9.1 Introduction 299
9.2 Best Linear Unbiased Estimation with Multiple Measurements 300
9.3 Modified Maximum Likelihood Estimation with Multiple Measurements 310
9.4 Modified Maximum Likelihood Estimation with Single Measurements 315

Chapter 10 A Sample Completion Technique for Censored Samples **329**

10.1 Introductory Remarks 329
10.2 Censored Samples from the Normal Distribution 330
10.3 Censored Samples from Skewed Distributions 331
10.4 Illustrative Examples 335

Bibliography 341
Author Index 367
Subject Index 373

Acknowledgments

The able and generous assistance of Edna Pathmanathan (McMaster University) and Gayle Rodriguez, Molly Rema, and Dawn Tolbert (University of Georgia) in typing the manuscript is acknowledged with thanks. Special thanks are extended to Mrs. Susan Gay (Assistant Editor, Academic Press, Boston) and Ms. Amy Strong (Production Editor) for their kind cooperation and prompt responses during the preparation of this book.

Thanks are offered to the American Statistical Association, Marcel Dekker, Inc., the American Society for Quality Control, Gordon and Breach Science Publishers, Inc., the Institute of Electrical and Electronics Engineers, Inc., the Biometrika Trustees, and the Editor of the *Journal of Statistical Planning and Inference*, for permission to reproduce previously published tables, charts, and examples.

The first author thanks the Natural Sciences and Engineering Research Council of Canada and the Science and Engineering Research Board of McMaster University for providing partial financial support (toward the typing and the computational costs) during the preparation of this book.

The second author extends thanks to Dr. Lynne Billard (former Head) and Dr. Robert L. Taylor (present Head) for encouragement and for making facilities of the Department of Statistics at the University of Georgia available during the preparation of this book, and to his colleague and former student, Dr. Betty Jones Whitten, for her assistance in computing, proofreading, and editing, and also for the encouragement she has given during the course of this project.

Preface

The subject of order statistics has been in the process of development for many years and recently has become increasingly important. Articles relating to this area have appeared in numerous different publications. In this volume, we have made a sincere effort to consolidate important developments in methods of estimation and to illustrate them with practical examples from various scientific disciplines. In addition to an account of classical moment and maximum likelihood estimation, emphasis has been placed on linear unbiased and optimal linear estimation based on complete as well as censored samples. Recent developments in simple approximate maximum likelihood estimation based on censored samples as well as the sample completion technique for censored data have been included and explained through several examples.

This volume is intended both as a text and as a reference source for researchers and practitioners in order statistics, estimation theory, life testing, quality control, and reliability. For this purpose, an up-to-date, comprehensive bibliography on this topic has been included. We have also incorporated numerous illustrative examples which will be of particular benefit to students and practitioners.

In reporting on various developments in this area, we have tried to the best of our knowledge to give credit and to recognize priority where appropriate for previously published results. We apologize for any oversights and will cheerfully accept corrections. We will be grateful for any information regarding omission of relevant references.

N. Balakrishnan
Hamilton, Ontario
Canada

A. Clifford Cohen
Athens, Georgia
U.S.A.

March 1990

List of Tables

4.4.1 The BLUEs of μ and σ and Their Standard Errors for the Data of D'agostino and Stephens 85

4.5.1 Days of Incubation among 10 Rabbits Following Inoculation with Graded Amounts of *Treponema pallidum*; BLUEs 93

4.6.1 Comparison of the Estimators $\tilde{\sigma}$, σ^*, and σ^{**} for the Normal Population 97

4.8.1 Coefficients $b_{ij}^{(s)}$ for Computing Elements of $\mathbf{\Omega}$ in the Normal Distribution 113

4.8.2 Efficiency of Range as an Estimate of σ in the Normal Distribution relative to

$$\sigma_{**} = C \sum_{i=1}^{n} (2i-n-1)X_{i:n}/\{n(n-1)\}\, [C = 1{\cdot}77245385]$$ 114

4.8.3 Coefficients $b_{ij}^{(s)}$ for Computing Elements of Ω in the Logistic Distribution 116

4.8.4 Values of Ω_{ij} for the Logistic Distribution When $n = 20$ 116

4.8.5 Coefficients, Variances, and Relative Efficiencies of the Estimators μ_{**} and σ_{**} When $n = 20$ 117

5.2.1 Variance-Covariance Factors for Maximum Likelihood Estimates of Weibull Parameters 125

5.3.1 Variance-Covariance Factors for Maximum Likelihood Estimates of Lognormal Parameters 130

5.4.1 Variance-Covariance Factors for Maximum Likelihood Estimates of Inverse Gaussian Parameters 134

5.5.1 Variance-Covariance Factors for Maximum Likelihood Estimates of Gamma Distribution Parameters 138

5.8.1 Auxiliary Estimation Function θ for Singly Truncated Normal Samples 150
5.8.2 Auxiliary Estimation Function $\lambda(h, \alpha)$ for Singly Censored Normal Samples 154
5.8.3 Variance Factors for Singly Truncated and Singly Censored Normal Samples 156
6.2.1 Bias, Variance, and Approximate Variance of $\hat{\sigma}$, and Variance of the BLUE of σ for the Rayleigh Distribution 166
6.3.1. Bias, Variances, and Covariance of Estimators $\hat{\mu}$ and $\hat{\sigma}$ for the Normal Distribution 172
6.3.2 Simulated Values of Bias, Variances, and Covariance of $\hat{\mu}$ and $\hat{\sigma}$ for the Normal Distribution 173
6.3.3 Variances of the Estimators of μ for the Normal Distribution When $r = s$ 174
6.4.1 Bias, Variances, and Covariance of Estimators $\hat{\mu}$ and $\hat{\sigma}$ for the Logistic Distribution 182
6.4.2 Simulated Values of Bias, Variances, and Covariance of $\hat{\mu}$ and $\hat{\sigma}$ for the Logistic Distribution 183
6.4.3 Variances of the Estimators of μ for the Logistic Distribution When $r = s$ 183
6.5.1 Values of Bias, Variances, and Covariance of $\hat{\mu}$ and $\hat{\sigma}$ for the Extreme Value Distribution When $n = 10$ 192
6.5.2 Values of Bias, Variances, and Covariance of $\hat{\mu}$ and $\hat{\sigma}$ for the Extreme Value Distribution When $n = 20$ 193
6.5.3 Comparison of the Bias and Mean Square Error of Various Estimators of μ and σ for the Extreme Value Distribution 194
6.6.1 Bias, Variances, and Covariance of $\hat{\mu}$ and $\hat{\sigma}$ for the Generalized Logistic Distribution When $n = 15$ and $b = 1.0$ 203
6.6.2 Bias, Variances, and Covariance of $\hat{\mu}$ and $\hat{\sigma}$ for the Generalized Logistic Distribution When $n = 15$ and $b = 2.0$ 204
6.6.3 Bias, Variances, and Covariance of $\hat{\mu}$ and $\hat{\sigma}$ for the Generalized Logistic Distribution When $n = 15$ and $b = 3.0$ 205
6.6.4 Comparison of $\hat{\mu}$ and $\hat{\sigma}$ with BLUEs μ^* and σ^* for the Generalized Logistic Distribution 206
6.6.5 Values of $\hat{\mu}$ and $\hat{\sigma}$ and Their Standard Errors for Censored Samples from the Generalised Logistic Distribution 207
6.7.1 Unbiasing Factor of the AMLE and Variance of the Unbiased AMLE, Unbiasing Factor of the MLE and Variance of the Unbiased MLE, and Variance of the BLUE of σ for the Half Logistic Distribution When $n = 10$ 210
6.7.2 Unbiasing Factor of the AMLE and Variance of the Unbiased AMLE, Unbiasing Factor of the MLE and Variance of the Unbiased MLE, and Variance of the BLUE of σ for the Half Logistic Distribution When $n = 15$ 211

6.7.3 Unbiasing Factor of the AMLE and Variance of the Unbiased AMLE, Unbiasing Factor of the MLE and Variance of the Unbiased MLE, and Variance of the BLUE of σ for the Half Logistic Distribution When $n = 20$ 212
7.3.1 Optimum Spacings, the Corresponding Coefficients, and Relative Efficiencies for $\tilde{\sigma}^*$ of the One-Parameter Exponential Distribution 241
7.3.2. Time Intervals in Days between Explosions in Mines Involving More Than 10 Men Killed from 6 December 1875 to 29 May 1951 250
7.4.1 Estimator in (7.4.1) for the Mean of the Normal Population and Values of Variance and Efficiency Relative to the Sample Mean 251
7.4.2 Estimator in (7.4.3) for the Standard Deviation of the Normal Population and Values of Variance and Relative Efficiencies 252
7.4.3 Estimator in (7.4.4) for the Standard Deviation of the Normal Population and Values of Variance and Relative Efficiencies 253
7.4.4 Estimator in (7.4.1) for the Mean of the Logistic Population and Values of Variance and Efficiency Relative to the BLUE 254
7.4.5 Estimator in (7.4.3) for the Standard Deviation of the Logistic Population and Values of Variance and Efficiency Relative to the BLUE . . 255
7.4.6 Estimator in (7.4.1) for the Mean of the Double Exponential Population and Values of Variance and Efficiency Relative to the BLUE . . 256
7.4.7 Estimator in (7.4.3) for σ of the Double Exponential Population and Values of Variance and Efficiency Relative to the BLUE 257
8.3.1 Expected Values of the First Order Statistics from the Standard Normal Distribution (0, 1) 281
8.7.1 Maximum Flood Levels of Susquehanna River at Harrisburg, PA over Four-Year periods (1890–1969) in Millions of Cubic Feet per Second. Example 8.7.1 291
8.7.2 Fatigue Life in Hours of a Certain Type of Bearing. Example 8.7.2 291
8.7.3 Random Sample from an Inverse Gaussian Distribution: $E(X) = 16$, $V(X) = 9$, $\alpha_3(X) = 1.5$. Example 8.7.3 292
8.7.4 Summary of Estimates for Example 8.7.1. $\bar{x} = 0.423125$, $s = 0.1252789$, $a_3 = 1.0673243$, $x_1 = 0.265$, $n = 20$, $s^2/(\bar{x} - x_1)^2 = 0.6277$, $z_1 = -1.2622$ 293
8.7.5 Summary of Estimates for Example 8.7.2. $\bar{x} = 220.48$, $s = 78.405638$, $a_3 = 1.8635835$, $x_1 = 152.7$, $n = 10$, $s^2/(\bar{x} - x_1)^2 = 1.338$, $z_1 = -0.8645$ 294
8.7.6 Summary of Estimates for Example 8.7.3. $\bar{x} = 15.7415$, $s = 3.1276$, $a_3 = 1.3230$, $x_1 = 11.3557$, $n = 100$, $s^2/(\bar{x} - x_1)^2 = 0.50854$, $z_1 = -1.4023$ 294
9.2.1 Sales (y) by Months on the Job (x) 305
9.2.2 BLUEs of α, β, and σ and Their Variances for Different Distributions 306
9.3.1 The Log Survival Times of the Specimens at Five Doses 313
9.3.2 Computational Tableau for the Evaluation of the MMLEs 313
9.4.1 Values of Bias and Variance of the MLEs and the MMLEs of α, β, and σ; $n = 10$ 323

9.4.2 Values of Bias and Variance of the MLEs and the MMLEs of α, β, and σ; $n=20$ 324
9.4.3 Normalized Residuals and Their Ranks for Daniel and Wood's Data 325
9.4.4 Normalized Residuals and Their Ranks for Daniel and Wood's Data with Contamination 327
9.4.5 Mendenhall's EPA 1980 Mileage Rating Data 328
10.4.1 Random Sample from a Normal Distribution; $N=100$, $\mu=60$, and $\sigma=10$ 336
10.4.2 Censored Sample Estimates of Normal Distribution Parameters 337
10.4.3 Successive Iterations of Pseudo-Complete Sample Estimates of Normal Distribution Parameters; $N=100$, $n=95$, $c=5$, $h=0.05$ 337
10.4.4 Random Sample from an Inverse Gaussian Distribution; $\gamma=10$, $\mu=6$, $\sigma=5$ ($\alpha_3=2.5$) 338
10.4.5 Summary of Estimates of Inverse Gaussian Distribution Parameters 338
10.4.6 Successive Iterations of Pseudo-Complete Sample Estimates of Inverse Gaussian Parameters; $\gamma=10$, $\mu=6$, $\sigma=5$, $E(X)=16$, $\alpha_3=2.5$, $N=100$, $h=0.05$ 339

List of Figures

5.6.1 Truncated sample estimating function for Rayleigh distribution 142

5.8.1 Estimation curve for singly truncated normal samples............ 149

8.2.1 Graphs of $C(\delta)$ and $D(\delta)$ for the Weibull distribution 276

8.2.2 Graphs of $W(n, \hat{\delta}) = s^2/(\bar{x} - x_1)^2$ for the Weibull distribution 277

8.3.1 Graphs of $\omega(\omega - 1)/[\sqrt{\omega} - \exp\{\sigma E(Z_{1:n})\}]^2$ for the lognormal distribution 282

8.4.1 Graphs of α_3 as a function of z_1 and n for the inverse Gaussian distribution 285

8.5.1 Graphs of α_3 as a function of z_1 and n for the gamma distribution 288

Chapter 1 Introduction

1.1. Introductory Remarks

The subject of order statistics deals with properties and applications of ordered random variables and of functions of these variables. If the random variables $\{X_i\}$, $i = 1, 2, \ldots, n$ are arranged in ascending order of magnitude and then written as

$$X_{1:n} \leq X_{2:n} \leq \cdots \leq X_{n:n},$$

then $X_{i:n}$ is said to be the ith-order statistic in a sample of size n. In the usual random sampling theory, the unordered X_i are assumed to be statistically independent and identically distributed. Because of the inequality relations among them, the order statistics $X_{i:n}$ are necessarily dependent. A distinction is made between the random variables $X_{i:n}$ and the corresponding sample observations $x_{i:n}$. As an illustration, consider a life test on a certain electrical component. A random selection of n specimens is made from the population of items at issue. This sample is placed "on test." If the test is continued until all sample specimens have failed, the sample is said to be complete, and it consists of the naturally ordered observations

$$x_{1:n} \leq x_{2:n} \leq \cdots \leq x_{n:n}.$$

If sample specimens are withdrawn prior to failure and/or testing is terminated with survivors, the sample is said to be censored. In some situations, the original sample observations x_i may be unordered with respect to magnitude, and a transformation — that is, a rearrangement — is required to produce a corresponding ordered sample.

Some frequently encountered functions of order statistics are the *extremes* $X_{1:n}$ and $X_{n:n}$, the *range* $W = X_{n:n} - X_{1:n}$, the *extreme deviate* from the sample mean, $X_{n:n} - \bar{X}$, and for a random sample from a normal distribution $N(\mu, \sigma^2)$, the *studentized range*, W/S_ν, where S_ν is a root-mean-square estimator of σ based on ν degrees of freedom. All of these statistics have important applications. The extremes arise in the statistical study of floods and droughts, as well as in breaking strength and fatigue failure studies. The range is widely employed in the field of quality control as a quick estimator of the standard deviation σ. The extreme deviate is a basic tool in procedures for detecting outliers. Large values of $(X_{n:n} - \bar{X})/\sigma$ suggest the presence of outliers. When outliers are not confined to one direction, the studentized range is also useful in the detection process. Furthermore, the studentized range is the basis of many quick tests in small samples, and it is of key importance in ranking "treatment" means in analysis of variance situations.

With regard to notation, random variables are designated by uppercase letters, and their realizations (observations) by corresponding lowercase letters, as shown below.

$X_1, X_2, \ldots, X_n$	unordered random variables
$x_1, x_2, \ldots, x_n$	unordered observations
$X_{1:n} \leq X_{2:n} \leq \cdots \leq X_{n:n}$	ordered random variables
$x_{1:n} \leq x_{2:n} \leq \cdots \leq x_{n:n}$	ordered observations.

By order statistics, we mean either ordered random variables or ordered observations. In some applications where only the first order statistic is involved, the notation will be abbreviated from $X_{1:n}$ to simply X_1, and from $x_{1:n}$ to x_1. This shortened notation will be employed only where no confusion is likely to result.

1.2. The Role of Order Statistics in Practical Applications

Order statistics and functions of these statistics play an important role in numerous practical applications. As previously mentioned, the range is widely used by quality control practitioners to provide quick estimates of σ in the normal distribution. The first- and the nth-order statistics are valuable tools for the detection of outliers. The studentized range is also useful in the detection of outliers. However, its principal use is as the basis of quick tests in small samples and in ranking procedures.

Sarhan and Greenberg (1962) employed linear functions of order statistics in conjunction with the Gauss–Markov theorem to systematically estimate location and scale parameters in both complete and censored samples. They provided tables of the coefficients necessary for the calculations of these estimates from samples varying in size from 2 to 20.

Other applications of order statistics arise in the study of system reliability. A system of n components is called a *k-out-of-n system* if it remains operational only if at least k components continue to function. For components with independent lifetime distributions, the time to failure of the system is thus the $(n-k+1)$th-order statistic. The special cases $k=1$ and $k=n$ correspond respectively to parallel and series systems.

A major impetus for the study of order statistics has been provided by the development of modern computers. Through their use it is feasible to make repeated examinations of the same data in many different ways. Tukey (1970), Mosteller and Tukey (1977), and others have employed various informal techniques in the analysis of data. It is thus possible to determine quickly if the data indeed are in accord with an assumed distribution and with an assumed model. A plot of the ordered observations against some simple function of their ranks, preferably on probability paper appropriate for the assumed distribution, will often prove helpful in making such determinations. Nelson (1972) has exploited this technique to obtain parameter estimates from progressively censored samples by plotting hazard functions. A straight-line fit to either a probability or a hazard function plot indicates that both the model and the parameter estimates are generally satisfactory, whereas serious departures from a straight line reveal either the presence of outliers or the selection of an incorrect model.

The first-order statistic is particularly useful in estimating the threshold parameter in skewed distributions. Cohen (1988), Cohen and Whitten (1980, 1981, 1982, 1985, 1986, 1988), and Cohen *et al.* (1984, 1985) have employed the first-order statistic to advantage in various modified moment and maximum likelihood estimators of parameters in the Weibull, lognormal, inverse Gaussian, gamma, generalized gamma, and other positively skewed distributions.

A somewhat more limited application of order statistics can be found in the area of data compression, in which large masses of data are replaced by a small number of selected order statistics. This application is of special importance in the space program where it is necessary to compress excessively large quantities of data accumulated on board an orbiting capsule in order to facilitate their transmission to a ground station.

An alternate approach to the problem of parameter estimation from censored samples was recently proposed by Whitten *et al.* (1988), whereby order statistics are employed to estimate expected values of the censored observations, and thus to restore censored samples to complete sample status.

1.3. Scope of This Volume

The primary objectives in writing this book have been to produce an up-to-date volume, with emphasis on applications, that will be suitable both as a text for a first course in order statistics and as a reference for practitioners whose analyses of sample data involve order statistics. Consideration is given to the various topics mentioned in the preceding sections and to many others as well. A full treatment of basic theory including the derivation of moments and the expected values of other functions of order statistics is also included. However, only brief mention will be made of rank-order statistics as exemplified by the work of Wilcoxon (1945, 1947). Although they involve an ordering of observations with respect to magnitude, rank-order statistics use only ranks of the ordered observations and not their actual values. This area is considered to be beyond the scope of this book.

Due recognition is given to the pioneering work of numerous previous authors. Among these are Wilks (1962), Sarhan and Greenberg (1962), Harter (1970a, b, 1978, 1983), David (1981), and many others. An extensive list of references is contained in the excellent multivolume chronological annotated bibliography of order statistics by Harter (1978, 1983). Recognition is also given to books primarily concerned with multiple comparisons, but which are also involved with order statistics, by Miller (1966) and by Gupta and Panchapakesan (1979). The book by Miller has been updated by an annotated bibliography (Miller, 1977). The asymptotic theory of extremes and of related statistics has been developed in a book by Galambos (1978), and a more applied account that continues to be relevant was given earlier by Gumbel (1958). A special edition of *Communications in Statistics—Theory and Methods (1988), Vol. 17 (7)* on *Order Statistics and Applications*, edited by Balakrishnan, contains a number of important recent results concerning order statistics. In addition, other important results on recurrence relations, bounds, and approximations for order statistics are included in a recent monograph by Arnold and Balakrishnan (1989).

It is recognized that the effective application of order statistic techniques requires numerous tables. Many of these are included in this volume, but space limitations do not permit the inclusion of all that might be desired. For additional tables, readers are referred to two volumes of general tables by Pearson and Hartley (1970, 1972), to specialized tables included by Sarhan and Greenberg (1962), and to two volumes of tables devoted to order statistics by Harter (1970a, b). Other references to tables that appear in original papers are given throughout this text.

Chapter 2 Basic Theory

2.1. Introduction

In this chapter we first start with the joint distribution of all the n order statistics, and then derive the joint distribution of two order statistics and the marginal distribution of a single order statistic. Next, we consider some systematic statistics, such as the sample range and the sample quasi-range, and derive their distributions through appropriate transformations. By considering the conditional distributions of order statistics, we also establish that the order statistics in a sample form a Markov chain. All the theoretical developments presented in this chapter are for the case in which the order statistics arise from the realizations of n absolutely continuous random variables, as this is the case that is essential for the rest of this volume. For similar results on order statistics from a discrete population, we refer interested readers to David (1981).

2.2. Joint Distribution of n Order Statistics

Let $X_1, X_2, \ldots, X_n$ be independent, absolutely continuous random variables with common pdf $f(x)$ and cdf $F(x)$, and let $X_{1:n} \leq X_{2:n} \leq \cdots \leq X_{n:n}$ be the corresponding order statistics. Then, given the realizations of these n order statistics to be $x_{1:n} \leq x_{2:n} \leq \cdots \leq x_{n:n}$, the original variables X_i $(i = 1, 2, \ldots, n)$ are constrained to take on the values $x_{i_k:n}$, which by symmetry carries equal probability for each of the $n!$ permutations $(i_1, i_2, \ldots, i_n)$

of $(1, 2, \ldots, n)$. As a result we derive the joint density of all n order statistics to be

$$f_{1,2,\ldots,n:n}(x_{1:n}, x_{2:n}, \ldots, x_{n:n}) = n! \prod_{i=1}^{n} f(x_{i:n}), \qquad -\infty < x_{1:n} < x_{2:n} < \cdots < x_{n:n} < \infty. \tag{2.2.1}$$

For example, if X_i's are i.i.d. random variables from a uniform $(0, 1)$ distribution, we have the joint density of all n order statistics as

$$f_{1,2,\ldots,n:n}(x_{1:n}, x_{2:n}, \ldots, x_{n:n}) = n!, \qquad 0 < x_{1:n} < x_{2:n} < \cdots < x_{n:n} < 1. \tag{2.2.2}$$

Similarly, when the random sample X_i $(i = 1, 2, \ldots, n)$ is from a standard exponential distribution with pdf $f(x) = e^{-x}$, $0 \le x < \infty$, we have the joint density of all n order statistics as

$$f_{1,2,\ldots,n:n}(x_{1:n}, x_{2:n}, \ldots, x_{n:n}) = n!\, e^{-\sum_{i=1}^{n} x_{i:n}}, \qquad 0 \le x_{1:n} < x_{2:n} < \cdots < x_{n:n} < \infty. \tag{2.2.3}$$

We will make use of the above joint density later, in Chapter 3, in order to show that the smallest order statistic $X_{1:n}$ of independent exponential random variables is again exponentially distributed, and also that the set of exponential spacings $X_{i:n} - X_{i-1:n}$ (with $X_{0:n} \equiv 0$) constitute a set of n independent exponential random variables.

2.3. Joint Distribution of Two Order Statistics

Let us consider the order statistics $X_{i:n}$ and $X_{j:n}$ $(1 \le i < j \le n)$ and derive their joint density from Eq. (2.2.1).

By considering the joint density of all n order statistics and integrating out the variables $(X_{1:n}, \ldots, X_{i-1:n})$, $(X_{i+1:n}, \ldots, X_{j-1:n})$, $(X_{j+1:n}, \ldots, X_{n:n})$, we derive the joint density function of $X_{i:n}$ and $X_{j:n}$ as

$$\begin{aligned} f_{i,j:n}(x_{i:n}, x_{j:n}) &= n!\, f(x_{i:n}) f(x_{j:n}) \\ &\times \int_{-\infty}^{x_{i:n}} \cdots \int_{-\infty}^{x_{2:n}} f(x_{1:n}) \cdots f(x_{i-1:n})\, dx_{1:n} \cdots dx_{i-1:n} \\ &\times \int_{x_{i:n}}^{x_{j:n}} \cdots \int_{x_{i:n}}^{x_{i+2:n}} f(x_{i+1:n}) \cdots f(x_{j-1:n})\, dx_{i+1:n} \cdots dx_{j-1:n} \\ &\times \int_{x_{j:n}}^{\infty} \cdots \int_{x_{j:n}}^{x_{j+2:n}} f(x_{j+1:n}) \cdots f(x_{n:n})\, dx_{j+1:n} \cdots dx_{n:n}. \end{aligned} \tag{2.3.1}$$

Direct integration yields

$$\int_{-\infty}^{x_{i:n}} \cdots \int_{-\infty}^{x_{2:n}} f(x_{1:n}) \cdots f(x_{i-1:n})\, dx_{1:n} \cdots dx_{i-1:n}$$

$$= \{F(x_{i:n})\}^{i-1}/(i-1)!, \tag{2.3.2}$$

$$\int_{x_{i:n}}^{x_{j:n}} \cdots \int_{x_{i:n}}^{x_{i+2:n}} f(x_{i+1:n}) \cdots f(x_{j-1:n})\, dx_{i+1:n} \cdots dx_{j-1:n}$$

$$= \{F(x_{j:n}) - F(x_{i:n})\}^{j-i-1}/(j-i-1)!, \tag{2.3.3}$$

and

$$\int_{x_{j:n}}^{\infty} \cdots \int_{x_{j:n}}^{x_{j+2:n}} f(x_{j+1:n}) \cdots f(x_{n:n})\, dx_{j+1:n} \cdots dx_{n:n}$$

$$= \{1 - F(x_{j:n})\}^{n-j}/(n-j)!. \tag{2.3.4}$$

Upon using the expressions (2.3.2)-(2.3.4) for the three sets of integrals in Eq. (2.3.1), we obtain the joint density function of $X_{i:n}$ and $X_{j:n}$ $(1 \le i < j \le n)$ as

$$f_{i,j:n}(x_{i:n}, x_{j:n})$$

$$= \frac{n!}{(i-1)!(j-i-1)!(n-j)!} \{F(x_{i:n})\}^{i-1} \{F(x_{j:n}) - F(x_{i:n})\}^{j-i-1}$$

$$\times \{1 - F(x_{j:n})\}^{n-j} f(x_{i:n}) f(x_{j:n}), \qquad -\infty < x_{i:n} < x_{j:n} < \infty. \tag{2.3.5}$$

For example, if X_i's are i.i.d. random variables from a uniform (0, 1) distribution, we have the joint density function of $X_{i:n}$ and $X_{j:n}$ $(1 \le i < j \le n)$ as

$$f_{i,j:n}(x_{i:n}, x_{j:n})$$

$$= \frac{n!}{(i-1)!(j-i-1)!(n-j)!} x_{i:n}^{i-1} (x_{j:n} - x_{i:n})^{j-i-1} (1 - x_{j:n})^{n-j},$$

$$0 < x_{i:n} < x_{j:n} < 1. \tag{2.3.6}$$

It may be noted that the above function is a bivariate beta (or Dirichlet) density function and that the constant term in (2.3.6) is simply $\{B(i, j-i, n-j+1)\}^{-1}$, where $B(a, b, c)$ is the three-parameter complete beta function defined by

$$B(a, b, c) = \Gamma(a)\Gamma(b)\Gamma(c)/\Gamma(a+b+c); \tag{2.3.7}$$

here, $\Gamma(\cdot)$ denotes the complete gamma function defined by

$$\Gamma(z)=\int_0^{\infty} e^{-t}t^{z-1}\,dt.$$

Similarly, when the random sample X_i $(i=1,2,\ldots,n)$ is from a standard exponential distribution with pdf $f(x)=e^{-x}$, $0\le x<\infty$, and cdf $F(x)=1-e^{-x}$, $0\le x<\infty$, we have the joint density function of $X_{i:n}$ and $X_{j:n}$ $(1\le i<j\le n)$ as

$$\begin{aligned} f_{i,j:n}(x_{i:n}, x_{j:n}) &= \frac{n!}{(i-1)!(j-i-1)!(n-j)!}(1-e^{-x_{i:n}})^{i-1}(e^{-x_{i:n}}-e^{-x_{j:n}})^{j-i-1} \\ &\quad\times e^{-x_{i:n}}(e^{-x_{j:n}})^{n-j+1}, \qquad 0<x_{i:n}<x_{j:n}<\infty. \end{aligned} \tag{2.3.8}$$

The joint density function of $X_{i:n}$ and $X_{j:n}$ in (2.3.5) has been derived from the joint density of all n order statistics by judicious integration. For the absolute continuous case that we are interested in, however, we present the following probabilistic derivation, which is algebraically simpler and also lends itself to further extensions; see David (1981), David and Shu (1978), and Arnold and Balakrishnan (1989). The event $(x_{i:n}<X_{i:n}\le x_{i:n}+\delta x_{i:n},\ x_{j:n}<X_{j:n}\le x_{j:n}+\delta x_{j:n})$ may be seen as follows:

$i-1$	1	$j-i-1$	1	$n-j$
$-\infty$	$x_{i:n}\quad x_{i:n}+\delta x_{i:n}$		$x_{j:n}\quad x_{j:n}+\delta x_{j:n}$	∞

$X_r\le x_{i:n}$ for $i-1$ of the X_r, $x_{i:n}<X_r\le x_{i:n}+\delta x_{i:n}$ for exactly one X_r, $x_{i:n}+\delta x_{i:n}<X_r\le x_{j:n}$ for $j-i-1$ of the X_r, $x_{j:n}<X_r\le x_{j:n}+\delta x_{j:n}$ for exactly one X_r, and $X_r>x_{j:n}+\delta x_{j:n}$ for the remaining $n-j$ of the X_r. Realizing now that

$$\begin{aligned} &\Pr(x_{i:n}<X_{i:n}\le x_{i:n}+\delta x_{i:n},\ x_{j:n}<X_{j:n}\le x_{j:n}+\delta x_{j:n}) \\ &= \frac{n!}{(i-1)!(j-i-1)!(n-j)!}\{F(x_{i:n})\}^{i-1}\{F(x_{j:n})-F(x_{i:n}+\delta x_{i:n})\}^{j-i-1} \\ &\quad\times\{1-F(x_{j:n}+\delta x_{j:n})\}^{n-j}\{F(x_{i:n}+\delta x_{i:n})-F(x_{i:n})\} \\ &\quad\times\{F(x_{j:n}+\delta x_{j:n})-F(x_{j:n})\} \\ &\quad+O((\delta x_{i:n})^2\delta x_{j:n})+O(\delta x_{i:n}(\delta x_{j:n})^2) \end{aligned} \tag{2.3.9}$$

upon regarding $\delta x_{i:n}$ and $\delta x_{j:n}$ as small, where $O((\delta x_{i:n})^2\delta x_{j:n})$ and $O(\delta x_{i:n}(\delta x_{j:n})^2)$ denote the probabilities of more than one X_r falling in the intervals $(x_{i:n}, x_{i:n}+\delta x_{i:n}]$ and $(x_{j:n}, x_{j:n}+\delta x_{j:n}]$, respectively, we derive the joint density function of $X_{i:n}$ and $X_{j:n}$ $(1 \le i < j \le n)$ as

$$
\begin{aligned}
f_{i,j:n}(x_{i:n}, x_{j:n})
&= \lim_{\substack{\delta x_{i:n}\to 0\\ \delta x_{j:n}\to 0}} \frac{\Pr(x_{i:n} < X_{i:n} \le x_{i:n}+\delta x_{i:n}, \quad x_{j:n} < X_{j:n} \le x_{j:n}+\delta x_{j:n})}{\delta x_{i:n}\,\delta x_{j:n}} \\
&= \frac{n!}{(i-1)!(j-i-1)!(n-j)!}\{F(x_{i:n})\}^{i-1}\{F(x_{j:n})-F(x_{i:n})\}^{j-i-1} \\
&\quad \times \{1-F(x_{j:n})\}^{n-j} f(x_{i:n}) f(x_{j:n}), \qquad -\infty < x_{i:n} < x_{j:n} < \infty.
\end{aligned}
$$

This is exactly the same expression as the one derived in (2.3.5).

2.4. Distribution of a Single Order Statistic

As done in the last section, we start with the joint density of all n order statistics and then integrate out the variables $(X_{1:n}, \ldots, X_{i-1:n})$, $(X_{i+1:n}, \ldots, X_{n:n})$ in order to derive the marginal density function of $X_{i:n}$ $(1 \le i \le n)$ as

$$
\begin{aligned}
f_{i:n}(x_{i:n}) &= n!\, f(x_{i:n}) \\
&\quad \times \int_{-\infty}^{x_{i:n}} \cdots \int_{-\infty}^{x_{2:n}} f(x_{1:n}) \cdots f(x_{i-1:n})\, dx_{1:n} \cdots dx_{i-1:n} \\
&\quad \times \int_{x_{i:n}}^{\infty} \cdots \int_{x_{i:n}}^{x_{i+2:n}} f(x_{i+1:n}) \cdots f(x_{n:n})\, dx_{i+1:n} \cdots dx_{n:n}. \qquad (2.4.1)
\end{aligned}
$$

As before, direct integration yields

$$
\int_{-\infty}^{x_{i:n}} \cdots \int_{-\infty}^{x_{2:n}} f(x_{1:n}) \cdots f(x_{i-1:n})\, dx_{1:n} \cdots dx_{i-1:n} = \{F(x_{i:n})\}^{i-1}/(i-1)! \qquad (2.4.2)
$$

and

$$
\begin{aligned}
&\int_{x_{i:n}}^{\infty} \cdots \int_{x_{i:n}}^{x_{i+2:n}} f(x_{i+1:n}) \cdots f(x_{n:n})\, dx_{i+1:n} \cdots dx_{n:n} \\
&\qquad = \{1-F(x_{i:n})\}^{n-i}/(n-i)!. \qquad (2.4.3)
\end{aligned}
$$

Upon substituting the expressions (2.4.2) and (2.4.3) for the two sets of integrals in Eq. (2.4.1), we obtain the marginal density function of $X_{i:n}$ $(1 \le i \le n)$ as

$$f_{i:n}(x_{i:n}) = \frac{n!}{(i-1)!(n-i)!}\{F(x_{i:n})\}^{i-1}\{1-F(x_{i:n})\}^{n-i}f(x_{i:n}),$$
$$-\infty < x_{i:n} < \infty. \tag{2.4.4}$$

The smallest and the largest order statistics in samples of size n are of special interest in numerous practical applications. The pdf of these statistics follow from (2.4.4) when $i=1$ and n, respectively, as

$$f_{1:n}(x_{1:n}) = n\{1-F(x_{1:n})\}^{n-1}f(x_{1:n}), \qquad -\infty < x_{1:n} < \infty$$

and

$$f_{n:n}(x_{n:n}) = n\{F(x_{n:n})\}^{n-1}f(x_{n:n}), \qquad -\infty < x_{n:n} < \infty.$$

As an example, if the X_i's are i.i.d. random variables from a uniform (0, 1) distribution, the density function of $X_{i:n}$ $(1 \le i \le n)$ becomes

$$f_{i:n}(x_{i:n}) = \frac{n!}{(i-1)!(n-i)!}x_{i:n}^{i-1}(1-x_{i:n})^{n-i}, \qquad 0 < x_{i:n} < 1, \tag{2.4.5}$$

which is the density function of a two-parameter beta distribution. We will make use of this density function later in Chapter 3 in order to derive the single moments of uniform order statistics.

Similarly, when the random sample X_i $(i=1, 2, \ldots, n)$ is from a standard exponential distribution with pdf $f(x)=e^{-x}$, $0 \le x < \infty$, and cdf $F(x)= 1-e^{-x}$, $0 \le x < \infty$, we have the density function of $X_{i:n}$ $(1 \le i \le n)$ as

$$f_{i:n}(x_{i:n}) = \frac{n!}{(i-1)!(n-i)!}(1-e^{-x_{i:n}})^{i-1}(e^{-x_{i:n}})^{n-i+1}, \qquad 0 \le x_{i:n} < \infty. \tag{2.4.6}$$

Even though we could use the above density function in order to derive the single moments of exponential order statistics, we present in Chapter 3 an alternative and simpler way of deriving moments through an interesting and convenient representation of the exponential order statistics themselves.

As in the last section we present the following probabilistic derivation of the marginal density function of a single order statistic. The event $x_{i:n} < X_{i:n} \le x_{i:n} + \delta x_{i:n}$ may be seen as follows:

$$\begin{array}{ccccc} & i-1 & 1 & n-i & \\ -\infty & & x_{i:n} \quad x_{i:n}+\delta x_{i:n} & & \infty \end{array}$$

$X_r \le x_{i:n}$ for $i-1$ of the X_r, $x_{i:n} < X_r \le x_{i:n} + \delta x_{i:n}$ for exactly one X_r, and $X_r > x_{i:n} + \delta x_{i:n}$ for the remaining $n-i$ of the X_r. By regarding $\delta x_{i:n}$ as small, we realize that

$$\begin{aligned}\Pr(x_{i:n} < X_{i:n} \le x_{i:n} + \delta x_{i:n})\\ &= \frac{n!}{(i-1)!(n-i)!}\{F(x_{i:n})\}^{i-1}\{1-F(x_{i:n}+\delta x_{i:n})\}^{n-i}\\ &\quad \times\{F(x_{i:n}+\delta x_{i:n}) - F(x_{i:n})\} + O((\delta x_{i:n})^2), \qquad (2.4.7)\end{aligned}$$

where $O((\delta x_{i:n})^2)$ denotes the probability of more than one X_r falling in the interval $(x_{i:n}, x_{i:n} + \delta x_{i:n}]$. From (2.4.7) we derive the density function of $X_{i:n}$ $(1 \le i \le n)$ to be

$$\begin{aligned}f_{i:n}(x_{i:n})\\ &= \lim_{\delta x_{i:n}\to 0} \frac{\Pr(x_{i:n} < X_{i:n} \le x_{i:n} + \delta x_{i:n})}{\delta x_{i:n}}\\ &= \frac{n!}{(i-1)!(n-i)!}\{F(x_{i:n})\}^{i-1}\{1-F(x_{i:n})\}^{n-i} f(x_{i:n}), \qquad -\infty < x_{i:n} < \infty,\end{aligned}$$

which is exactly the same as the expression derived in (2.4.6).

The cumulative distribution function of an individual order statistic $X_{i:n}$ may be derived without any difficulty in the following manner. For any $i = 1, 2, \ldots, n$ and any $x \in \mathbb{R}$, we have

$$\begin{aligned}F_{i:n}(x) &= \Pr(X_{i:n} \le x)\\ &= \Pr(\text{at least } i \text{ of the } X_r\text{'s are } \le x)\\ &= \sum_{r=i}^{n} \binom{n}{r} \{F(x)\}^r \{1-F(x)\}^{n-r}. \qquad (2.4.8)\end{aligned}$$

We thus find the cdf of $X_{i:n}$ to be simply the tail probability (starting from i) of a binomial distribution with number of trials n and probability of success $F(x)$. From (2.4.8) we obtain the cdf of the largest order statistic $X_{n:n}$ as

$$F_{n:n}(x) = \{F(x)\}^n, \qquad (2.4.9)$$

and that of the smallest order statistic $X_{1:n}$ as

$$\begin{aligned}F_{1:n}(x) &= \sum_{r=1}^{n} \binom{n}{r} \{F(x)\}^r \{1-F(x)\}^{n-r}\\ &= 1 - \{1-F(x)\}^n. \qquad (2.4.10)\end{aligned}$$

Let us now denote a random sample of size n from a uniform $(0,1)$ distribution by $U_1, U_2, \ldots, U_n$, and their order statistics by $U_{1:n}$, $U_{2:n}, \ldots, U_{n:n}$. Further, let $X_1, X_2, \ldots, X_n$ denote a random sample of size n from any distribution function $F(x)$, and let their order statistics be denoted by $X_{1:n}, X_{2:n}, \ldots, X_{n:n}$. Then if the inverse distribution function F^{-1} is defined by

$$F^{-1}(u) = \sup\{x\colon F(x) \leq u\}, \tag{2.4.11}$$

it can be easily verified that

$$F^{-1}(U_i) \overset{d}{=} X_i \tag{2.4.12}$$

and

$$F^{-1}(U_{i:n}) \overset{d}{=} X_{i:n}. \tag{2.4.13}$$

From (2.4.8) we simply have the cdf of $U_{i:n}$ $(1 \leq i \leq n)$ as

$$F_{U_{i:n}}(u) = \sum_{r=i}^{n} \binom{n}{r} u^r (1-u)^{n-r}, \tag{2.4.14}$$

which, upon differentiation, yields the density function of $U_{i:n}$ as

$$f_{U_{i:n}}(u) = \frac{n!}{(i-1)!(n-i)!} u^{i-1}(1-u)^{n-i}, \qquad 0 < u < 1.$$

This is the same density that has already been derived in (2.4.5). From the above density function, we may also write the cdf of $U_{i:n}$ $(1 \leq i \leq n)$ as

$$F_{U_{i:n}}(u) = \int_0^u \frac{n!}{(i-1)!(n-i)!} t^{i-1}(1-t)^{n-i}\, dt. \tag{2.4.15}$$

By comparing Eqs. (2.4.8), (2.4.14), and (2.4.15), we derive the expression for the cdf of $X_{i:n}$ $(1 \leq i \leq n)$ as

$$F_{i:n}(x) = \int_0^{F(x)} \frac{n!}{(i-1)!(n-i)!} t^{i-1}(1-t)^{n-i}\, dt, \qquad -\infty < x < \infty. \tag{2.4.16}$$

This is just Karl Pearson's (1934) incomplete beta function. It is of interest to mention here that the above expression for the cdf of $X_{i:n}$ is valid for any common parent distribution $F(x)$ of the X_r's (whether discrete or continuous). Moreover, for the case in which the X_r's are absolutely continuous random variables, we may derive the density function of $X_{i:n}$ upon differentiating the expression of the cdf in (2.4.16) by employing the chain rule. The density thus derived is identical to the form of $f_{i:n}(x_{i:n})$ given in (2.4.4).

From the marginal density function of $X_{i:n}$ in (2.4.4) and the joint density function of $X_{i:n}$ and $X_{j:n}$ in (2.3.6), we obtain the conditional density function of $X_{j:n}$, given $X_{i:n} = x_{i:n}$, to be

$$f_{j:n}(x_{j:n} \mid x_{i:n}) = \frac{(n-i)!}{(j-i-1)!(n-j)!} \left\{ \frac{F(x_{j:n}) - F(x_{i:n})}{1 - F(x_{i:n})} \right\}^{j-i-1} \left\{ \frac{1 - F(x_{j:n})}{1 - F(x_{i:n})} \right\}^{n-j} \times \frac{f(x_{j:n})}{1 - F(x_{i:n})}, \qquad x_{i:n} \leq x_{j:n} < \infty. \tag{2.4.17}$$

Since $\{F(x_{j:n}) - F(x_{i:n})\}/\{1 - F(x_{i:n})\}$ and $f(x_{j:n})/\{1 - F(x_{i:n})\}$ are the cdf and the pdf, respectively, of the parent distribution truncated on the left at $x_{i:n}$, we may state the following theorem.

Theorem 2.4.1. *Let $X_1, X_2, \ldots, X_n$ be i.i.d. random variables from a population with cdf $F(x)$ and pdf $f(x)$, and let $X_{1:n} \leq X_{2:n} \leq \cdots \leq X_{n:n}$ be the corresponding order statistics. Then the conditional distribution of $X_{j:n}$, given $X_{i:n} = x_{i:n}$, is the same as the distribution of the $(j-i)$th order statistic in a sample of size $n-i$ from a population with distribution $F(x)$ truncated on the left at $x_{i:n}$.*

Similarly, from (2.4.4) and (2.3.6) we obtain the conditional density function of $X_{i:n}$, given $X_{j:n} = x_{j:n}$, to be

$$f_{i:n}(x_{i:n} \mid x_{j:n}) = \frac{(j-1)!}{(i-1)!(j-i-1)!} \left\{ \frac{F(x_{i:n})}{F(x_{j:n})} \right\}^{i-1} \left\{ \frac{F(x_{j:n}) - F(x_{i:n})}{F(x_{j:n})} \right\}^{j-i-1} \frac{f(x_{i:n})}{F(x_{j:n})}, \qquad -\infty < x_{i:n} \leq x_{j:n}. \tag{2.4.18}$$

Since $F(x_{i:n})/F(x_{j:n})$ and $f(x_{i:n})/F(x_{j:n})$ are the cdf and the pdf of the parent distribution truncated on the right at $x_{j:n}$, respectively, we may state the following theorem.

Theorem 2.4.2. *Let $X_1, X_2, \ldots, X_n$ be i.i.d. random variables from a population with cdf $F(x)$ and pdf $f(x)$, and let $X_{1:n} \leq X_{2:n} \leq \cdots \leq X_{n:n}$ be the corresponding order statistics. Then the conditional distribution of $X_{i:n}$, given $X_{j:n} = x_{j:n}$, is the same as the distribution of the ith order statistic in a sample of size $j-1$ from a population with distribution $F(x)$ truncated on the right at $x_{j:n}$.*

Furthermore, from the joint density of all n order statistics given in (2.2.1) we obtain the joint density of $X_{1:n}, X_{2:n}, \ldots, X_{i:n}$ to be

$$\begin{aligned} f_{1,2,\ldots,i:n}&(x_{1:n}, x_{2:n}, \ldots, x_{i:n}) \\ &= n!\, f(x_{1:n}) f(x_{2:n}) \cdots f(x_{i:n}) \\ &\quad \times \int_{x_{i:n}}^{\infty} \cdots \int_{x_{i:n}}^{x_{i+2:n}} f(x_{i+1:n}) \cdots f(x_{n:n})\, dx_{i+1:n} \cdots dx_{n:n} \\ &= \frac{n!}{(n-i)!} \{1 - F(x_{i:n})\}^{n-i} f(x_{1:n}) f(x_{2:n}) \cdots f(x_{i:n}), \\ &\qquad -\infty < x_{1:n} < x_{2:n} < \cdots < x_{i:n} < \infty. \end{aligned} \tag{2.4.19}$$

Proceeding similarly, we obtain from (2.2.1) the joint density of $X_{1:n}, X_{2:n}, \ldots, X_{i:n}, X_{j:n}$ $(1 \le i < j \le n)$ to be

$$\begin{aligned} f_{1,\ldots,i,j:n}&(x_{1:n}, \ldots, x_{i:n}, x_{j:n}) \\ &= n!\, f(x_{1:n}) f(x_{2:n}) \cdots f(x_{i:n}) f(x_{j:n}) \\ &\quad \times \int_{x_{i:n}}^{x_{j:n}} \cdots \int_{x_{i:n}}^{x_{i+2:n}} f(x_{i+1:n}) \cdots f(x_{j-1:n})\, dx_{i+1:n} \cdots dx_{j-1:n} \\ &\quad \times \int_{x_{j:n}}^{\infty} \cdots \int_{x_{j:n}}^{x_{j+2:n}} f(x_{j+1:n}) \cdots f(x_{n:n})\, dx_{j+1:n} \cdots dx_{n:n} \\ &= \frac{n!}{(j-i-1)!(n-j)!} \{F(x_{j:n}) - F(x_{i:n})\}^{j-i-1} \{1 - F(x_{j:n})\}^{n-j} \\ &\quad \times f(x_{1:n}) f(x_{2:n}) \cdots f(x_{i:n}) f(x_{j:n}), \\ &\qquad -\infty < x_{1:n} < \cdots < x_{i:n} < x_{j:n} < \infty. \end{aligned} \tag{2.4.20}$$

From (2.4.19) and (2.4.20) we derive the conditional density of $X_{j:n}$, given $X_{1:n} = x_{1:n}$, $X_{2:n} = x_{2:n}, \ldots, X_{i:n} = x_{i:n}$, to be

$$\begin{aligned} f_{j:n}&(x_{j:n} \mid x_{1:n}, x_{2:n}, \ldots, x_{i:n}) \\ &= \frac{(n-i)!}{(j-i-1)!(n-j)!} \left\{ \frac{F(x_{j:n}) - F(x_{i:n})}{1 - F(x_{i:n})} \right\}^{j-i-1} \left\{ \frac{1 - F(x_{j:n})}{1 - F(x_{i:n})} \right\}^{n-j} \\ &\quad \times \frac{f(x_{j:n})}{1 - F(x_{i:n})}, \qquad x_{i:n} \le x_{j:n} < \infty. \end{aligned} \tag{2.4.21}$$

We realize that this is the same as the conditional density of $X_{j:n}$, given $X_{i:n} = x_{i:n}$, derived in (2.4.17). Thus, we observe that the order statistics in a sample from an absolutely continuous population form a Markov chain.

2.5. Distribution of Range and Some Other Statistics

From (2.3.5) we have the joint density function of $X_{1:n}$ and $X_{n:n}$ to be

$$f_{1,n:n}(x_{1:n}, x_{n:n}) = n(n-1)\{F(x_{n:n}) - F(x_{1:n})\}^{n-2} f(x_{1:n}) f(x_{n:n}), \quad -\infty < x_{1:n} < x_{n:n} < \infty. \tag{2.5.1}$$

Now denoting $W = X_{n:n} - X_{1:n}$ for the sample range, we obtain from (2.5.1) the joint density function of $X_{1:n}$ and W to be

$$g(x_{1:n}, w) = n(n-1)\{F(x_{1:n}+w) - F(x_{1:n})\}^{n-2} f(x_{1:n}) f(x_{1:n}+w), \quad -\infty < x_{1:n} < \infty, \quad 0 < w < \infty. \tag{2.5.2}$$

By integrating out $x_{1:n}$ in (2.5.2), we derive the density function of the sample range W as

$$g_W(w) = n(n-1)\int_{-\infty}^{\infty} \{F(x_{1:n}+w) - F(x_{1:n})\}^{n-2} f(x_{1:n}) f(x_{1:n}+w)\, dx_{1:n}, \quad 0 < w < \infty. \tag{2.5.3}$$

The integration to be performed in (2.5.3) does not assume a manageable form in most cases. Yet, the cdf of the range W takes on a simpler form. From (2.5.3) we immediately have the cdf of W as

$$G_W(w_0) = \Pr(W \le w_0) = n(n-1)\int_0^{w_0}\int_{-\infty}^{\infty} \{F(x_{1:n}+w) - F(x_{1:n})\}^{n-2} \times f(x_{1:n}) f(x_{1:n}+w)\, dx_{1:n}\, dw, \quad 0 \le w_0 < \infty. \tag{2.5.4}$$

By applying Fubini's theorem and changing the order of integration, we derive the cdf of W as

$$\begin{aligned} G_W(w_0) &= n\int_{-\infty}^{\infty} f(x_{1:n}) \left[(n-1)\int_0^{w_0} \{F(x_{1:n}+w) - F(x_{1:n})\}^{n-2} \times f(x_{1:n}+w)\, dw \right] dx_{1:n}, \quad 0 \le w_0 < \infty. \\ &= n\int_{-\infty}^{\infty} \{F(x_{1:n}+w_0) - F(x_{1:n})\}^{n-1} f(x_{1:n})\, dx_{1:n}, \quad 0 \le w_0 < \infty. \end{aligned} \tag{2.5.5}$$

Some similar expressions have been derived for the discrete case by Abdel-Aty (1954), Burr (1955), and Siotani (1957).

In deriving (2.5.3) we have assumed that the population distribution has an infinite support. In the case in which the population distribution has a finite support, however, we have to fix the limits of integration carefully. For example, in the case of the uniform (0, 1) distribution, by realizing that $f(x_{1:n}+w)\equiv 0$ whenever $x_{1:n}+w>1$, we obtain the density function of W from (2.5.3) to be

$$\begin{aligned} g_W(w) &= n(n-1)\int_0^{1-w} w^{n-2}\,dx_{1:n} \\ &= n(n-1)w^{n-2}(1-w), \qquad 0<w<1, \end{aligned} \tag{2.5.6}$$

and the cdf of W from (2.5.6) to be

$$\begin{aligned} G_W(w_0) &= \int_0^{w_0} n(n-1)(w^{n-2}-w^{n-1})\,dw \\ &= nw_0^{n-1}-(n-1)w_0^n, \qquad 0<w_0<1. \end{aligned} \tag{2.5.7}$$

Also, the formula for the cdf of W in (2.5.5) yields in this case

$$\begin{aligned} G_W(w_0) &= n\int_0^1 \{F(x_{1:n}+w_0)-F(x_{1:n})\}^{n-1} f(x_{1:n})\,dx_{1:n} \\ &= n\int_0^{1-w_0} w_0^{n-1}\,dx_{1:n} + n\int_{1-w_0}^1 (1-x_{1:n})^{n-1}\,dx_{1:n} \\ &= nw_0^{n-1}(1-w_0)+w_0^n, \qquad 0\le w_0\le 1, \end{aligned}$$

which is exactly the same as the expression derived in (2.5.7).

Next, for the case of the standard exponential distribution with pdf $f(x)=e^{-x}$, $0\le x<\infty$, and cdf $F(x)=1-e^{-x}$, $0\le x<\infty$, we obtain the density function of the sample range W from (2.5.3) to be

$$\begin{aligned} g_W(w) &= n(n-1)\int_0^\infty \{e^{-x_{1:n}}-e^{-x_{1:n}-w}\}^{n-2}\, e^{-x_{1:n}}\, e^{-x_{1:n}-w}\,dx_{1:n} \\ &= n(n-1)\, e^{-w}(1-e^{-w})^{n-2}\int_0^\infty e^{-nx_{1:n}}\,dx_{1:n} \\ &= (n-1)(1-e^{-w})^{n-2}\, e^{-w}, \qquad 0\le w<\infty. \end{aligned} \tag{2.5.8}$$

The cdf of W may be obtained from (2.5.8) to be

$$\begin{aligned} G_W(w_0) &= \int_0^{w_0} (n-1)(1-e^{-w})^{n-2}\, e^{-w}\,dw \\ &= (1-e^{-w_0})^{n-1}, \qquad 0\le w_0<\infty. \end{aligned} \tag{2.5.9}$$

This expression could have been derived directly from (2.5.5) as well.

The result on the sample range presented in the beginning of this section may be generalized for the subrange $W_{i,j} = X_{j:n} - X_{i:n}$, $1 \le i < j \le n$. From the joint density function of $X_{i:n}$ and $X_{j:n}$ given in (2.3.5), we find the joint density function of $X_{i:n}$ and $W_{i,j}$ to be

$$g(x_{i:n}, w_{i,j}) = \frac{n!}{(i-1)!(j-i-1)!(n-j)!} \{F(x_{i:n})\}^{i-1} \{F(x_{i:n} + w_{i,j}) - F(x_{i:n})\}^{j-i-1} \{1 - F(x_{i:n} + w_{i,j})\}^{n-j} f(x_{i:n}) f(x_{i:n} + w_{i,j}),$$

$$-\infty < x_{i:n} < \infty, \quad 0 < w_{i,j} < \infty. \tag{2.5.10}$$

By integrating out $x_{i:n}$ from the above equation, we derive the density function of the subrange $W_{i,j}$ to be

$$g_{W_{i,j}}(w_{i,j}) = \frac{n!}{(i-1)!(j-i-1)!(n-j)!} \int_{-\infty}^{\infty} \{F(x_{i:n})\}^{i-1} \{F(x_{i:n} + w_{i,j}) - F(x_{i:n})\}^{j-i-1} \times \{1 - F(x_{i:n} + w_{i,j})\}^{n-j} f(x_{i:n}) f(x_{i:n} + w_{ij})\, dx_{i:n}, \qquad 0 < w_{i,j} < \infty. \tag{2.5.11}$$

As before, for the uniform (0, 1) distribution, we obtain the density function of $W_{i,j}$ from (2.5.11) to be

$$g_{W_{i,j}}(w_{i,j}) = \frac{n!}{(i-1)!(j-i-1)!(n-j)!} \int_0^{1-w_{i,j}} x_{i:n}^{i-1} w_{i,j}^{j-i-1} \times (1 - w_{i,j} - x_{i:n})^{n-j}\, dx_{i:n}$$

$$= \frac{n!}{(j-i-1)!(n-j+i)!} w_{i,j}^{j-i-1} (1 - w_{i,j})^{n-j+i}, \qquad 0 < w_{i,j} < 1. \tag{2.5.12}$$

Thus, $W_{i,j}$ has a beta $(j-i, n-j+i+1)$ distribution, and hence depends only on $j-i$ and not on j and i individually. It is also of interest to note here that, for the uniform (0, 1) distribution, the subrange $W_{i,j}$ is distributed exactly as the $(j-i)$th order statistic in a sample of size n from the uniform (0, 1) distribution. Some more interesting distributional results of this sort will be presented in the next chapter.

Furthermore, for the standard exponential distribution, we obtain from Eq. (2.5.11) the density function of $W_{i,j}$ to be

$$\begin{aligned} g_{W_{i,j}}(w_{i,j}) &= \frac{n!}{(i-1)!(j-i-1)!(n-j)!} \\ &\quad \times \int_0^\infty (1-e^{-x_{i:n}})^{i-1}(e^{-x_{i:n}} - e^{-x_{i:n}-w})^{j-i-1} \\ &\quad \times (e^{-x_{i:n}-w})^{n-j}\, e^{-x_{i:n}}\, e^{-x_{i:n}-w}\, dx_{i:n} \\ &= \frac{(n-i)!}{(j-i-1)!(n-j)!} \\ &\quad \times (1-e^{-w})^{j-i-1}(e^{-w})^{n-j+1}, \qquad 0 < w < \infty. \qquad (2.5.13) \end{aligned}$$

From (2.5.13) we observe a very interesting result that, for the standard exponential distribution, the subrange $W_{i,j}$ is distributed exactly as the $(j-i)$th order statistic in a sample of size $n-i$ from the standard exponential distribution. Some more interesting distributional properties of exponential order statistics of this nature will be presented in the next chapter.

Chapter 3 Moments and Other Expected Values

3.1. Introduction

In the last chapter we gave the derivation of the distribution of an order statistic, the joint distribution of two order statistics, and also the distributions of some systematic statistics, such as the sample range, the sample quasi-range, etc. By making use of these results, we first present in this chapter some fundamental formulas that are necessary for the computation of single and product moments of order statistics. Next, we present some recurrence relations and identities satisfied by these moments of order statistics. Further, we consider some specific distributions such as the uniform, exponential, logistic, Weibull, gamma, normal, and half logistic, and discuss the evaluation of moments of order statistics in these cases. We also present some interesting distributional results involving order statistics from the uniform, exponential, and normal populations. Some recurrence relations for the single and product moments of order statistics are given for the exponential, logistic, gamma, normal, and half logistic cases and their uses are illustrated. Finally, we explain the David-Johnson (1954) approximation for the moments of order statistics from an arbitrary continuous distribution. This method of approximation, based on the probability integral transformation and the explicit expressions of the moments of uniform order statistics, provides a simple and practical method of approximating the single and the product moments of order statistics for large sample sizes. A detailed account of this and some other methods of approximation for the moments of order statistics may be found in the recent monograph on this topic by Arnold and Balakrishnan (1989).

3.2. Some Basic Formulas

Let $X_1, X_2, \ldots, X_n$ be a random sample from a population with pdf $f(x)$ and cdf $F(x)$, and let $X_{1:n} \le X_{2:n} \le \cdots \le X_{n:n}$ be the order statistics obtained from the above sample. Let us denote the single moment $E(X_{i:n}^k)$ by $\alpha_{i:n}^{(k)}$. Then, from the density function of $X_{i:n}$ in (2.4.4), we have

$$\alpha_{i:n}^{(k)} = E(X_{i:n}^k) = \int_{-\infty}^{\infty} x^k f_{i:n}(x)\,dx$$
$$= \frac{n!}{(i-1)!(n-i)!} \int_{-\infty}^{\infty} x^k \{F(x)\}^{i-1}\{1-F(x)\}^{n-i} f(x)\,dx,$$
$$i = 1, 2, \ldots, n,\ k \ge 1. \tag{3.2.1}$$

For convenience, let us denote $\alpha_{i:n}^{(1)}$ simply by $\alpha_{i:n}$. Then from the first two single moments of $X_{i:n}$, we may compute the variance as

$$\beta_{i,i:n} = \text{Var}(X_{i:n}) = \alpha_{i:n}^{(2)} - \alpha_{i:n}^2, \qquad 1 \le i \le n. \tag{3.2.2}$$

Next, let us denote the product moment $E(X_{i:n}X_{j:n})$ by $\alpha_{i,j:n}$. Then, from the joint density function of $X_{i:n}$ and $X_{j:n}$ in (2.3.5), we have

$$\alpha_{i,j:n} = E(X_{i:n}X_{j:n})$$
$$= \iint_{-\infty<x<y<\infty} xy f_{i,j:n}(x, y)\,dy\,dx$$
$$= \frac{n!}{(i-1)!(j-i-1)!(n-j)!} \iint_{-\infty<x<y<\infty} xy\{F(x)\}^{i-1}\{F(y)-F(x)\}^{j-i-1}$$
$$\times\{1-F(y)\}^{n-j} f(x) f(y)\,dy\,dx, \qquad 1 \le i < j \le n. \tag{3.2.3}$$

For convenience, we may also sometimes denote $\alpha_{i:n}^{(2)}$ by $\alpha_{i,i:n}$. From the first two single moments and the above product moments, we may compute the covariance of $X_{i:n}$ and $X_{j:n}$ as

$$\beta_{i,j:n} = \text{Cov}(X_{i:n}, X_{j:n}) = \alpha_{i,j:n} - \alpha_{i:n}\alpha_{j:n}, \qquad 1 \le i < j \le n. \tag{3.2.4}$$

The formulas (3.2.1) and (3.2.3) will enable us to derive exact and explicit expressions for the single and product moments of order statistics, respectively, in most cases. When such an explicit algebraic integration is not possible, the formulas (3.2.1) and (3.2.3) may be used for purposes of numerical integration. For example, one may refer to Tietjen *et al.* (1977) for the case of the normal distribution.

3.3. Recurrence Relations and Identities

In this section we present some basic recurrence relations and identities satisfied by the single and the product moments, and also covariances of order statistics. All the results presented in this section for moments of order statistics from an arbitrary continuous distribution and many others have been listed and analyzed by Malik *et al.* (1988) in their recent expository review article on this topic, and similar results for some specific continuous distributions have been reviewed by Balakrishnan *et al.* (1988). Interested readers may also refer to Arnold and Balakrishnan (1989) for a compendium of all these results.

By using the identities

$$\sum_{i=1}^{n} X_{i:n}^{k} = \sum_{i=1}^{n} X_{i}^{k} \tag{3.3.1}$$

and

$$\sum_{i=1}^{n} \sum_{j=1}^{n} X_{i:n} X_{j:n} = \sum_{i=1}^{n} \sum_{j=1}^{n} X_i X_j \tag{3.3.2}$$

and taking expectation on both sides, we derive the identities

$$\sum_{i=1}^{n} \alpha_{i:n}^{(k)} = n\alpha_{1:1}^{(k)} = nE(X^k) \tag{3.3.3}$$

and

$$\sum_{i=1}^{n} \sum_{j=1}^{n} \alpha_{i,j:n} = n\alpha_{1:1}^{(2)} + n(n-1)\alpha_{1:1}^{2}. \tag{3.3.4}$$

In particular, for $k = 1$ and 2, (3.3.3) yields the identities

$$\sum_{i=1}^{n} \alpha_{i:n} = n\alpha_{1:1} = nE(X) \tag{3.3.5}$$

and

$$\sum_{i=1}^{n} \alpha_{i:n}^{(2)} = n\alpha_{1:1}^{(2)} = nE(X^2); \tag{3.3.6}$$

see also Hoeffding (1953). Upon using (3.3.6) in (3.3.4), we simply derive the identity

$$\sum_{i=1}^{n}\sum_{j=i+1}^{n} \alpha_{i,j:n} = \binom{n}{2}\alpha_{1:1}^{2}. \tag{3.3.7}$$

Now, by making use of the fact that

$$\alpha_{1,2:2} = E(X_{1:2}X_{2:2}) = E(X_1X_2) = \alpha_{1:1}^{2}, \tag{3.3.8}$$

we may also rewrite the identity in (3.3.7) as (Govindarajulu, 1963)

$$\sum_{i=1}^{n-1}\sum_{j=i+1}^{n} \alpha_{i,j:n} = \binom{n}{2}\alpha_{1,2:2}. \tag{3.3.9}$$

From (3.3.4) and (3.3.5), we also obtain the identity

$$\begin{aligned}\sum_{i=1}^{n}\sum_{j=1}^{n} \beta_{i,j:n} &= \sum_{i=1}^{n}\sum_{j=1}^{n} \alpha_{i,j:n} - \left(\sum_{i=1}^{n}\alpha_{i:n}\right)\left(\sum_{j=1}^{n}\alpha_{j:n}\right)\\ &= n(\alpha_{1:1}^{(2)} - \alpha_{1:1}^{2})\\ &= n\beta_{1,1:1} = n\ \mathrm{Var}(X).\end{aligned} \tag{3.3.10}$$

The identities in (3.3.3), (3.3.4), and (3.3.10) are very simple to use and, hence, can be applied effectively to check the accuracy of the computation of the single moments, product moments, and covariances of order statistics, respectively.

In addition to the above identities, the following recurrence relations are also satisfied by these quantities.

Relation 3.3.1. *For $i = 1, 2, \ldots, n-1$ and $k \geq 1$,*

$$i\alpha_{i+1:n}^{(k)} + (n-i)\alpha_{i:n}^{(k)} = n\alpha_{i:n-1}^{(k)}. \tag{3.3.11}$$

Proof. From Eq. (3.2.1) we have for $i = 1, 2, \ldots, n-1$

$$\begin{aligned}n\alpha_{i:n-1}^{(k)} &= \frac{n!}{(i-1)!(n-i-1)!}\int_{-\infty}^{\infty} x^k\{F(x)\}^{i-1}\{1-F(x)\}^{n-i-1}f(x)\,dx\\ &= \frac{n!}{(i-1)!(n-i-1)!}\int_{-\infty}^{\infty} x^k\{F(x)\}^{i-1}\{1-F(x)\}^{n-i-1}\\ &\quad\times[F(x)+1-F(x)]f(x)\,dx\\ &= \frac{n!}{(i-1)!(n-i-1)!}\left[\int_{-\infty}^{\infty} x^k\{F(x)\}^{i}\{1-F(x)\}^{n-i-1}f(x)\,dx\right.\\ &\quad\left.+\int_{-\infty}^{\infty} x^k\{F(x)\}^{i-1}\{1-F(x)\}^{n-i}f(x)\,dx\right]\\ &= i\alpha_{i+1:n}^{(k)} + (n-i)\alpha_{i:n}^{(k)}.\end{aligned}$$

It may be noted that Relation 3.3.1 just requires the value of the kth moment of a single order statistic in a sample of size n for the evaluation of the kth moment of all n order statistics, assuming that these moments in samples of size less than n are known. This relation was first derived by Cole (1951) and was proved for the case of discrete populations by Abdel-Aty (1954), Melnick (1964), and Balakrishnan (1986). Arnold and Meeden (1975) called Relation 3.3.1 "the triangle rule" and discussed some applications of it in the area of characterization of distributions. While Arnold (1977) has given extensions of this relation, Balakrishnan (1988b) has recently generalized it to the case in which the order statistics arise from n independent and nonidentically distributed random variables.

For even values of n, say $n=2m$, by setting $i=m$ in Relation 3.3.1 we immediately obtain the relation

$$\tfrac{1}{2}\{\alpha_{m+1:2m}^{(k)}+\alpha_{m:2m}^{(k)}\}=\alpha_{m:2m-1}^{(k)}, \tag{3.3.12}$$

which, for the case $k=1$, simply implies that the expected value of the median in a sample of even size $(2m)$ is exactly equal to the expected value of the median in a sample of odd size $(2m-1)$.

Relation 3.3.2. *For $i=1,2,\ldots,n-1$ and $k\geq 1$,*

$$\alpha_{i:n}^{(k)}=\sum_{j=i}^{n}(-1)^{j-i}\binom{n}{j}\binom{j-1}{i-1}\alpha_{j:j}^{(k)}. \tag{3.3.13}$$

Proof. By considering the expression of $\alpha_{i:n}^{(k)}$ in (3.2.1) and expanding the term $\{1-F(x)\}^{n-i}$ in the integrand binomially in powers of $F(x)$, we get

$$\alpha_{i:n}^{(k)}=\frac{n!}{(i-1)!(n-i)!}\sum_{j=0}^{n-i}(-1)^{j}\binom{n-i}{j}\int_{-\infty}^{\infty}x^{k}\{F(x)\}^{i+j-1}f(x)\,dx,$$

which, when simplified, yields the relation in (3.3.13).

Relation 3.3.3. *For $i=2,3,\ldots,n$ and $k\geq 1$,*

$$\alpha_{i:n}^{(k)}=\sum_{j=n-i+1}^{n}(-1)^{j-n+i-1}\binom{n}{j}\binom{j-1}{n-i}\alpha_{1:j}^{(k)}. \tag{3.3.14}$$

Proof. By considering the expression of $\alpha_{i:n}^{(k)}$ in (3.2.1) and writing the term $\{F(x)\}^{i-1}$ in the integrand as $[1-\{1-F(x)\}]^{i-1}$, and then expanding it binomially in powers of $1-F(x)$, we obtain

$$\alpha_{i:n}^{(k)}=\frac{n!}{(i-1)!(n-i)!}\sum_{j=0}^{i-1}(-1)^{j}\binom{i-1}{j}\int_{-\infty}^{\infty}x^{k}\{1-F(x)\}^{n-i+j}f(x)\,dx,$$

which, when simplified, yields the relation in (3.3.14).

Relations 3.3.2 and 3.3.3 have been derived by Srikantan (1962) and Govindarajulu (1963), and both are quite useful as they express the kth moment of the ith order statistic in a sample of size n in terms of the kth moment of the largest and the smallest order statistics in samples of size n and less, respectively. These relations could have also been obtained by repeated application of Relation 3.3.1. Hence, as one would expect, we note from Relations 3.3.2 and 3.3.3 that we just require the value of the kth moment of a single order statistic (either the largest or the smallest) in a sample of size n for the evaluation of the kth moment of all n order statistics, given these moments in samples of size less than n. One needs to be somewhat careful, however, in applying Relations 3.3.2 and 3.3.3, as increasing values of n result in large combinatorial terms and hence in an error of large magnitude; Srikantan (1962) has investigated the cumulative rounding error propagated by the use of these two relations. Recently, Balakrishnan (1988b) has generalized Relations 3.3.2 and 3.3.3 to the case in which the order statistics arise from n independent and nonidentically distributed random variables.

In the case of certain specific distributions such as the gamma, Weibull, extreme value, and geometric, by noting that the moments $\alpha_{n:n}^{(k)}$ or $\alpha_{1:n}^{(k)}$ may be derived exactly, some authors have employed one of Relations 3.3.1-3.3.3 for computing the single moments of all the remaining order statistics; see, for example, Kimball (1947), Lieblein (1955), Gupta (1960), Margolin and Winokur (1967), and Saleh *et al.* (1975). These computations have often been checked by the identity in (3.3.3). This is not meaningful, since Balakrishnan and Malik (1986) have shown that the identity in (3.3.3) will be automatically satisfied if one of Relations 3.3.1-3.3.3 is used in the computational procedure. Let us demonstrate this point by considering Relation 3.3.2. By setting $i=1, i=2, \ldots, i=n-1$ in Relation 3.3.2 and then adding the resulting $n-1$ equations, we obtain

$$\sum_{i=1}^{n-1} \alpha_{i:n}^{(k)} = \binom{n}{1} \alpha_{1:1}^{(k)} + \sum_{r=2}^{n-1} (-1)^{r-1} \binom{n}{r} \sum_{j=0}^{r-1} (-1)^j \binom{r-1}{j} \alpha_{r:r}^{(k)}$$
$$+(-1)^{n-1} \sum_{r=0}^{n-2} (-1)^r \binom{n-1}{r} \alpha_{n:n}^{(k)}. \tag{3.3.15}$$

By making use of the combinatorial identities

$$\sum_{j=0}^{r-1} (-1)^j \binom{r-1}{j} = 0 \quad \text{and} \quad \sum_{r=0}^{n-2} (-1)^r \binom{n-1}{r} = (-1)^n$$

in (3.3.15), we get

$$\sum_{i=1}^{n-1} \alpha_{i:n}^{(k)} = n\alpha_{1:1}^{(k)} - \alpha_{n:n}^{(k)},$$

which simply reveals that the identity in (3.3.3) will be satisfied automatically. In other words, when we use the identity in (3.3.3) as a check, it will not discover any error in the starting calculations, i.e., in the value of $\alpha_{i:i}^{(k)}$ or $\alpha_{1:i}^{(k)}$ for $i=1,2,\ldots,n$. However, as rightly mentioned by Balakrishnan and Malik (1986), the identity in (3.3.3) is so simple to use that one might still apply it to check, at least partially, whether the other single moments have been computed correctly from the values of $\alpha_{i:i}^{(k)}$ or $\alpha_{1:i}^{(k)}$.

Relation 3.3.4. *For* $2 \le i < j \le n$,

$$(i-1)\alpha_{i,j:n} + (j-i)\alpha_{i-1,j:n} + (n-j+1)\alpha_{i-1,j-1:n} = n\alpha_{i-1,j-1:n-1}. \tag{3.3.16}$$

Proof. From Eq. (3.2.3) we have for $2 \le i < j \le n$

$$n\alpha_{i-1,j-1:n-1}$$

$$= \frac{n!}{(i-2)!(j-i-1)!(n-j)!} \iint\limits_{-\infty<x<y<\infty} xy\{F(x)\}^{i-2}\{F(y)-F(x)\}^{j-i-1} \times\{1-F(y)\}^{n-j} f(x)f(y)\,dy\,dx$$

$$= \frac{n!}{(i-2)!(j-i-1)!(n-j)!} \iint\limits_{-\infty<x<y<\infty} xy\{F(x)\}^{i-2}\{F(y)-F(x)\}^{j-i-1} \times\{1-F(y)\}^{n-j}[F(x)+\{F(y)-F(x)\}+\{1-F(y)\}] f(x)f(y)\,dy\,dx$$

$$= \frac{n!}{(i-2)!(j-i-1)!(n-j)!}\Bigg[\iint\limits_{-\infty<x<y<\infty} xy\{F(x)\}^{i-1} \times\{F(y)-F(x)\}^{j-i-1}\{1-F(y)\}^{n-j} f(x)f(y)\,dy\,dx$$

$$+ \iint\limits_{-\infty<x<y<\infty} xy\{F(x)\}^{i-2}\{F(y)-F(x)\}^{j-i} \times\{1-F(y)\}^{n-j} f(x)f(y)\,dy\,dx$$

$$+ \iint\limits_{-\infty<x<y<\infty} xy\{F(x)\}^{i-2}\{F(y)-F(x)\}^{j-i-1} \times\{1-F(y)\}^{n-j+1} f(x)f(y)\,dy\,dx\Bigg]$$

$$= (i-1)\alpha_{i,j:n} + (j-i)\alpha_{i-1,j:n} + (n-j+1)\alpha_{i-1,j-1:n}.$$

This relation was originally pointed out for the case of normal order statistics by Teichroew (1956) and was proved in general by Govindarajulu (1963). It may be noted that Relation 3.3.4 will enable us to compute all the product moments $\alpha_{i,j:n}$ $(1 \le i < j \le n)$ if we know $n-1$ suitably chosen moments; for example, the knowledge of the $n-1$ immediate upper-diagonal product moments, viz., $\alpha_{i,i+1:n}$ $(1 \le i \le n-1)$, will suffice. This important application of Relation 3.3.4 has been realized and utilized extensively for reducing the amount of numerical computation by several authors including Gupta (1960), Shah (1966), Saleh *et al.* (1975), Joshi (1982), Balakrishnan and Joshi (1984), Balakrishnan (1985), Balakrishnan and Puthenpura (1986), Ragab and Green (1984), and Balakrishnan *et al.* (1987). Relation 3.3.4 has been generalized by Balakrishnan (1988b) to the case in which the order statistics arise from n independent and nonidentically distributed random variables.

It is important to mention here that Balakrishnan and Malik (1986) have shown that the identity in (3.3.7) will be automatically satisfied if Relation 3.3.4 is used in the computational procedure. To see this, by setting $n=3$, $i=2$, and $j=3$ in (3.3.16) and then using (3.3.8), we obtain

$$\alpha_{1,2:3} + \alpha_{1,3:3} + \alpha_{2,3:3} = 3\alpha_{1,2:2} = 3\alpha_{1:1}^2,$$

which is precisely the identity in (3.3.7) for $n=3$. For a general n, by setting $i = 2, 3, \ldots, n-1$ and $j = i+1, i+2, \ldots, n$, and then adding the resulting $(n-2)(n-1)/2$ equations, we derive

$$(n-2) \sum_{i=1}^{n-1} \sum_{j=i+1}^{n} \alpha_{i,j:n} = n \sum_{i=1}^{n-2} \sum_{j=i+1}^{n-1} \alpha_{i,j:n-1}, \tag{3.3.17}$$

which shows immediately that the identity in (3.3.7) will be automatically satisfied.

Relation 3.3.5. *For* $2 \le i < j \le n$,

$$(i-1)\beta_{i,j:n} + (j-i)\beta_{i-1,j:n} + (n-j+1)\beta_{i-1,j-1:n} = n\{\beta_{i-1,j-1:n-1} + (\mu_{i-1:n-1} - \mu_{i-1:n})(\mu_{j-1:n-1} - \mu_{j:n})\}. \tag{3.3.18}$$

Proof. From Relation 3.3.4 we have for $2 \le i < j \le n$

$$(i-1)\beta_{i,j:n} + (j-i)\beta_{i-1,j:n} + (n-j+1)\beta_{i-1,j-1:n} = n\beta_{i-1,j-1:n-1} + n\alpha_{i-1:n-1}\alpha_{j-1:n-1} - (i-1)\alpha_{i:n}\alpha_{j:n} - (j-i)\alpha_{i-1:n}\alpha_{j:n} - (n-j+1)\alpha_{i-1:n}\alpha_{j-1:n}. \tag{3.3.19}$$

Now consider

$$\begin{aligned}(i-1)\alpha_{i:n}\alpha_{j:n}&+(j-i)\alpha_{i-1:n}\alpha_{j:n}+(n-j+1)\alpha_{i-1:n}\alpha_{j-1:n}\\ &=\alpha_{j:n}\{(i-1)\alpha_{i:n}+(n-i+1)\alpha_{i-1:n}\}+(n-j+1)\alpha_{i-1:n}(\alpha_{j-1:n}-\alpha_{j:n})\\ &=n\alpha_{i-1:n-1}\alpha_{j:n}+n\alpha_{i-1:n}(\alpha_{j-1:n-1}-\alpha_{j:n})\end{aligned}$$

by using (3.3.11). Upon substituting for this in the RHS of (3.3.19) and simplifying the resulting expression, we derive the relation in (3.3.18).

Relation 3.3.5 has been proved recently by Balakrishnan (1989c) and has also been generalized by him to the case in which the order statistics arise from n independent and nonidentically distributed random variables. In addition to the above results, several recurrence relations and identities are available in the literature for these moments of order statistics. A detailed description of all these results may be found in the recent monograph by Arnold and Balakrishnan (1989).

By a systematic application of several recurrence relations satisfied by the single and product moments of order statistics, Joshi and Balakrishnan (1982) have obtained upper bounds for the number of single and double integrals to be evaluated for the calculation of all means, variances, and covariances in a sample of size n, provided these are available in samples of sizes $n-1$ and less. Their result is summarized in the following theorem.

Theorem 3.3.1. *In order to determine the means, variances, and covariances of order statistics in a sample of size n drawn from an arbitrary distribution, given these quantities in samples of sizes $n-1$ and less, one has to evaluate at most two single integrals and $(n-2)/2$ double integrals if n is even, and two single integrals and $(n-1)/2$ double integrals if n is odd.*

For distributions that are symmetric about zero, it may be easily noted from Eq. (2.4.13) that

$$X_{i:n} \stackrel{d}{=} (-X)_{n-i+1:n}, \qquad 1 \le i \le n, \tag{3.3.20}$$

and

$$(X_{i:n}, X_{j:n}) \stackrel{d}{=} ((-X)_{n-j+1:n}, (-X)_{n-i+1:n}), \qquad 1 \le i < j \le n. \tag{3.3.21}$$

Therefore, for distributions which are symmetric about zero, we have the results

$$\alpha_{n-i+1:n}^{(k)} = (-1)^k \alpha_{i:n}^{(k)}, \qquad 1 \le i \le n,\ k \ge 1, \tag{3.3.22}$$

$$\alpha_{n-j+1,n-i+1:n} = \alpha_{i,j:n}, \qquad 1 \le i < j \le n, \tag{3.3.23}$$

and

$$\begin{aligned}\beta_{n-j+1,n-i+1:n} &= \alpha_{n-j+1,n-i+1:n} - \alpha_{n-j+1:n}\alpha_{n-i+1:n} \\ &= \alpha_{i,j:n} - \alpha_{j:n}\alpha_{i:n} \\ &= \beta_{i,j:n}, \qquad 1 \le i \le j \le n. \end{aligned} \tag{3.3.24}$$

The above results help reduce the number of computations to be done for the case in which the parent population distribution is symmetric about zero. These results also help us reduce the bounds given in Theorem 3.3.1 for the number of single and double integrals to be evaluated in order to compute the means, variances, and covariances of order statistics in a sample of size n. This result, due to Joshi (1971), is summarized in the following theorem.

Theorem 3.3.2. *In order to determine the means, variances, and covariances of order statistics in a sample of size n drawn from an arbitrary distribution symmetric about zero, given these quantities for all sample sizes less than n, one has to evaluate at most one single integral if n is even, and one single integral and $(n-1)/2$ double integrals if n is odd.*

The important point to be noted from Theorem 3.3.2 is that for even values of n there is no need to evaluate any double integral. This is quite apparent from a result of Joshi and Balakrishnan (1982), wherein they expressed the product moment $\alpha_{i,j:n}$, for the case in which the population distribution is symmetric about zero and n is even, explicitly in terms of the first single moments and the product moments of order statistics in samples of sizes $n-1$ and less.

3.4. Results for the Uniform Distribution

Let $U_{1:n} \le U_{2:n} \le \cdots \le U_{n:n}$ be the order statistics from a uniform $(0, 1)$ distribution. Then, from (3.2.1), we get for $1 \le i \le n$ and $k \ge 1$

$$\begin{aligned}\alpha_{i:n}^{(k)} &= E(U_{i:n}^k) \\ &= \frac{n!}{(i-1)!(n-i)!} \int_0^1 x^{k+i-1}(1-x)^{n-i}\,dx \\ &= \frac{n!}{(i-1)!} \frac{(k+i-1)!}{(n+k)!} \\ &= \frac{i(i+1)\cdots(i+k-1)}{(n+1)(n+2)\cdots(n+k)}. \end{aligned} \tag{3.4.1}$$

Thus, by setting $p_i = i/(n+1)$ and $q_i = 1 - p_i$, we find from (3.4.1) that

$$\alpha_{i:n} = E(U_{i:n}) = p_i \tag{3.4.2}$$

and

$$\begin{aligned}\beta_{i,i:n} &= \mathrm{Var}(U_{i:n}) \\ &= \frac{i(i+1)}{(n+1)(n+2)} - \frac{i^2}{(n+1)^2} \\ &= p_i q_i/(n+2).\end{aligned} \tag{3.4.3}$$

Similarly, from (3.2.3), we get for $1 \le i < j \le n$ and $k_1, k_2 \ge 1$

$$\begin{aligned}\alpha_{i,j:n}^{(k_1,k_2)} &= E(U_{i:n}^{k_1} U_{j:n}^{k_2}) \\ &= \frac{n!}{(i-1)!(j-i-1)!(n-j)!} \iint_{0<x<y<1} x^{k_1+i-1}(y-x)^{j-i-1} \\ &\quad \times y^{k_2}(1-y)^{n-j}\,dx\,dy \\ &= \frac{n!}{(i-1)!}\,\frac{(k_1+i-1)!}{(k_1+j-1)!}\,\frac{(k_1+k_2+j-1)!}{(n+k_1+k_2)!} \\ &= \frac{i(i+1)\cdots(i+k_1-1)(j+k_1)(j+k_1+1)\cdots(j+k_1+k_2-1)}{(n+1)(n+2)\cdots(n+k_1+k_2)}.\end{aligned} \tag{3.4.4}$$

By setting $k_1 = k_2 = 1$ in (3.4.4), we immediately obtain

$$\alpha_{i,j:n} = \frac{i(j+1)}{(n+1)(n+2)}, \tag{3.4.5}$$

which, when used with (3.4.2), yields

$$\begin{aligned}\beta_{i,j:n} &= \mathrm{Cov}(U_{i:n}, U_{j:n}) \\ &= \frac{i(j+1)}{(n+1)(n+2)} - \frac{ij}{(n+1)^2} \\ &= p_i q_j/(n+2).\end{aligned} \tag{3.4.6}$$

Proceeding similarly, we derive for $1 \le i_1 < i_2 < i_3 < i_4 \le n$

$$\begin{aligned}&E(U_{i_1:n}^{k_1} U_{i_2:n}^{k_2} U_{i_3:n}^{k_3} U_{i_4:n}^{k_4}) \\ &= \frac{n!}{(i_1-1)!}\,\frac{(k_1+i_1-1)!}{(k_1+i_2-1)!}\,\frac{(k_1+k_2+i_2-1)!}{(k_1+k_2+i_3-1)!} \\ &\quad \times \frac{(k_1+k_2+k_3+i_3-1)!}{(k_1+k_2+k_3+i_4-1)!}\,\frac{(k_1+k_2+k_3+k_4+i_4-1)!}{(n+k_1+k_2+k_3+k_4)!}.\end{aligned} \tag{3.4.7}$$

The first four cumulants and cross-cumulants of uniform order statistics derived from (3.4.7) have been used by David and Johnson (1954) in developing some approximations for the first four cumulants and cross-cumulants of order statistics from an arbitrary continuous distribution. This method of approximation is discussed in detail in Section 3.10.

For $1 \le i_1 < i_2 < \cdots < i_\ell \le n$, David and Johnson (1954) have also derived the general expression

$$E\left(\prod_{j=1}^{\ell} U_{i_j:n}^{k_j}\right) = \frac{n!}{\left(n + \sum_{j=1}^{\ell} k_j\right)!} \prod_{j=1}^{\ell} \left\{ \frac{(i_j + k_1 + k_2 + \cdots + k_j - 1)!}{(i_j + k_1 + k_2 + \cdots + k_{j-1} - 1)!} \right\}, \tag{3.4.8}$$

with k_0 taken as 0.

Instead of employing the above direct integration approach to evaluate the single moments and product moments of uniform order statistics, one may use the following theorem and simplify the process considerably.

Theorem 3.4.1. *For the uniform* $(0, 1)$ *distribution, the variables* $V_1 = U_{i:n}/U_{j:n}$ *and* $V_2 = U_{j:n}$, *for* $1 \le i < j \le n$, *are statistically independent, with* V_1 *and* V_2 *having* $\text{beta}(i, j-i)$ *and* $\text{beta}(j, n-j+1)$ *distributions, respectively.*

Proof. From Eq. (2.3.6), we have the joint density function of $U_{i:n}$ and $U_{j:n}$ $(1 \le i < j \le n)$ as

$$f_{i,j:n}(u_{i:n}, u_{j:n}) = \frac{n!}{(i-1)!(j-i-1)!(n-j)!} u_{i:n}^{i-1} (u_{j:n} - u_{i:n})^{j-i-1} (1 - u_{j:n})^{n-j},$$
$$0 < u_{i:n} < u_{j:n} < 1. \tag{3.4.9}$$

By making the transformation

$$V_1 = \frac{U_{i:n}}{U_{j:n}}, \quad V_2 = U_{j:n} \quad \Rightarrow \quad U_{i:n} = V_1 V_2, \qquad U_{j:n} = V_2$$

and realizing that the Jacobian of this transformation is V_2, we obtain the joint density function of V_1 and V_2 as

$$g(v_1, v_2) = \frac{(j-1)!}{(i-1)!(j-i-1)!} v_1^{i-1} (1 - v_1)^{j-i-1} \frac{n!}{(j-1)!(n-j)!} v_2^{j-1} (1 - v_2)^{n-j},$$
$$0 \le v_1, v_2 \le 1. \tag{3.4.10}$$

By a factorization theorem (see Rao, 1973), we observe from (3.4.10) that the variables V_1 and V_2 are statistically independent. Furthermore, we also immediately note that V_1 and V_2 have beta$(i, j-i)$ and beta$(j, n-j+1)$ distributions, respectively.

As mentioned earlier, we may use Theorem 3.4.1 to simplify the evaluation of product moments. For example, we may find

$$\begin{aligned}\alpha_{i,j:n}^{(k_1,k_2)} &= E(V_1^{k_1} V_2^{k_1+k_2})\\ &= E(V_1^{k_1}) E(V_2^{k_1+k_2})\\ &= \frac{(j-1)!(k_1+i-1)!}{(i-1)!(k_1+j-1)!} \frac{n!(k_1+k_2+j-1)!}{(j-1)!(n+k_1+k_2)!},\end{aligned}$$

which is exactly the same as the expression derived in (3.4.4) by direct integration.

Theorem 3.4.1 can, in a straightforward manner, be generalized as follows.

Theorem 3.4.2. *For the uniform* (0, 1) *distribution, the variables*

$$V_1 = \frac{U_{i_1:n}}{U_{i_2:n}}, \; V_2 = \frac{U_{i_2:n}}{U_{i_3:n}}, \ldots, V_{\ell-1} = \frac{U_{i_{\ell-1}:n}}{U_{i_\ell:n}}, \; V_\ell = U_{i_\ell:n}$$

for $1 \le i_1 < i_2 < \cdots < i_\ell \le n$ *are all statistically independent, having* beta$(i_1, i_2 - i_1)$,beta$(i_2, i_3 - i_2), \ldots$, beta$(i_{\ell-1}, i_\ell - i_{\ell-1})$,*and*beta$(i_\ell, n - i_\ell + 1)$ *distributions, respectively.*

Now, by using Theorem 3.4.2 and writing

$$\begin{aligned}E\left(\prod_{j=1}^{\ell} U_{i_j:n}^{k_j}\right) &= E\left(\prod_{j=1}^{\ell} V_j^{k_1+k_2+\cdots+k_j}\right)\\ &= \prod_{j=1}^{\ell} E(V_j^{k_1+k_2+\cdots+k_j}),\end{aligned}$$

we derive the formula given in (3.4.8).

It is quite important to realize here that if $X_{1:n} \le X_{2:n} \le \cdots \le X_{n:n}$ are the order statistics from an arbitrary continuous distribution with cdf $F(x)$, then all the formulas given in this section for the uniform order statistics $U_{i:n}$ continue to hold for the order statistics $F(X_{i:n})$. This is obvious from the fact that (see Section 2.4) $F(X_{i:n}) \stackrel{d}{=} U_{i:n}$ for $i = 1, 2, \ldots, n$. Thus, for example, by combining this fact with the formula in (3.4.1), we simply obtain

$$E\{F(X_{i:n})\}^k = \frac{i(i+1)\cdots(i+k-1)}{(n+1)(n+2)\cdots(n+k)}. \tag{3.4.11}$$

In particular, by setting $i = 1$ in (3.4.11), we get

$$E\{F(X_{1:n})\}^k = k!/\{(n+1)(n+2)\cdots(n+k)\}. \tag{3.4.12}$$

3.5. Results for the Exponential Distribution

Let $X_{1:n} \le X_{2:n} \le \cdots \le X_{n:n}$ be the order statistics from the standard exponential distribution with pdf

$$f(x) = e^{-x}, \qquad 0 \le x < \infty \tag{3.5.1}$$

and cdf

$$F(x) = 1 - e^{-x}, \qquad 0 \le x < \infty. \tag{3.5.2}$$

Then, from (2.2.1) and (3.5.1), we have the joint density function of $X_{i:n}$ $(i = 1, 2, \ldots, n)$ as

$$\begin{aligned} &f_{1,2,\ldots,n:n}(x_{1:n}, x_{2:n}, \ldots, x_{n:n}) \\ &\quad = n!\, e^{-\sum_{i=1}^{n} x_{i:n}}, \qquad 0 \le x_{1:n} < x_{2:n} < \cdots < x_{n:n} < \infty. \end{aligned} \tag{3.5.3}$$

Let us now consider the transformation

$$Y_1 = nX_{1:n},\ Y_2 = (n-1)(X_{2:n} - X_{1:n}), \ldots,\ Y_n = X_{n:n} - X_{n-1:n},$$

or equivalently,

$$X_{1:n} = \frac{Y_1}{n},\ X_{2:n} = \frac{Y_1}{n} + \frac{Y_2}{n-1}, \ldots,\ X_{n:n} = \frac{Y_1}{n} + \frac{Y_2}{n-1} + \cdots + Y_n. \tag{3.5.4}$$

By noting that the Jacobian of the transformation in (3.5.4) is $1/n!$, we derive the joint density function of $Y_1, Y_2, \ldots, Y_n$ from (3.5.3) to be

$$g(y_1, y_2, \ldots, y_n) = e^{-\sum_{i=1}^{n} y_i}, \qquad 0 \le y_1, y_2, \ldots, y_n < \infty. \tag{3.5.5}$$

By using the factorization theorem (Rao, 1973), we immediately observe that the variables $Y_1, Y_2, \ldots, Y_n$ are all statistically independent, and also that they all have standard exponential distributions. This result is due to Sukhatme (1937). (3.5.4) enables us, therefore, to write

$$X_{i:n} = \sum_{\ell=1}^{i} Y_\ell/(n - \ell + 1). \tag{3.5.6}$$

Thus, in (3.5.6), we have expressed the ith order statistic in a sample of size n from the standard exponential distribution in (3.5.1) as a linear function of i independent standard exponential random variables. As a result, we see immediately that the exponential order statistics form an additive Markov chain; for example, see Rényi (1953) or Karlin and Taylor (1975).

From the representation of the ith exponential order statistic given in (3.5.6), we obtain at once

$$\alpha_{i:n} = \sum_{\ell=1}^{i} E(Y_\ell)/(n-\ell+1) = \sum_{\ell=1}^{i} 1/(n-\ell+1), \qquad 1 \le i \le n, \quad (3.5.7)$$

$$\beta_{i,i:n} = \sum_{\ell=1}^{i} \mathrm{Var}(Y_\ell)/(n-\ell+1)^2 = \sum_{\ell=1}^{i} 1/(n-\ell+1)^2, \qquad 1 \le i \le n, \quad (3.5.8)$$

and for $1 \le i < j \le n$

$$\beta_{i,j:n} = \sum_{\ell=1}^{i} \mathrm{Var}(Y_\ell)/(n-\ell+1)^2$$

$$= \sum_{\ell=1}^{i} 1/(n-\ell+1)^2 = \beta_{i,i:n}. \quad (3.5.9)$$

Higher order moments of $X_{i:n}$ may also be derived similarly from (3.5.6).

The following theorem, due to Malmquist (1950), gives a distributional result satisfied by the uniform order statistics by making use of the representation of the exponential order statistic given in (3.5.6).

Theorem 3.5.1. *Let* $U_{1:n} \le U_{2:n} \le \cdots \le U_{n:n}$ *be the order statistics from a uniform* (0, 1) *distribution. Then, with* $U_{n+1:n} \equiv 1$, *the random variables*

$$V_i^* = \left\{ \frac{U_{i:n}}{U_{i+1:n}} \right\}^i, \qquad i = 1, 2, \ldots, n,$$

are all statistically independent, each having a uniform (0, 1) *distribution.*

Proof. With $X_{1:n} \le X_{2:n} \le \cdots \le X_{n:n}$ denoting the exponential order statistics, since the random variable $X = -\ln U$ has a standard exponential distribution with density as in (3.5.1) and also that $-\ln u$ is a monotonically decreasing function in u, we have the relation

$$X_{i:n} = -\ln U_{n-i+1:n}, \qquad 1 \le i \le n. \quad (3.5.10)$$

Hence, we obtain from (3.5.10) that for $1 \le i \le n$

$$V_i^* = \left\{ \frac{U_{i:n}}{U_{i+1:n}} \right\}^i = \left\{ \frac{e^{-X_{n-i+1:n}}}{e^{-X_{n-i:n}}} \right\}^i$$

$$= e^{-i(X_{n-i+1:n} - X_{n-i:n})}$$

$$= e^{-Y_{n-i+1}} \quad (3.5.11)$$

upon using (3.5.6). As a result, we see that the variables V_i^*, $i = 1, 2, \ldots, n$, are all statistically independent. Further, since the Y_i's are standard exponential random variables, we also observe from (3.5.11) that the variables V_i^* all have a uniform (0, 1) distribution. Hence, the theorem.

Several characterization results are available for the exponential distribution based on order statistics. Interested readers may refer to the monograph on this topic by Galambos and Kotz (1978).

In the following two theorems, we present some simple recurrence relations satisfied by the single and the product moments of order statistics from the exponential distribution. As pointed out by Joshi (1978), these recurrence relations are so easy to use that one can write a simple computer program to evaluate the first k single moments and the product moments of all order statistics without introducing serious rounding errors, at least up to moderately large sample sizes. Also, the method of derivation of these relations may be extended to the truncated exponential distributions; see, for example, Joshi (1979a), Joshi (1982), and Balakrishnan and Joshi (1984).

Theorem 3.5.2. *For the standard exponential distribution, we have*

$$\alpha_{1:n}^{(k)} = \frac{k}{n}\,\alpha_{1:n}^{(k-1)}, \qquad n \geq 1,\ k \geq 1 \tag{3.5.12}$$

and

$$\alpha_{i:n}^{(k)} = \alpha_{i-1:n-1}^{(k)} + \frac{k}{n}\,\alpha_{i:n}^{(k-1)}, \qquad 2 \leq i \leq n,\ k \geq 1, \tag{3.5.13}$$

with $\alpha_{i:n}^{(0)} \equiv 1$ *for* $1 \leq i \leq n$.

Proof. Let us consider the expression of $\alpha_{i:n}^{(k-1)}$ in (3.2.1). By noting that $f(x) = 1 - F(x)$ for the standard exponential distribution, we may write for $1 \leq i \leq n$ and $k \geq 1$

$$\alpha_{i:n}^{(k-1)} = \frac{n!}{(i-1)!(n-i)!} \int_0^\infty x^{k-1}\{F(x)\}^{i-1}\{1-F(x)\}^{n-i+1}\,dx. \tag{3.5.14}$$

Integrating the RHS of (3.5.14) by parts, treating x^{k-1} for integration and the rest of the integrand for differentiation, we get for $1 \leq i \leq n$ and $k \geq 1$

$$\alpha_{i:n}^{(k-1)} = \frac{n!}{(i-1)!(n-i)!k}\left[(n-i+1)\int_0^\infty x^k\{F(x)\}^{i-1}\{1-F(x)\}^{n-i}f(x)\,dx \right.$$
$$\left. -(i-1)\int_0^\infty x^k\{F(x)\}^{i-2}\{1-F(x)\}^{n-i+1}f(x)\,dx\right]. \tag{3.5.15}$$

The recurrence relation in (3.5.12) follows readily from (3.5.15) if i is set equal to 1. If we split the first integral on the RHS of (3.5.15) into two and combine with the second integral, (3.5.15) may be rewritten as

$$\alpha_{i:n}^{(k-1)} = \frac{n!}{(i-1)!(n-i)!k}\left[n\int_0^\infty x^k\{F(x)\}^{i-1}\{1-F(x)\}^{n-i}f(x)\,dx\right.$$

$$\left.-(i-1)\int_0^\infty x^k\{F(x)\}^{i-2}\{1-F(x)\}^{n-i}f(x)\,dx\right]. \qquad (3.5.16)$$

Eq. (3.5.16), when simplified, yields the relation in (3.5.13).

Theorem 3.5.3. *For the standard exponential distribution, we have*

$$\alpha_{i,i+1:n} = \alpha_{i:n}^{(2)} + \frac{1}{n-i}\,\alpha_{i:n}, \qquad 1 \le i \le n-1, \qquad (3.5.17)$$

and

$$\alpha_{i,j:n} = \alpha_{i,j-1:n} + \frac{1}{n-j+1}\,\alpha_{i:n}, \qquad 1 \le i < j \le n,\, j-i \ge 2. \qquad (3.5.18)$$

Proof. Let us first consider the expression of $f_{i,j:n}(x, y)$ in (2.3.5). As before, by using the fact that $f(y) = 1 - F(y)$ for the standard exponential distribution, we may write for $1 \le i < j \le n$

$$\alpha_{i:n} = E(X_{i:n}X_{j:n}^0)$$

$$= \frac{n!}{(i-1)!(j-i-1)!(n-j)!}\int_0^\infty x\{F(x)\}^{i-1}I(x)f(x)\,dx, \qquad (3.5.19)$$

where

$$I(x) = \int_x^\infty \{F(y)-F(x)\}^{j-i-1}\{1-F(y)\}^{n-j+1}\,dy. \qquad (3.5.20)$$

Integrating the RHS of (3.5.20) by parts, treating dy for integration and the rest of the integrand for differentiation, we get for $j = i+1$

$$I(x) = (n-i)\int_x^\infty y\{1-F(y)\}^{n-i-1}f(y)\,dy - x\{1-F(x)\}^{n-i}, \qquad (3.5.21)$$

and for $1 \le i < j \le n$ and $j - i \ge 2$

$$I(x) = (n-j+1)\int_x^\infty y\{F(y)-F(x)\}^{j-i-1}\{1-F(y)\}^{n-j}f(y)\,dy$$

$$-(j-i-1)\int_x^\infty y\{F(y)-F(x)\}^{j-i-2}\{1-F(y)\}^{n-j+1}f(y)\,dy. \qquad (3.5.22)$$

By substituting the expressions of $I(x)$ in (3.5.21) and (3.5.22) in Eq. (3.5.19) and simplifying the resulting equation, we derive the recurrence relations given in (3.5.17) and (3.5.18).

It should be pointed out here that the relation in (3.5.17) itself is sufficient for the evaluation of all the product moments of order statistics, because the remaining product moments, viz., $\alpha_{i,j:n}$ for $1 \le i < j \le n$ and $j-i \ge 2$, can all be computed from the product moments $\alpha_{i,i+1:n}$ $(1 \le i \le n-1)$ by using Relation 3.3.4.

3.6. Results for the Logistic Distribution

Let $X_{1:n} \le X_{2:n} \le \cdots \le X_{n:n}$ be the order statistics from a logistic population with pdf

$$f(x) = e^{-x}/(1+e^{-x})^2, \qquad -\infty < x < \infty \tag{3.6.1}$$

and cdf

$$F(x) = 1/(1+e^{-x}), \qquad -\infty < x < \infty. \tag{3.6.2}$$

Then, from the expression of $f_{i:n}(x)$ in (2.4.4) we obtain the moment generating function of $X_{i:n}$ as

$$m_{i:n}(t) = E(e^{tX_{i:n}})$$
$$= \frac{n!}{(i-1)!(n-i)!} \int_{-\infty}^{\infty} e^{tx} \left\{\frac{1}{1+e^{-x}}\right\}^{i-1} \left\{\frac{e^{-x}}{1+e^{-x}}\right\}^{n-i} \frac{e^{-x}}{(1+e^{-x})^2}\, dx. \tag{3.6.3}$$

By making the substitution $u = 1/(1+e^{-x})$ in the integral on the RHS of (3.6.3), we get

$$\begin{aligned} m_{i:n}(t) &= \frac{n!}{(i-1)!(n-i)!} \int_0^1 \left(\frac{u}{1-u}\right)^t u^{i-1}(1-u)^{n-i}\, du \\ &= \frac{n!}{(i-1)!(n-i)!} \int_0^1 u^{i+t-1}(1-u)^{n-i-t}\, du \\ &= \Gamma(i+t)\Gamma(n-i+1-t)/\{\Gamma(i)\Gamma(n-i+1)\}. \end{aligned} \tag{3.6.4}$$

From the moment generating function of $X_{i:n}$ in (3.6.4), we derive

$$\begin{aligned} \alpha_{i:n} &= \frac{d}{dt} m_{i:n}(t)\big|_{t=0} \\ &= \frac{\Gamma'(i)}{\Gamma(i)} - \frac{\Gamma'(n-i+1)}{\Gamma(n-i+1)} \\ &= \psi(i) - \psi(n-i+1), \end{aligned} \tag{3.6.5}$$

where $\psi(z)=d/dz \ln\Gamma(z)=\Gamma'(z)/\Gamma(z)$ is the psi (or digamma) function. Similarly, we derive

$$\begin{aligned}\alpha_{i:n}^{(2)}&=\frac{d^2}{dt^2}m_{i:n}(t)|_{t=0}\\&=\frac{\Gamma''(i)}{\Gamma(i)}-2\frac{\Gamma'(i)}{\Gamma(i)}\frac{\Gamma'(n-i+1)}{\Gamma(n-i+1)}+\frac{\Gamma''(n-i+1)}{\Gamma(n-i+1)}\\&=[\psi'(i)+\{\psi(i)\}^2]-2\psi(i)\psi(n-i+1)+[\psi'(n-i+1)+\{\psi(n-i+1)\}^2]\\&=\psi'(i)+\psi'(n-i+1)+[\psi(i)-\psi(n-i+1)]^2, \qquad (3.6.6)\end{aligned}$$

where $\psi'(z)=d/dz\,\psi(z)=d^2/dz^2 \ln\Gamma(z)$ is the derivative of the psi function and commonly termed as the trigamma function. From (3.6.6) and (3.6.5), we immediately derive the variance of $X_{i:n}$ as

$$\begin{aligned}\beta_{i,i:n}&=\alpha_{i:n}^{(2)}-\alpha_{i:n}^2\\&=\psi'(i)+\psi'(n-i+1). \qquad (3.6.7)\end{aligned}$$

These expressions have been derived by Birnbaum and Dudman (1963), Gupta and Shah (1965), and Tarter and Clark (1965). From the formulas of the mean and the variance of $X_{i:n}$ in (3.6.5) and (3.6.7), respectively, we observe that

$$\alpha_{i:n}=-\alpha_{n-i+1:n} \quad \text{and} \quad \beta_{i,i:n}=\beta_{n-i+1,n-i+1:n}.$$

This is because of the symmetry of the logistic distribution (see Section 3.3).

Proceeding similarly, Gupta and Shah (1965) and Shah (1966) have derived an expression for the product moment of $X_{i:n}$ and $X_{j:n}$ as

$$\begin{aligned}\alpha_{i,j:n}=\alpha_{j:n}^{(2)}+\sum_{r=i}^{j-1}\sum_{s=1}^{r-1}&\left\{(-1)^{r+i}\binom{r-1}{i-1}\binom{n}{r}\binom{j-i-r+s}{s}B(s,n-r+1)\right.\\&\left.\times\alpha_{j+s-r:n+s-r}\right\}+\binom{n}{r}\sum_{r=0}^{j-i-1}\left[(-1)^r\binom{n-i}{r}\frac{1}{i+r}\{-\psi'(n-j+1)\right.\\&\left.+(\psi(n-j+1)-\psi(n-i-r+1))(\psi(j-i-r)-\psi(n-j+1))\}\right]. \qquad (3.6.8)\end{aligned}$$

The digamma and trigamma functions involved in the formulas in (3.6.5)–(3.6.8) have been computed rather extensively; for example, see Abramowitz and Stegun (1965) or Davis (1935). Reference may also be made to Bernardo (1976) and Schneider (1978) for Fortran programs for the computation of these functions. By following an exactly similar approach, Balakrishnan and Leung (1988a) have derived expressions for the means, variances, and covariances of order statistics from a Type I generalized logistic distribution.

In the following two theorems, we present several recurrence relations satisfied by the single and the product moments of logistic order statistics. These relations, established by Shah (1966, 1970), will enable one to compute the first k single moments and the product moments of all order statistics in a very simple recursive way without accumulating serious rounding errors. Recently, Balakrishnan and Malik (1990) have successfully employed this recursive procedure and tabulated the means, variances, and covariances of logistic order statistics for sample sizes up to 50.

Theorem 3.6.1. *For the logistic population with pdf as in* (3.6.1), *we have*

$$\alpha_{i+1:n+1}^{(k)} = \alpha_{i:n}^{(k)} + \frac{k}{i}\,\alpha_{i:n}^{(k-1)}, \qquad 1 \le i \le n,\ k \ge 1, \tag{3.6.9}$$

with $\alpha_{i:n}^{(0)} \equiv 1$ *for* $1 \le i \le n$.

Proof. Let us first consider the expression of $\alpha_{i:n}^{(k-1)}$ in (3.2.1). By realizing that $f(x) = F(x)\{1 - F(x)\}$ for the logistic population, we may write for $1 \le i \le n$ and $k \ge 1$

$$\alpha_{i:n}^{(k-1)} = \frac{n!}{(i-1)!(n-i)!} \int_{-\infty}^{\infty} x^{k-1}\{F(x)\}^{i}\{1 - F(x)\}^{n-i+1}\, dx. \tag{3.6.10}$$

Integrating the RHS of (3.6.10) by parts, treating x^{k-1} for integration and the rest of the integrand for differentiation, we obtain for $1 \le i \le n$ and $k \ge 1$

$$\alpha_{i:n}^{(k-1)} = \frac{n!}{(i-1)!(n-i)!k}\left[(n-i+1)\int_{-\infty}^{\infty} x^{k}\{F(x)\}^{i}\{1-F(x)\}^{n-i} f(x)\, dx \right.$$
$$\left. - i\int_{-\infty}^{\infty} x^{k}\{F(x)\}^{i-1}\{1-F(x)\}^{n-i+1} f(x)\, dx\right]. \tag{3.6.11}$$

If we split the first integral on the RHS of (3.6.11) into two and combine with the second integral, (3.6.11) may be rewritten as

$$\alpha_{i:n}^{(k-1)} = \frac{n!}{(i-1)!(n-i)!k}\left[(n+1)\int_{-\infty}^{\infty} x^{k}\{F(x)\}^{i}\{1-F(x)\}^{n-i} f(x)\, dx \right.$$
$$\left. - i\int_{-\infty}^{\infty} x^{k}\{F(x)\}^{i-1}\{1-F(x)\}^{n-i} f(x)\, dx\right]. \tag{3.6.12}$$

Eq. (3.6.12), when simplified, yields the relation in (3.6.9).

Theorem 3.6.1 will allow one to calculate all the moments $\alpha_{i:n+1}^{(k)}$ $(2 \le i \le n+1)$ starting from $\alpha_{1:1}^{(k)}$ in a very simple recursive manner. Note that because of the symmetry of the logistic distribution about zero, we simply have $\alpha_{1:n+1}^{(k)} = (-1)^k \alpha_{n+1:n+1}^{(k)}$ and hence $\alpha_{1:n+1}^{(k)}$ will also be known. Thus, for example, starting with $\alpha_{1:1} = 0$ and $\alpha_{1:1}^{(2)} = \pi^2/3$, one can employ Theorem 3.6.1 to compute the means and variances of all order statistics.

Theorem 3.6.2. *For the logistic population with pdf as in* (3.6.1), *we have*

$$\alpha_{i,i+1:n+1} = \frac{n+1}{n-i+1}\left[\alpha_{i,i+1:n} - \frac{i}{n+1}\alpha_{i+1:n+1}^{(2)} - \frac{1}{n-i}\alpha_{i:n}\right],$$

$$1 \le i \le n-1, \tag{3.6.13}$$

and

$$\alpha_{i,j:n+1} = \frac{n+1}{n-j+2}\left[\alpha_{i,j:n} - \alpha_{i,j-1:n} + \frac{n-j+2}{n+1}\alpha_{i,j-1:n+1} - \frac{1}{n-j+1}\alpha_{i:n}\right],$$

$$1 \le i < j \le n,\ j-i \ge 2. \tag{3.6.14}$$

Proof. By considering the expression of $f_{i,j:n}(x, y)$ in (2.3.5) and using the fact that $f(y) = F(y)\{1 - F(y)\}$ for the logistic population with pdf as in (3.6.1), we may write for $1 \le i < j \le n$

$$\begin{aligned} \alpha_{i:n} &= E(X_{i:n}X_{j:n}^{0}) \\ &= \frac{n!}{(i-1)!(j-i-1)!(n-j)!}\int_{-\infty}^{\infty} x\{F(x)\}^{i-1} I(x) f(x)\, dx, \end{aligned} \tag{3.6.15}$$

where

$$I(x) = \int_{x}^{\infty} \{F(y) - F(x)\}^{j-i-1}\{1 - F(y)\}^{n-j+1} F(y)\, dy. \tag{3.6.16}$$

By writing $F(y)$ as $1 - \{1 - F(y)\}$ and splitting the integral in (3.6.16) accordingly into two, we get

$$\begin{aligned} I(x) = &\int_{x}^{\infty} \{F(y) - F(x)\}^{j-i-1}\{1 - F(y)\}^{n-j+1}\, dy \\ &- \int_{x}^{\infty} \{F(y) - F(x)\}^{j-i-1}\{1 - F(y)\}^{n-j+2}\, dy. \end{aligned} \tag{3.6.17}$$

Integrating the RHS of (3.6.17) by parts treating dy for integration, we obtain for $j=i+1$

$$I(x)=\left[(n-i)\int_x^\infty y\{1-F(y)\}^{n-i-1}f(y)\,dy-x\{1-F(x)\}^{n-i}\right]$$
$$-\left[(n-i+1)\int_x^\infty y\{1-F(y)\}^{n-i}f(y)\,dy-x\{1-F(x)\}^{n-i+1}\right], \tag{3.6.18}$$

and for $j-i\geq 2$

$$I(x)=\left[(n-j+1)\int_x^\infty y\{F(y)-F(x)\}^{j-i-1}\{1-F(y)\}^{n-j}f(y)\,dy\right.$$
$$\left.-(j-i-1)\int_x^\infty y\{F(y)-F(x)\}^{j-i-2}\{1-F(y)\}^{n-j+1}f(y)\,dy\right]$$
$$-\left[(n-j+2)\int_x^\infty y\{F(y)-F(x)\}^{j-i-1}\{1-F(y)\}^{n-j+1}f(y)\,dy\right.$$
$$\left.-(j-i-1)\int_x^\infty y\{F(y)-F(x)\}^{j-i-2}\{1-F(y)\}^{n-j+2}f(y)\,dy\right]. \tag{3.6.19}$$

Substitution of the expression for $I(x)$ from (3.6.18) in Eq. (3.6.15) gives for $1\leq i\leq n-1$

$$\alpha_{i:n}=(n-i)\left[\alpha_{i,i+1:n}-\alpha_{i:n}^{(2)}+\frac{(n-i+1)}{(n+1)}\alpha_{i:n+1}^{(2)}-\frac{(n-i+1)}{(n+1)}\alpha_{i,i+1:n+1}\right]. \tag{3.6.20}$$

Now, by using the result

$$(n-i+1)\alpha_{i:n+1}^{(2)}-(n+1)\alpha_{i:n}^{(2)}=-i\alpha_{i+1:n+1}^{(2)}$$

obtained from Relation 3.3.1 in Eq. (3.6.20) and simplifying the resulting expression, we derive the recurrence relation in (3.6.13). Similarly, by substituting the expression for $I(x)$ from (3.6.19) in Eq. (3.6.15) and simplifying the resulting expression, we derive the recurrence relation in (3.6.14).

It should be mentioned here that the relation in (3.6.13) itself is sufficient for the evaluation of all the product moments of order statistics. To see this, we first note from (3.3.8) that

$$\alpha_{1,2:2} = \alpha_{1:1}^2 = 0 \tag{3.6.21}$$

and that

$$\alpha_{n,n+1:n+1} = \alpha_{1,2:n+1} \tag{3.6.22}$$

because of the symmetry of the logistic distribution. Thus, the relation in (3.6.13), when used along with (3.6.21) and (3.6.22), will enable one to calculate the product moments $\alpha_{i,i+1:n+1}$ $(1 \le i \le n)$. This is sufficient for the evaluation of all the product moments, because the remaining product moments, viz., $\alpha_{i,j:n+1}$ for $1 \le i < j \le n+1$ and $j-i \ge 2$, can all be systematically computed by using Relation 3.3.4.

By proceeding on exactly similar lines, Balakrishnan and Joshi (1983) and Balakrishnan and Kocherlakota (1986) have derived several recurrence relations satisfied by the single and the product moments of order statistics from the truncated logistic distributions. Explicit finite series expressions for the moments of order statistics have been derived in this case by Tarter (1966).

3.7. Results for the Gamma Distribution

Let $X_{1:n} \le X_{2:n} \le \cdots \le X_{n:n}$ be the order statistics from a gamma population with pdf

$$f(x) = \frac{1}{\Gamma(\rho)} e^{-x} x^{\rho-1}, \qquad 0 \le x < \infty, \rho > 0. \tag{3.7.1}$$

The standard exponential distribution considered in Section 3.5 is observed to be a particular case (when $\rho = 1$) of the gamma distribution in (3.7.1).

Let us now consider the case when ρ is an integer and present the formulas that have been derived by Gupta (1960, 1962) for the computation of the single and the product moments of order statistics. First of all, we note easily that when ρ is an integer, the cumulative distribution function $F(x)$ of the gamma population can be written as a partial sum of the probabilities in a Poisson distribution, viz.,

$$F(x) = \sum_{\ell=\rho}^{\infty} e^{-x} x^{\ell} / \ell!, \qquad x \ge 0. \tag{3.7.2}$$

Now, from (3.2.1) we have for $1 \le i \le n$ and $k \ge 1$

$$\alpha_{i:n}^{(k)} = \frac{n!}{(i-1)!(n-i)!} \int_0^\infty x^k \{F(x)\}^{i-1} \{1-F(x)\}^{n-i} f(x)\, dx$$
$$= \frac{n!}{(i-1)!(n-i)!} \sum_{r=0}^{i-1} (-1)^r \binom{i-1}{r} \int_0^\infty x^k \{1-F(x)\}^{n-i+r} f(x)\, dx. \tag{3.7.3}$$

By using the expressions of $f(x)$ and $F(x)$ in (3.7.1) and (3.7.2), respectively, in Eq. (3.7.3), we obtain for $1 \le i \le n$ and $k \ge 1$

$$\alpha_{i:n}^{(k)} = \frac{n!}{(i-1)!(n-i)!} \sum_{r=0}^{i-1} (-1)^r \binom{i-1}{r} \int_0^\infty x^k \left\{ \sum_{\ell=0}^{\rho-1} e^{-x} x^\ell / \ell! \right\}^{n-i+r}$$
$$\times \frac{1}{\Gamma(\rho)} e^{-x} x^{\rho-1}\, dx$$
$$= \frac{n!}{(i-1)!(n-i)!} \sum_{r=0}^{i-1} (-1)^r \binom{i-1}{r} \frac{1}{\Gamma(\rho)} \int_0^\infty e^{-(n-i+r+1)x}$$
$$\times x^{k+\rho-1} \left\{ \sum_{\ell=0}^{\rho-1} x^\ell / \ell! \right\}^{n-i+r} dx$$
$$= \frac{n!}{(i-1)!(n-i)!\Gamma(\rho)} \sum_{r=0}^{i-1} (-1)^r \binom{i-1}{r} \sum_{s=0}^{(\rho-1)(n-i+r)} a_s(\rho, n-i+r)$$
$$\times \int_0^\infty e^{-(n-i+r+1)x} x^{k+\rho+s-1}\, dx$$
$$= \frac{n!}{(i-1)!(n-i)!\Gamma(\rho)} \sum_{r=0}^{i-1} (-1)^r \binom{i-1}{r} \sum_{s=0}^{(\rho-1)(n-i+r)} a_s(\rho, n-i+r)$$
$$\times \Gamma(k+\rho+s)/(n-i+r+1)^{k+\rho+s}, \tag{3.7.4}$$

where $a_s(\rho, n-i+r)$ is the coefficient of x^s in the expansion of

$$\left\{ \sum_{\ell=0}^{\rho-1} x^\ell / \ell! \right\}^{n-i+r}.$$

The kth moment of the smallest order statistic may be obtained from (3.7.4) by setting $i = 1$ to be

$$\alpha_{1:n}^{(k)} = \frac{n}{\Gamma(\rho)} \sum_{s=0}^{(\rho-1)(n-1)} a_s(\rho, n-1) \Gamma(k+\rho+s) n^{-(k+\rho+s)}. \tag{3.7.5}$$

The coefficients $a_s(\rho, n-1)$ can be generated very easily as follows. First of all, we see that

$$a_0(\rho, 1) = 1,\ a_1(\rho, 1) = 1,\ a_2(\rho, 1) = \frac{1}{2!}, \ldots, a_{\rho-1}(\rho, 1) = \frac{1}{(\rho-1)!}. \tag{3.7.6}$$

Next, let us consider $a_s(\rho, m)$ for $m \geq 2$, given by

$$a_s(\rho, m) = \text{coefficient of } x^s \text{ in } \left\{ \sum_{\ell=0}^{\rho-1} x^\ell/\ell! \right\}^m$$

$$= \sum_{\ell=0}^{\rho-1} \left[\text{coefficient of } x^\ell \text{ in } \left\{ \sum_{\ell=0}^{\rho-1} x^\ell/\ell! \right\} \right]$$

$$\times \left[\text{coefficient of } x^{s-\ell} \text{ in } \left\{ \sum_{\ell=0}^{\rho-1} x^\ell/\ell! \right\}^{m-1} \right]$$

$$= \sum_{\ell=0}^{\rho-1} \frac{1}{\ell!} a_{s-\ell}(\rho, m-1). \tag{3.7.7}$$

Thus, by starting with the values of $a_s(\rho, 1)$ given in (3.7.6), we can compute the coefficients $a_s(\rho, m)$ for any value of m by repeated application of the recurrence relation in (3.7.7). After computing the coefficients $a_s(\rho, m)$ this way, one may either directly use the formula in (3.7.4) to compute the single moments $\alpha_{i:n}^{(k)}$, or simply compute the moments $\alpha_{1:n}^{(k)}$ from (3.7.5) and then employ Relation 3.3.3 to compute the single moments of other order statistics.

By using the above formulas, Gupta (1960, 1962) has tabulated the first four moments of all order statistics for sample sizes up to 10 when $\rho = 1(1)5$ and for the smallest order statistic up to sample size 15. Breiter and Krishnaiah (1968) have provided similar tables for sample sizes up to 9 when $\rho = 0.5(1)10.5$. Harter (1970b) has tabulated just the means of all order statistics for sample sizes 40 and less when $\rho = 0.5(0.5)4.0$.

From (3.2.3) we have for $1 \leq i < j \leq n$

$$\alpha_{i,j:n} = \frac{n!}{(i-1)!(j-i-1)!(n-j)!} \int_0^\infty \int_x^\infty xy\{F(x)\}^{i-1}\{F(y) - F(x)\}^{j-i-1}$$

$$\times \{1 - F(y)\}^{n-j} f(x) f(y) \, dy \, dx$$

$$= \frac{n!}{(i-1)!(j-i-1)!(n-j)!} \sum_{r=0}^{i-1} \sum_{s=0}^{j-i-1} (-1)^{r+s} \binom{i-1}{r} \binom{j-i-1}{s}$$

$$\times \int_0^\infty \int_x^\infty xy\{1 - F(x)\}^{j-i-1-s+r}\{1 - F(y)\}^{n-j+s} f(x) f(y) \, dy \, dx. \tag{3.7.8}$$

By using the expressions of the pdf and the cdf in (3.7.1) and (3.7.2), respectively, in Eq. (3.7.8), we obtain for integer values of ρ and $1 \le i < j \le n$

$$\alpha_{i,j:n} = \frac{n!}{(i-1)!(j-i-1)!(n-j)!} \sum_{r=0}^{i-1} \sum_{s=0}^{j-i-1} (-1)^{r+s} \binom{i-1}{r} \binom{j-i-1}{s}$$
$$\times \int_0^\infty x \left\{ \sum_{\ell=0}^{\rho-1} e^{-x} x^\ell / \ell! \right\}^{j-i-1-s+r} \frac{1}{\Gamma(\rho)} e^{-x} x^{\rho-1}$$
$$\times \left[\int_x^\infty y \left\{ \sum_{\ell=0}^{\rho-1} e^{-y} y^\ell / \ell! \right\}^{n-j+s} \frac{1}{\Gamma(\rho)} e^{-y} y^{\rho-1} \, dy \right] dx.$$
$$= \frac{n!}{(i-1)!(j-i-1)!(n-j)!\{\Gamma(\rho)\}^2} \sum_{r=0}^{i-1} \sum_{s=0}^{j-i-1} (-1)^{r+s} \binom{i-1}{r} \binom{j-i-1}{s}$$
$$\times \sum_{t_1=0}^{(\rho-1)(j-i-1-s+r)} a_{t_1}(\rho, j-i-1-s+r) \sum_{t_2=0}^{(\rho-1)(n-j+s)} a_{t_2}(\rho, n-j+s)$$
$$\times \int_0^\infty e^{-(j-i-s+r)x} x^{t_1+\rho} \left[\int_x^\infty e^{-(n-j+s+1)y} y^{t_2+\rho} \, dy \right] dx. \qquad (3.7.9)$$

Let us consider the integral

$$I(x) = \int_x^\infty e^{-(n-j+s+1)y} y^{t_2+\rho} \, dy.$$

By making the transformation $u = (n-j+s+1)y$, we get

$$I(x) = \frac{1}{(n-j+s+1)^{t_2+\rho+1}} \int_{(n-j+s+1)x}^\infty e^{-u} u^{(t_2+\rho+1)-1} \, du,$$

which, when used with (3.7.2), gives

$$I(x) = \frac{\Gamma(t_2+\rho+1)}{(n-j+s+1)^{t_2+\rho+1}} \sum_{\ell=0}^{t_2+\rho} e^{-(n-j+s+1)x} \frac{(n-j+s+1)^\ell x^\ell}{\ell!}$$
$$= \Gamma(t_2+\rho+1) \sum_{\ell=0}^{t_2+\rho} e^{-(n-j+s+1)x} x^\ell / \{(n-j+s+1)^{t_2+\rho+1-\ell} \ell!\}.$$
$$(3.7.10)$$

Upon substituting the above expression of $I(x)$ in (3.7.9), we derive for integer values of ρ and $1 \le i < j \le n$

$$\alpha_{i,j:n} = \frac{n!}{(i-1)!(j-i-1)!(n-j)!\{\Gamma(\rho)\}^2} \sum_{r=0}^{i-1} \sum_{s=0}^{j-i-1} (-1)^{r+s} \binom{i-1}{r} \binom{j-i-1}{s}$$
$$\times \sum_{t_1=0}^{(\rho-1)(j-i-1-s+r)} a_{t_1}(\rho, j-i-1-s+r)$$
$$\times \sum_{t_2=0}^{(\rho-1)(n-j+s)} a_{t_2}(\rho, n-j+s) \Gamma(t_2+\rho+1)$$
$$\times \sum_{\ell=0}^{t_2+\rho} \int_0^\infty e^{-(n-i+r+1)x} x^{t_1+\rho+\ell} \, dx / \{(n-j+s+1)^{t_2+\rho+1-\ell} \ell!\}$$

$$= \frac{n!}{(i-1)!(j-i-1)!(n-j)!\{\Gamma(\rho)\}^2} \sum_{r=0}^{i-1} \sum_{s=0}^{j-i-1} (-1)^{r+s} \binom{i-1}{r} \binom{j-i-1}{s}$$

$$\times \sum_{t_1=0}^{(\rho-1)(j-i-1-s+r)} a_{t_1}(\rho, j-i-1-s+r)$$

$$\times \sum_{t_2=0}^{(\rho-1)(n-j+s)} a_{t_2}(\rho, n-j+s)\Gamma(t_2+\rho+1)$$

$$\times \sum_{\ell=0}^{t_2+\rho} \Gamma(t_1+\rho+\ell+1)/\{(n-i+r+1)^{t_1+\rho+\ell+1}(n-j+s+1)^{t_2+\rho-\ell+1}\ell!\}. \tag{3.7.11}$$

The formula in (3.7.11), derived by Gupta (1960, 1962), may be used to compute the covariances of gamma order statistics for integer values of ρ. Prescott (1974) has tabulated the covariances of order statistics for sample sizes up to 10 when $\rho = 2(1)5$. Also, Young (1971) has derived some expressions for the moments of gamma order statistics for the limiting case, i.e., when $\rho \to \infty$. In addition, for the case in which ρ is an integer, Joshi (1979b) has established the recurrence relations

$$\alpha_{1:n}^{(k)} = \frac{k}{n}\Gamma(\rho) \sum_{j=0}^{\rho-1} \alpha_{1:n}^{(k+j-\rho)}/j!, \qquad k \geq 1 \tag{3.7.12}$$

and

$$\alpha_{i:n}^{(k)} = \alpha_{i-1:n-1}^{(k)} + \frac{k}{n}\Gamma(\rho) \sum_{j=0}^{\rho-1} \alpha_{i:n}^{(k+j-\rho)}/j!, \qquad 2 \leq i \leq n, k \geq 1, \tag{3.7.13}$$

and has successfully applied them to compute the existing negative moments of all order statistics by using the table of first four positive moments prepared by Gupta (1960, 1962). Reference may also be made to Tadikamalla (1977), who has developed some approximation to the moments of gamma order statistics via a four-parameter Burr distribution approximation to the gamma distribution in (3.7.1), and to Tiku and Malik (1972) for a three-moment chi square and t approximations for the distributions of gamma order statistics.

3.8. Results for the Weibull Distribution

Let $X_{1:n} \leq X_{2:n} \leq \cdots \leq X_{n:n}$ be the order statistics from a Weibull population with pdf

$$f(x) = e^{-x^\delta} \delta x^{\delta-1}, \qquad 0 \leq x < \infty, \delta > 0, \tag{3.8.1}$$

and cdf

$$F(x)=1-e^{-x^{\delta}}, \qquad 0\le x<\infty,\ \delta>0. \tag{3.8.2}$$

Then, from (3.2.1) we obtain for $1\le i\le n$ and $k\ge 1$

$$\begin{aligned}\alpha_{i:n}^{(k)}&=\frac{n!}{(i-1)!(n-i)!}\int_0^{\infty} x^k(1-e^{-x^{\delta}})^{i-1}(e^{-x^{\delta}})^{n-i}\, e^{-x^{\delta}}\,\delta x^{\delta-1}\,dx\\&=\frac{n!}{(i-1)!(n-i)!}\sum_{r=0}^{i-1}(-1)^r\binom{i-1}{r}\int_0^{\infty} e^{-(n-i+r+1)x^{\delta}}\,x^k\,\delta x^{\delta-1}\,dx.\end{aligned} \tag{3.8.3}$$

By setting $u=x^{\delta}$ in the integral in (3.8.3), we get for $1\le i\le n$ and $k\ge 1$

$$\begin{aligned}\alpha_{i:n}^{(k)}&=\frac{n!}{(i-1)!(n-i)!}\sum_{r=0}^{i-1}(-1)^r\binom{i-1}{r}\int_0^{\infty} e^{-(n-i+r+1)u}\,u^{k/\delta}\,du\\&=\frac{n!}{(i-1)!(n-i)!}\Gamma\left(1+\frac{k}{\delta}\right)\sum_{r=0}^{i-1}(-1)^r\binom{i-1}{r}\Big/(n-i+r+1)^{1+(k/\delta)}.\end{aligned} \tag{3.8.4}$$

By making use of the formula in (3.8.4), first derived by Lieblein (1955), Govindarajulu and Joshi (1968) have tabulated the means and variances of all order statistics for sample sizes up to 10 when $\delta=1, 2, 2.5, 3(1)10$. Harter (1970b) has prepared a more extensive table of means of order statistics for samples of sizes 40 and less when $\delta=0.5(0.5)4(1)8$.

Next, from the expression of the product moment in (3.2.3) we have for $1\le i<j\le n$

$$\begin{aligned}\alpha_{i,j:n}&=\frac{n!}{(i-1)!(j-i-1)!(n-j)!}\int_0^{\infty}\int_0^{y} xy(1-e^{-x^{\delta}})^{i-1}(e^{-x^{\delta}}-e^{-y^{\delta}})^{j-i-1}\\&\quad\times(e^{-y^{\delta}})^{n-j}\,e^{-x^{\delta}}\delta x^{\delta-1}\,e^{-y^{\delta}}\,\delta y^{\delta-1}\,dx\,dy\\&=\frac{n!}{(i-1)!(j-i-1)!(n-j)!}\delta^2\sum_{r=0}^{i-1}\sum_{s=0}^{j-i-1}(-1)^{j-i-1-s+r}\binom{i-1}{r}\binom{j-i-1}{s}\\&\quad\times\int_0^{\infty}\int_0^{y} e^{-(r+s+1)x^{\delta}}\,e^{-(n-i-s)y^{\delta}}x^{\delta}y^{\delta}\,dx\,dy\\&=\frac{n!}{(i-1)!(j-i-1)!(n-j)!}\sum_{r=0}^{i-1}\sum_{s=0}^{j-i-1}(-1)^{j-i-1-s+r}\binom{i-1}{r}\binom{j-i-1}{s}\\&\quad\times\phi_{\delta}(r+s+1,\,n-i-s),\end{aligned} \tag{3.8.5}$$

where $\phi_{\delta}(a, b)$ is Lieblein's ϕ-function defined by

$$\phi_{\delta}(a,b)=\delta^2\int_0^{\infty}\int_0^{y} e^{-ax^{\delta}-by^{\delta}}x^{\delta}y^{\delta}\,dx\,dy. \tag{3.8.6}$$

In the following we present an algebraic evaluation of the above ϕ-function, as derived by Lieblein (1955) and also explained in detail by Balakrishnan and Kocherlakota (1985), while studying the moments of order statistics from the double Weibull distribution.

By considering the definition of $\phi_\delta(a, b)$ in (3.8.6) and integrating x by parts, we get

$$\phi_\delta(a, b) = -\frac{\Gamma\left(1+\frac{2}{\delta}\right)}{a(a+b)^{1+(2/\delta)}} + \frac{1}{a}\int_0^\infty \int_0^y y\, e^{-ax^\delta}\, e^{-by^\delta}\, \delta y^{\delta-1}\, dx\, dy. \quad (3.8.7)$$

Differentiating Eq. (3.8.7) partially with respect to a, we obtain

$$\frac{\partial \phi_\delta(a, b)}{\partial a} = \frac{\Gamma\left(2+\frac{2}{\delta}\right)}{a(a+b)^{2+(2/\delta)}} - \frac{1}{a}\left(1+\frac{1}{\delta}\right)\phi_\delta(a, b),$$

which, when rewritten, yields the differential equation

$$\frac{\partial \phi_\delta(a, b)}{\partial a} + \frac{1}{a}\left(1+\frac{1}{\delta}\right)\phi_\delta(a, b) = \frac{\Gamma\left(2+\frac{2}{\delta}\right)}{a(a+b)^{2+(2/\delta)}}. \quad (3.8.8)$$

Now by treating (3.8.8) as a linear differential equation in $\phi_\delta(a, b)$ and solving it, we get for $a \geq b$

$$\phi_\delta(a, b) = \frac{1}{a^{1+(1/\delta)}}\left[\Gamma\left(2+\frac{2}{\delta}\right)\int_b^a \frac{x^{1/\delta}}{(b+x)^{2+(2/\delta)}}\, dx + K(b)\right], \quad (3.8.9)$$

where $K(b)$ is the constant of integration to be determined. By setting $w = x/(b+x)$ in the integral on the RHS of (3.8.9), we may rewrite (3.8.9) as

$$\phi_\delta(a, b) = \frac{1}{a^{1+(1/\delta)}}\left[\frac{\Gamma\left(2+\frac{2}{\delta}\right)}{b^{1+(1/\delta)}}\int_{1/2}^{a/(a+b)} w^{1/\delta}(1-w)^{1/\delta}\, dw + K(b)\right]. \quad (3.8.10)$$

In order to determine the constant $K(b)$, let us set $a = b$ in (3.8.10); then, we get

$$K(b) = b^{1+(1/\delta)}\phi_\delta(b, b). \quad (3.8.11)$$

Now, from (3.8.6) we have

$$\phi_\delta(b, b) = \delta^2 \int_0^\infty \int_0^y e^{-bx^\delta}\, e^{-by^\delta}\, x^\delta y^\delta\, dx\, dy,$$

which, because of the symmetry in x and y, may be written as

$$\begin{aligned}\phi_\delta(b, b) &= \frac{1}{2}\delta^2\left\{\int_0^\infty e^{-bx^\delta}x^\delta\,dx\right\}^2 \\ &= \frac{1}{2}\left\{\Gamma\left(1+\frac{1}{\delta}\right)\Big/ b^{1+(1/\delta)}\right\}^2 \\ &= \left\{\Gamma\left(2+\frac{2}{\delta}\right)\Big/ b^{2+(2/\delta)}\right\}\frac{1}{2}B\left(1+\frac{1}{\delta}, 1+\frac{1}{\delta}\right). \qquad (3.8.12)\end{aligned}$$

In (3.8.12), $B(1+(1/\delta), 1+(1/\delta))$ denotes the complete beta integral

$$\int_0^1 w^{1/\delta}(1-w)^{1/\delta}\,dw,$$

which, by its symmetry around $\frac{1}{2}$, may be written equivalently as

$$2\int_0^{1/2} w^{1/\delta}(1-w)^{1/\delta}\,dw.$$

By making use of this expression in (3.8.12), we obtain

$$\phi_\delta(b, b) = \frac{\Gamma\left(2+\dfrac{2}{\delta}\right)}{b^{2+(2/\delta)}}\int_0^{1/2} w^{1/\delta}(1-w)^{1/\delta}\,dw, \qquad (3.8.13)$$

which, when used in (3.8.11), immediately yields

$$K(b) = \frac{\Gamma\left(2+\dfrac{2}{\delta}\right)}{b^{1+(1/\delta)}}\int_0^{1/2} w^{1/\delta}(1-w)^{1/\delta}\,dw. \qquad (3.8.14)$$

Upon substituting the expression of $K(b)$ from (3.8.14) in Eq. (3.8.10), we derive an explicit algebraic formula for the ϕ-function to be

$$\begin{aligned}\phi_\delta(a, b) &= \frac{\Gamma\left(2+\dfrac{2}{\delta}\right)}{(ab)^{1+(1/\delta)}}\int_0^{a/(a+b)} w^{1/\delta}(1-w)^{1/\delta}\,dw \\ &= \frac{\left\{\Gamma\left(1+\dfrac{1}{\delta}\right)\right\}^2}{(ab)^{1+(1/\delta)}}\,\mathrm{IB}_{a/(a+b)}\left(1+\frac{1}{\delta}, 1+\frac{1}{\delta}\right), \qquad a \geq b, \qquad (3.8.15)\end{aligned}$$

where $\mathrm{IB}_p(d_1, d_2)$ is Karl Pearson's (1934) incomplete beta function defined as

$$\mathrm{IB}_p(d_1, d_2) = \frac{1}{B(d_1, d_2)}\int_0^p w^{d_1-1}(1-w)^{d_2-1}\,dw, \qquad 0 \leq p \leq 1. \quad (3.8.16)$$

Furthermore, we also have from (3.8.6) that

$$\phi_\delta(a, b)+\phi_\delta(b, a)=\delta^2 \int_0^\infty e^{-ax^\delta} x^\delta \, dx \int_0^\infty e^{-by^\delta} y^\delta \, dy$$
$$=\left\{\Gamma\left(1+\frac{1}{\delta}\right)\right\}^2 \Big/ (ab)^{1+(1/\delta)}. \tag{3.8.17}$$

So, after computing the function $\phi_\delta(a, b)$ from (3.8.15) for $a \geq b$, one can use the relation in (3.8.17) to compute the function $\phi_\delta(a, b)$ for $a < b$. These formulas for the ϕ-function will then enable one to compute the product moments of Weibull order statistics from (3.8.5). Govindarajulu and Joshi (1968) have used this procedure and tabulated the covariances of order statistics for sample sizes up to 10 when $\delta = 1, 2, 2.5, 3(1)10$.

3.9. Results for the Normal Distribution

Let us consider the case when the order statistics $X_{1:n} \leq X_{2:n} \leq \cdots \leq X_{n:n}$ are from the standard normal population with cdf $F(x)$ and pdf

$$f(x)=\frac{1}{\sqrt{2\pi}} e^{-x^2/2}, \qquad -\infty < x < \infty. \tag{3.9.1}$$

Derivations of the single and the product moments of order statistics explicitly in terms of some elementary functions have been attempted by several authors, including Jones (1948), Godwin (1949), Ruben (1954, 1956), Watanabe *et al.* (1957, 1958), Bose and Gupta (1959), and David (1963). These authors have been successful in achieving this for small sample sizes at least. To explain the method of derivation of single moments, let us follow the lines of Bose and Gupta (1959) and denote

$$I_n(a)=\int_{-\infty}^\infty \{F(ax)\}^n \frac{1}{\sqrt{\pi}} e^{-x^2} \, dx, \qquad n = 0, 1, 2, \ldots. \tag{3.9.2}$$

Setting $n = 0$ in (3.9.2), we immediately obtain

$$I_0(a) = 1. \tag{3.9.3}$$

Let us now consider the expression of $\alpha_{2:2}$ obtained from (3.2.1) as

$$\alpha_{2:2} = 2 \int_{-\infty}^\infty xF(x)f(x) \, dx. \tag{3.9.4}$$

By using the property that $f'(x) = -xf(x)$ for the standard normal distribution in the integral on the RHS of (3.9.4), and then integrating by parts treating $f'(x)$ for integration and the rest of the integrand for differentiation, we obtain

$$\alpha_{2:2} = \int_{-\infty}^{\infty} \frac{1}{\pi} e^{-x^2}\, dx = \frac{1}{\sqrt{\pi}} I_0(a) = \frac{1}{\sqrt{\pi}} = 0.5641895835$$

from (3.9.3). Also, because of the symmetry of $f(x)$, we have $\alpha_{1:2} = -\alpha_{2:2} = -0.5641895835$.

Next, let us consider the integral

$$\int_{-\infty}^{\infty} \left\{F(ax) - \frac{1}{2}\right\}^{2m+1} \frac{1}{\sqrt{\pi}} e^{-x^2}\, dx \tag{3.9.5}$$

for $m = 0, 1, 2, \ldots$. Since the integrand in (3.9.5) is an odd function of x, we immediately have

$$\int_{-\infty}^{\infty} \left\{F(ax) - \frac{1}{2}\right\}^{2m+1} \frac{1}{\sqrt{\pi}} e^{-x^2}\, dx = 0, \qquad m \geq 0. \tag{3.9.6}$$

Upon expanding the term $\{F(ax) - \frac{1}{2}\}^{2m+1}$ binomially in the above integral and using (3.9.2) to express the integrals in terms of $I_n(a)$, we get

$$\sum_{i=0}^{2m+1} (-1)^i \binom{2m+1}{i} \frac{1}{2^i} I_{2m+1-i}(a) = 0, \qquad m \geq 0,$$

which yields the relation

$$I_{2m+1}(a) = \sum_{i=0}^{2m+1} (-1)^{i+1} \binom{2m+1}{i} \frac{1}{2^i} I_{2m+1-i}(a), \qquad m \geq 0. \tag{3.9.7}$$

Thus, for example, setting $m = 0$ in (3.9.7), we obtain

$$I_1(a) = \frac{1}{2} I_0(a) = \frac{1}{2}. \tag{3.9.8}$$

Now, from (3.2.1) let us consider the expression of $\alpha_{3:3}$ and integrate by parts as before by using $f'(x) = -xf(x)$. We then obtain

$$\alpha_{3:3} = 3 \int_{-\infty}^{\infty} F(x) \frac{1}{\pi} e^{-x^2}\, dx = \frac{3}{\sqrt{\pi}} I_1(1) = \frac{1.5}{\sqrt{\pi}} = 0.8462843753$$

from (3.9.8). Because of the symmetry of $f(x)$, we also have $\alpha_{2:3} = 0$ and $\alpha_{1:3} = -\alpha_{3:3} = -0.8462843753$.

By differentiating the expression of $I_2(a)$ from (3.9.2) with respect to a, and by applying Fubini's theorem, we obtain

$$I_2'(a)=\frac{\sqrt{2}}{\pi}\int_{-\infty}^{\infty} F(ax)x\,e^{-1/2x^2(2+a^2)}\,dx$$

$$=-\frac{\sqrt{2}}{\pi(a^2+2)}\int_{-\infty}^{\infty} F(ax)\frac{d}{dx}\{e^{-1/2x^2(2+a^2)}\},$$

which, upon integration by parts, yields

$$I_2'(a)=\frac{\sqrt{2}}{\pi(a^2+2)}\frac{a}{\sqrt{2\pi}}\int_{-\infty}^{\infty} e^{-1/2x^2(a^2+1)2}\,dx$$

$$=\frac{a}{\pi(a^2+2)(a^2+1)^{1/2}}. \quad (3.9.9)$$

By solving (3.9.9), we derive

$$I_2(a)=\frac{1}{\pi}\tan^{-1}(a^2+1)^{1/2}. \quad (3.9.10)$$

From (3.2.1) let us now consider the expression of $\alpha_{4:4}$ and integrate by parts as before. Since $f'(x)=-xf(x)$, we obtain

$$\alpha_{4:4}=6\int_{-\infty}^{\infty}\{F(x)\}^2\frac{1}{\pi}e^{-x^2}\,dx=\frac{6}{\sqrt{\pi}}I_2(1)$$

$$=\frac{6}{\pi\sqrt{\pi}}\tan^{-1}\sqrt{2}=1.0293753730$$

from (3.9.10). From Relation 3.3.1 we get

$$\alpha_{3:4}=4\alpha_{3:3}-3\alpha_{4:4}=\frac{6}{\sqrt{\pi}}-\frac{18}{\pi\sqrt{\pi}}\tan^{-1}\sqrt{2}=0.2970113823,$$

and because of the symmetry of $f(x)$ we have $\alpha_{2:4}=-\alpha_{3:4}=-0.2970113823$ and $\alpha_{1:4}=-\alpha_{4:4}=-1.0293753730$.

By using the relation in (3.9.7), we get by setting $m=1$

$$I_3(a)=\frac{3}{2}I_2(a)-\frac{3}{4}I_1(a)+\frac{1}{8}I_0(a)$$

$$=\frac{3}{2\pi}\tan^{-1}(a^2+1)^{1/2}-\frac{1}{4}. \quad (3.9.11)$$

Let us now consider the expression of $\alpha_{5:5}$ from (3.2.1) and integrate by parts by using $f'(x) = -xf(x)$. We then get

$$\alpha_{5:5} = 10 \int_{-\infty}^{\infty} \{F(x)\}^3 \frac{1}{\pi} e^{-x^2}\, dx = \frac{10}{\sqrt{\pi}} I_3(1) = 1.1629644736,$$

and from Relation 3.3.1 we get

$$\alpha_{4:5} = 5\alpha_{4:4} - 4\alpha_{5:5} = 0.4950189705.$$

Also, because of the symmetry of $f(x)$ we have $\alpha_{3:5} = 0$, $\alpha_{2:5} = -\alpha_{4:5} = -0.4950189705$ and $\alpha_{1:5} = -\alpha_{5:5} = -1.1629644736$. However, because $I_4(a)$ cannot be expressed in terms of elementary functions, the above method fails for $\alpha_{6:6}$. Moreover, Ruben (1954), who has shown that the moments of order statistics can be expressed as linear functions of the contents of certain hyperspherical simplices, has noted that for dimension greater than three these contents cannot be expressed in terms of elementary functions; this (as pointed out above) explains why the method fails for $\alpha_{6:6}$.

By using a similar but somewhat cumbersome approach, the product moments of order statistics may also be determined in terms of elementary functions for small sample sizes. First of all, we know that

$$\alpha_{1,2:2} = \alpha_{1:1}^2 = 0.$$

Next, let us consider the expression of $\alpha_{1,2:3}$ from (3.2.3) given by

$$\alpha_{1,2:3} = 6 \int_{-\infty}^{\infty} \int_{-\infty}^{y} xy\{1 - F(y)\} f(x) f(y)\, dx\, dy. \tag{3.9.12}$$

By making use of the property that $f'(x) = -xf(x)$, we have

$$\int_{-\infty}^{y} xf(x)\, dx = -f(y), \tag{3.9.13}$$

which, when used in (3.9.12), gives

$$\alpha_{1,2:3} = -6 \int_{-\infty}^{\infty} y\{1 - F(y)\} f^2(y)\, dy. \tag{3.9.14}$$

By writing $yf(y)$ as $-f'(y)$ and integrating by parts, we get from (3.9.14)

$$\begin{aligned}\alpha_{1,2:3} &= 6 \int_{-\infty}^{\infty} f^3(y)\, dy + 6 \int_{-\infty}^{\infty} y\{1 - F(y)\} f^2(y)\, dy \\ &= 6 \int_{-\infty}^{\infty} f^3(y)\, dy - \alpha_{1,2:3},\end{aligned}$$

which immediately yields

$$\alpha_{1,2:3} = 3\int_{-\infty}^{\infty} f^3(y)\,dy = \frac{\sqrt{3}}{2\pi}\,I_0(a) = \frac{\sqrt{3}}{2\pi} = 0.2756644477$$

from (3.9.3). Because of the symmetry of $f(x)$ we have $\alpha_{2,3:3} = \alpha_{1,2:3} =$ 0.2756644477. Also, by using Relation 3.3.4 we get

$$\alpha_{1,3:3} = 3\alpha_{1,2:2} - \alpha_{1,2:3} - \alpha_{2,3:3} = -\frac{\sqrt{3}}{\pi} = -0.5513288954.$$

Next, let us consider the expression of $\alpha_{1,2:4}$ from (3.2.3) given by

$$\alpha_{1,2:4} = 12\int_{-\infty}^{\infty}\int_{-\infty}^{y} xy\{1-F(y)\}^2 f(x)f(y)\,dx\,dy. \tag{3.9.15}$$

By applying (3.9.13) to the above equation we get

$$\alpha_{1,2:4} = -12\int_{-\infty}^{\infty} y\{1-F(y)\}^2 f^2(y)\,dy. \tag{3.9.16}$$

By writing $yf(y)$ as $-f'(y)$ and integrating by parts, we get from (3.9.16)

$$\begin{aligned}\alpha_{1,2:4} &= 24\int_{-\infty}^{\infty}\{1-F(y)\}f^3(y)\,dy + 12\int_{-\infty}^{\infty} y\{1-F(y)\}^2 f^2(y)\,dy\\ &= 24\int_{-\infty}^{\infty}\{1-F(y)\}f^3(y)\,dy - \alpha_{1,2:4},\end{aligned}$$

which immediately yields

$$\begin{aligned}\alpha_{1,2:4} &= 12\int_{-\infty}^{\infty} f^3(y)\,dy - 12\int_{-\infty}^{\infty} F(y)f^3(y)\,dy\\ &= \frac{2\sqrt{3}}{\pi}\left\{I_0(a) - I_1\left(\sqrt{\frac{2}{3}}\right)\right\}\\ &= \frac{\sqrt{3}}{\pi} = 0.5513288954.\end{aligned}$$

Similarly, let us now start with the expression of $\alpha_{2,3:4}$ from (3.2.3) given by

$$\alpha_{2,3:4} = 24\int_{-\infty}^{\infty}\int_{-\infty}^{y} xyF(x)\{1-F(y)\}f(x)f(y)\,dx\,dy. \tag{3.9.17}$$

By making use of the property that $f'(x) = -xf(x)$, we have

$$\int_{-\infty}^{y} xF(x)f(x)\,dx = -F(y)f(y) + \frac{1}{2\sqrt{\pi}}\,F(\sqrt{2}y), \tag{3.9.18}$$

which, when used in (3.9.17), gives

$$\alpha_{2,3:4} = -24K_1 + \frac{24}{2\sqrt{\pi}} K_2, \tag{3.9.19}$$

where

$$K_1 = \int_{-\infty}^{\infty} y\{1 - F(y)\}F(y)f^2(y)\,dy \tag{3.9.20}$$

and

$$K_2 = \int_{-\infty}^{\infty} y\{1 - F(y)\}F(\sqrt{2}y)f(y)\,dy. \tag{3.9.21}$$

Since the integrand in the expression for K_1 in (3.9.20) is an odd function in y, we immediately have $K_1 = 0$. Next, by writing $yf(y)$ as $-f'(y)$ in the expression for K_2 in (3.9.21), integrating by parts, and then simplifying the resulting expression, we get

$$K_2 = \frac{1}{\sqrt{3}\pi} I_0(a) - \frac{1}{2\sqrt{\pi}} I_1(\sqrt{2}) - \frac{1}{\sqrt{3}\pi} I_1\left(\sqrt{\frac{2}{3}}\right) = \frac{2-\sqrt{3}}{4\sqrt{3}\pi}.$$

Upon substituting for K_1 and K_2 in (3.9.19), we get

$$\alpha_{2,3:4} = \frac{\sqrt{3}(2-\sqrt{3})}{\pi} = 0.1477281323.$$

By using Relation 3.3.4 we get

$$\begin{aligned}\alpha_{1,3:4} &= 4\alpha_{1,2:3} - 2\alpha_{1,2:4} - \alpha_{2,3:4} \\ &= -\frac{\sqrt{3}(2-\sqrt{3})}{\pi} = -0.1477281323,\end{aligned}$$

and by symmetry of $f(x)$ we also have $\alpha_{3,4:4} = \alpha_{1,2:4} = 0.5513288954$ and $\alpha_{2,4:4} = \alpha_{1,3:4} = -0.1477281323$. Finally, $\alpha_{1,4:4}$ may be derived by using Relation 3.3.4 as

$$\begin{aligned}\alpha_{1,4:4} &= 2\alpha_{1,3:3} - \frac{1}{2}(\alpha_{1,3:4} + \alpha_{2,4:4}) \\ &= -\frac{3}{\pi} = -0.9549296586.\end{aligned}$$

A general approach, given by Godwin (1949a), is to express the product moments in terms of integrals of the form

$$J_n = \int_0^{\infty}\int_0^{\infty}\cdots\int_0^{\infty} e^{-Q(x_1,\ldots,x_n)}\,dx_1\,dx_2\cdots dx_n,$$

where $Q(x_1, x_2, \ldots, x_n)$ is a quadratic form in the x_i's. For $n = 1, 2, 3$, J_n can be written down explicitly in terms of elementary functions as follows:

$n = 1$

$$Q(x_1) = a_{11}x_1^2$$

$$J_1 = \sqrt{\pi}/2a_{11}$$

$n = 2$

$$Q(x_1, x_2) = a_{11}x_1^2 + a_{22}x_2^2 + 2a_{12}x_1x_2$$

$$J_2 = \frac{1}{\sqrt{\Delta_2}}\left\{\frac{\pi}{2} - \tan^{-1}\left(\frac{a_{12}}{\sqrt{\Delta_2}}\right)\right\},$$

$$\text{where } \Delta_2 = \begin{vmatrix} a_{11} & a_{12} \\ a_{12} & a_{22} \end{vmatrix} = a_{11}a_{22} - a_{12}^2$$

$n = 3$

$$Q(x_1, x_2, x_3) = a_{11}x_1^2 + a_{22}x_2^2 + a_{33}x_3^2 + 2a_{12}x_1x_2 + 2a_{13}x_1x_3 + 2a_{23}x_2x_3$$

$$J_3 = \frac{\sqrt{\pi}}{4\sqrt{\Delta_3}}\left\{\frac{\pi}{2} + \tan^{-1}\left(\frac{a_{12}a_{13} - a_{11}a_{23}}{\sqrt{a_{11}\Delta_3}}\right) + \tan^{-1}\left(\frac{a_{12}a_{23} - a_{13}a_{22}}{\sqrt{a_{22}\Delta_3}}\right) + \tan^{-1}\left(\frac{a_{13}a_{23} - a_{12}a_{33}}{\sqrt{a_{33}\Delta_3}}\right)\right\},$$

$$\text{where } \Delta_3 = \begin{vmatrix} a_{11} & a_{12} & a_{13} \\ a_{12} & a_{22} & a_{23} \\ a_{13} & a_{23} & a_{33} \end{vmatrix} = a_{11}a_{22}a_{33} - a_{11}a_{23}^2 - a_{22}a_{13}^2 - a_{33}a_{12}^2 + 2a_{12}a_{13}a_{23}.$$

The values of $\alpha_{i:n}$ for all i and for $n = 2(1)100(25)250(50)400$ have been tabulated to five decimals by Harter (1961a), and also by Harter (1970) for some more choices of n. The mean and variance of the ith quasi-range have been tabulated by Harter (1959) for n up to 100. For the sample range, Tippett (1925) has computed the expected value for $n \leq 1000$, whereas Harter (1960) has tabulated the mean, variance, and the coefficients of skewness and kurtosis for $n \leq 100$. By using the tables of means and product moments of order statistics for sample sizes up to 20 prepared by Teichroew (1956), Sarhan and Greenberg (1956) have tabulated the variances and covariances to ten decimals for $n \leq 20$. These tables have been extended to $n \leq 50$ by Tietjen *et al.* (1977). The values of the mean and standard deviation prepared by Yamauti (1972) for sample sizes up to 50 are contained in the tables of Tietjen *et al.* (1977). For the largest order statistic $X_{n:n}$, Ruben (1954) has tabulated the first 10 moments for $n \leq 50$ and Borenius (1966), the first two moments for $n \leq 120$.

In addition to the above results, the moments of normal order statistics also satisfy some interesting recurrence relations and identities. We present a few of them in the following theorems and also illustrate some of their applications.

Theorem 3.9.1. *If $g(x)$ is any differentiable function such that differentiation of $g(x)$ with respect to x and $E\{g(X)\}$ with respect to an absolutely continuous distribution are interchangeable, then for $1 \le i \le n$*

$$E\{g'(X_{i:n})\} = -\sum_{j=1}^{n} E\{g(X_{i:n})f'(X_{j:n})/f(X_{j:n})\}, \tag{3.9.22}$$

where $f(x)$ is the pdf of the population.

Proof. By adopting a method of proof originally given by Seal (1956), Govindarajulu (1963) has proved this theorem as follows. For all real t we have for $1 \le i \le n$

$$\begin{aligned} E\{g(X_{i:n}+t)\} &= n! \underset{-\infty<x_{1:n}<\cdots<x_{n:n}<\infty}{\iint\cdots\int} g(x_{i:n}+t)\prod_{j=1}^{n} f(x_{j:n})\,dx_{j:n} \\ &= n! \underset{-\infty<x_{1:n}<\cdots<x_{n:n}<\infty}{\iint\cdots\int} g(x_{i:n})\prod_{j=1}^{n} f(x_{j:n}-t)\,dx_{j:n}. \end{aligned} \tag{3.9.23}$$

By differentiating both sides of (3.9.23) with respect to t and setting $t=0$, we derive

$$E\{g'(X_{i:n})\} = n! \underset{-\infty<x_{1:n}<\cdots<x_{n:n}<\infty}{\iint\cdots\int} g(x_{i:n})\left\{-\sum_{j=1}^{n}\frac{f'(x_{j:n})}{f(x_{j:n})}\right\}\prod_{k=1}^{n} f(x_{k:n})\,dx_{k:n},$$

which yields the relation in (3.9.22).

Corollary 3.9.1. *For $1 \le i \le n$, we have*

$$\sum_{j=1}^{n} E\{X_{i:n}f'(X_{j:n})/f(X_{j:n})\} = -1. \tag{3.9.24}$$

Proof. This follows directly from Theorem 3.9.1 if we take $g(x)=x$.

Corollary 3.9.2. *For the standard normal population with pdf $f(x)$ as in (3.9.1), we have for $1 \le i \le n$*

$$\sum_{j=1}^{n} \alpha_{i,j:n} = 1 \tag{3.9.25}$$

and

$$\sum_{j=1}^{n} \beta_{i,j:n} = 1. \tag{3.9.26}$$

Proof. (3.9.25) is derived from Corollary 3.9.1 simply by noting that $f'(x) = -xf(x)$ for the standard normal distribution. The identity in (3.9.26) then follows from (3.9.25) by using the fact that $\sum_{j=1}^{n} \alpha_{j:n} = n\alpha_{1:1} = 0$.

The two identities given in Corollary 3.9.2 may also be proved by using the independence of $\bar{X}$ and $X_{i:n} - \bar{X}$; see McKay (1935) and Daly (1946).

Theorem 3.9.2. *For the standard normal population, we have for $1 \le i \le n$*

$$\alpha_{i:n}^{(2)} = 1 + n \binom{n-1}{i-1} \sum_{j=0}^{n-i} (-1)^j \binom{n-i}{j} \frac{1}{i+j} \alpha_{1,2:i+j}. \tag{3.9.27}$$

Proof. From (3.2.1), we have for $1 \le i \le n$

$$\alpha_{i:n}^{(2)} = \frac{n!}{(i-1)!(n-i)!} \int_{-\infty}^{\infty} x^2 \{F(x)\}^{i-1} \{1 - F(x)\}^{n-i} f(x)\, dx. \tag{3.9.28}$$

By writing $xf(x)$ as $-f'(x)$ in (3.9.28) and integrating by parts, we obtain

$$\begin{aligned} \alpha_{i:n}^{(2)} = 1 &+ \frac{n!}{(i-1)!(n-i)!} (i-1) \sum_{j=0}^{n-i} (-1)^j \binom{n-i}{j} \int_{-\infty}^{\infty} x \{F(x)\}^{i+j-2} f^2(x)\, dx \\ &+ \frac{n!}{(i-1)!(n-i)!} (n-i) \sum_{j=0}^{n-i-1} (-1)^{j+1} \binom{n-i-1}{j} \\ &\times \int_{-\infty}^{\infty} x \{F(x)\}^{i+j-1} f^2(x)\, dx. \end{aligned} \tag{3.9.29}$$

Since

$$\int_x^{\infty} y f(y)\, dy = -\int_x^{\infty} f'(y)\, dy = f(x),$$

we have

$$\int_{-\infty}^{\infty} x\{F(x)\}^k f^2(x)\,dx = \int_{-\infty}^{\infty} x\{F(x)\}^k f(x)\left\{\int_x^{\infty} yf(y)\,dy\right\} dx$$

$$= \iint\limits_{-\infty<x<y<\infty} xy\{F(x)\}^k f(x)f(y)\,dy\,dx$$

$$= \alpha_{k+1,k+2:k+2}/\{(k+1)(k+2)\}$$

$$= \alpha_{1,2:k+2}/\{(k+1)(k+2)\}, \qquad (3.9.30)$$

because of the symmetry of the standard normal distribution. By using the formula in (3.9.30) for the two integrals on the RHS of (3.9.29) and simplifying the resulting expression, we derive the relation in (3.9.27).

In particular, by setting $i = n$ in (3.9.27) we obtain the relation

$$\alpha_{n:n}^{(2)} = \alpha_{1:n}^{(2)} = 1 + \alpha_{1,2:n}. \qquad (3.9.31)$$

The results given in Theorems 3.9.1 and 3.9.2 have been used for checking the computations of $\alpha_{i,j:n}$ and $\beta_{i,j:n}$. In addition, Davis and Stephens (1977, 1978) have applied (3.9.26) and (3.9.31) to improve the David–Johnson approximation of the variance-covariance matrix of normal order statistics. A detailed discussion of the David–Johnson approximation is presented in the next section.

By noting that the condition $f'(x) = -xf(x)$ is satisfied by both the standard normal and the half normal (or chi) distributions, and then proceeding exactly on the same lines as in the proof of Theorem 3.9.2, Joshi and Balakrishnan (1981) have established the following theorem.

Theorem 3.9.3. *For the standard normal and the half normal distributions, we have*

$$\sum_{j=i}^{n} \alpha_{i,j:n} = 1 + \sum_{j=i}^{n} \alpha_{i-1,j:n}, \qquad 1 \le i \le n, \qquad (3.9.32)$$

and

$$\sum_{j=i+1}^{n} \alpha_{i,j:n} = \sum_{j=i+1}^{n} \alpha_{j:n}^{(2)} - (n-i), \qquad 1 \le i \le n-1, \qquad (3.9.33)$$

with $\alpha_{0,j:n} = 0$ *for* $j \ge 1$.

For the case $i = n-1$, *the relation in* (3.9.33) *simply reduces to* (3.9.31).

Theorem 3.9.4. *For the standard normal and the half normal distributions, we have for* $1 \le i \le n$

$$\sum_{j=1}^{n} \alpha_{i,j:n} = 1 + n\alpha_{1:1}\alpha_{i-1:n-1} \tag{3.9.34}$$

and

$$\sum_{j=1}^{n} \beta_{i,j:n} = 1 - (n-i+1)\alpha_{1:1}(\alpha_{i:n} - \alpha_{i-1:n}), \tag{3.9.35}$$

with $\alpha_{0:j} = 0$ *for* $j \ge 1$.

Proof. For the case $i = 1$, (3.9.34) is the same as (3.9.32). For $2 \le i \le n$, it is known that for any arbitrary distribution (Joshi and Balakrishnan, 1982)

$$\sum_{j=i}^{n} \alpha_{i-1,j:n} + \sum_{j=1}^{i-1} \alpha_{j,i:n} = n\alpha_{1:1}\alpha_{i-1:n-1}. \tag{3.9.36}$$

Upon adding Eqs. (3.9.32) and (3.9.36), we derive the relation in (3.9.34). The relation in (3.9.35) readily follows from (3.9.34).

Joshi and Balakrishnan (1981) have used the results of Theorems 3.9.3 and 3.9.4 to find a convenient expression for the variance of the selection differential or reach statistic, defined as $\Delta_k = \bar{X}_k - \bar{X}_n$, where $\bar{X}_n$ is the sample mean and $\bar{X}_k$ is the average of the k largest order statistics. Some properties of $\bar{X}_k$ have been investigated by several authors including Schaeffer *et al.* (1970) and Burrows (1972, 1975). They have observed that $\nu_k = k\,\mathrm{Var}(\bar{X}_k)$ remains almost constant for the selected fraction k/n. While Schaeffer *et al.* (1970) tabulated ν_k for $n \le 20$ and all choices of k by making use of Sarhan and Greenberg's (1956) tables, Burrows (1972, 1975) has provided approximations to $E(\bar{X}_k)$ and ν_k for large values of n. By following Joshi and Balakrishnan (1981), we now have

$$\begin{aligned} k^2 E(\bar{X}_k^2) &= \sum_{i=n-k+1}^{n} \sum_{j=n-k+1}^{n} \alpha_{i,j:n} \\ &= \sum_{i=n-k+1}^{n} \alpha_{i:n}^{(2)} + 2 \sum_{i=n-k+1}^{n-1} \sum_{j=i+1}^{n} \alpha_{i,j:n} \\ &= \sum_{i=n-k+1}^{n} \alpha_{i:n}^{(2)} + 2 \sum_{i=n-k+1}^{n-1} \left\{ \sum_{j=i+1}^{n} \alpha_{j:n}^{(2)} - (n-i) \right\} \end{aligned} \tag{3.9.37}$$

upon using (3.9.33). By rearranging the terms in (3.9.37) and then simplifying, we derive

$$k^2 E(\bar{X}_k^2) = \sum_{i=n-k+1}^{n} (2i - 2n + 2k - 1)\alpha_{i:n}^{(2)} - k(k-1). \tag{3.9.38}$$

As a result, the mean and the variance of $\bar{X}_k$ can be calculated from the tables of $\alpha_{i:n}$ and $\alpha_{i:n}^{(2)}$ alone; Joshi and Balakrishnan (1981) have tabulated them for n up to 50 and have pointed out that Burrows' (1975) approximation for ν_k is not satisfactory for small values of k even when $n=50$, and that the approximation improves with increasing values of k.

For the purpose of testing the presence of outliers in normal samples, Murphy (1951) has proposed the internally studentized version of the selection differential defined by

$$D_k = k(\bar{X}_k - \bar{X}_n)/s = k\Delta_k/s, \tag{3.9.39}$$

where $s^2 = \sum_{i=1}^{n} (X_{i:n} - \bar{X}_n)^2/(n-1)$ is the sample variance. Statistics related to D_k for small values of k have been studied in great detail by several authors. Interested readers may refer to Barnett and Lewis (1978) and Hawkins (1979) for more information on this topic.

Let us now consider the case when $X_1, X_2, \ldots, X_n$ is a random sample from a normal $N(\mu, \sigma^2)$ population, and $X_{1:n} \le X_{2:n} \le \cdots \le X_{n:n}$ are the order statistics obtained from this sample. First of all, it is quite well known that $(\bar{X}, s)$ form a complete sufficient statistic for (μ, σ). Next, we note that the statistic D_k defined in (3.9.39) is invariant with respect to both location and scale, and hence its distribution does not involve (μ, σ). Then, by invoking Basu's (1955) theorem, we immediately find the variables D_k and s to be statistically independent. Consequently, we get from (3.9.39) that

$$\begin{aligned} E(D_k^\ell) &= k^\ell E(\Delta_k^\ell)/E(s^\ell) \\ &= \left\{ \frac{k^\ell \left(\frac{n-1}{2}\right)^{\ell/2} \Gamma\left(\frac{n-1}{2}\right)}{\Gamma\left(\frac{n-1+\ell}{2}\right)} \right\} E(\Delta_k^\ell). \end{aligned} \tag{3.9.40}$$

The mean and the variance of the internally studentized selection differential, D_k, can now be computed from (3.9.40) by using the results that

$$E(\Delta_k) = E(\bar{X}_k)$$

and

$$E(\Delta_k^2) = \mathrm{Var}(\bar{X}_k) - \frac{1}{n} + \{E(\bar{X}_k)\}^2.$$

These quantities have been tabulated for the case $k=1$ by Borenius (1966) for n up to 120.

3.10. Results for the Half Logistic Distribution

Let $X_{1:n} \leq X_{2:n} \leq \cdots \leq X_{n:n}$ be the order statistics from a half logistic population with pdf

$$f(x) = 2e^{-x}/(1+e^{-x})^2, \qquad 0 \leq x < \infty, \tag{3.10.1}$$

and cdf

$$F(x) = (1-e^{-x})/(1+e^{-x}), \qquad 0 \leq x < \infty. \tag{3.10.2}$$

The density function $f(x)$ in (3.10.1), obtained by folding the logistic density function in (3.6.1) at $x=0$, is a monotonic decreasing function of x in the interval $[0, \infty)$ and has an increasing hazard rate. It, therefore, serves as a good failure-time model in life-testing problems, as is shown later in Chapters 4 and 6.

From (3.10.1) and (3.10.2), we immediately observe the relations

$$f(x) = F(x)\{1-F(x)\} + \frac{1}{2}\{1-F(x)\}^2, \tag{3.10.3}$$

$$f(x) = \{1-F(x)\} - \frac{1}{2}\{1-F(x)\}^2 \tag{3.10.4}$$

and

$$f(x) = \frac{1}{2}\{1-F^2(x)\}. \tag{3.10.5}$$

By making use of the relations in (3.10.3)-(3.10.5), Balakrishnan (1985) has derived several recurrence relations satisfied by the single and the product moments of order statistics. These relations, when applied in a simple and systematic recursive way, will enable one to compute the means, variances, and covariances of all order statistics for all sample sizes simply by starting with the values of $\alpha_{1:1} = E(X) = \ln 4$ and $\alpha_{1:1}^{(2)} = E(X^2) = \pi^2/3$.

Theorem 3.10.1. *For the half logistic population with pdf as in* (3.10.1), *we have*

$$\alpha_{1:n+1}^{(k+1)} = 2\left\{\alpha_{1:n}^{(k+1)} - \frac{k+1}{n}\alpha_{1:n}^{(k)}\right\}, \qquad n \geq 1,\ k \geq 0, \tag{3.10.6}$$

with $\alpha_{1:n}^{(0)} \equiv 1$.

Proof. For $n \geq 1$, we have from (2.4.4) that

$$\alpha_{1:n}^{(k)} = n\int_0^\infty x^k\{1-F(x)\}^{n-1}f(x)\,dx.$$

Upon using the relation in (3.10.4) and splitting the above integral accordingly into two, we get

$$\alpha_{1:n}^{(k)} = n\left[\int_0^\infty x^k\{1-F(x)\}^n\,dx - \frac{1}{2}\int_0^\infty x^k\{1-F(x)\}^{n+1}\,dx\right]. \quad (3.10.7)$$

Integrating the RHS of (3.10.7) by parts, treating x^k for integration and the rest of the integrand for differentiation, we obtain for $n \geq 1$ and $k \geq 0$

$$\alpha_{1:n}^{(k)} = \frac{n}{k+1}\left[n\int_0^\infty x^{k+1}\{1-F(x)\}^{n-1}f(x)\,dx\right.$$

$$\left. - \frac{n+1}{2}\int_0^\infty x^{k+1}\{1-F(x)\}^n f(x)\,dx\right]. \quad (3.10.8)$$

Equation (3.10.8), when simplified, yields the relation in (3.10.6).

Theorem 3.10.2. *For the half logistic population with pdf as in* (3.10.1), *we have*

$$\alpha_{i+1:n+1}^{(k+1)} = \frac{1}{i}\left[\frac{(n+1)(k+1)}{n-i+1}\alpha_{i:n}^{(k)} + \frac{n+1}{2}\alpha_{i-1:n}^{(k+1)} - \frac{n-2i+1}{2}\alpha_{i:n+1}^{(k+1)}\right],$$

$$1 \leq i \leq n,\ k \geq 0, \quad (3.10.9)$$

with $\alpha_{0:t}^{(k)} \equiv 0$ *for* $t \geq 1$ *and* $k \geq 0$, *and* $\alpha_{i:t}^{(0)} \equiv 1$ *for* $1 \leq i \leq t$.

Proof. From (2.4.4), we have for $1 \leq i \leq n$ and $k \geq 0$

$$\alpha_{i:n}^{(k)} = \frac{n!}{(i-1)!(n-i)!}\int_0^\infty x^k\{F(x)\}^{i-1}\{1-F(x)\}^{n-i}f(x)\,dx.$$

Upon using the relation in (3.10.3) and splitting the integral accordingly into two, we get

$$\alpha_{i:n}^{(k)} = \frac{n!}{(i-1)!(n-i)!}\left[\int_0^\infty x^k\{F(x)\}^i\{1-F(x)\}^{n-i+1}\,dx\right.$$

$$\left. + \int_0^\infty x^k\{F(x)\}^{i-1}\{1-F(x)\}^{n-i+2}\,dx\right]. \quad (3.10.10)$$

Upon integrating the first integral on the RHS of (3.10.10) by parts, treating x^k for integration and the rest of the integrand for differentiation, we obtain

$$\int_0^\infty x^k\{F(x)\}^i\{1-F(x)\}^{n-i+1}\,dx$$

$$
= \frac{1}{k+1}\left[(n-i+1)\int_0^\infty x^{k+1}\{F(x)\}^i\{1-F(x)\}^{n-i}f(x)\,dx\right.
$$

$$
\left.-i\int_0^\infty x^{k+1}\{F(x)\}^{i-1}\{1-F(x)\}^{n-i+1}f(x)\,dx\right]
$$

$$
= \frac{i!(n-i+1)!}{(n+1)!(k+1)}\{\alpha_{i+1:n+1}^{(k+1)} - \alpha_{i:n+1}^{(k+1)}\}. \tag{3.10.11}
$$

By simply changing i to $i-1$ in the above equation, we obtain an expression for the second integral on the RHS of (3.10.10):

$$
\int_0^\infty x^k\{F(x)\}^{i-1}\{1-F(x)\}^{n-i+2}\,dx = \frac{(i-1)!(n-i+2)!}{(n+1)!(k+1)}\{\alpha_{i:n+1}^{(k+1)} - \alpha_{i-1:n+1}^{(k+1)}\}. \tag{3.10.12}
$$

By substituting the expressions of the two integrals in (3.10.11) and (3.10.12) in Eq. (3.10.10) and simplifying the resulting equation, we get

$$
\alpha_{i:n}^{(k)} = \frac{(n-i+1)}{(n+1)(k+1)}[i\{\alpha_{i+1:n+1}^{(k+1)} - \alpha_{i:n+1}^{(k+1)}\} + (n-i+2)\{\alpha_{i:n+1}^{(k+1)} - \alpha_{i-1:n+1}^{(k+1)}\}]. \tag{3.10.13}
$$

From Relation 3.3.1, we have

$$
(n-i+2)\{\alpha_{i:n+1}^{(k+1)} - \alpha_{i-1:n+1}^{(k+1)}\} = (n+1)\{\alpha_{i:n+1}^{(k+1)} - \alpha_{i-1:n}^{(k+1)}\},
$$

which, when used in Eq. (3.10.13) and simplified, yields the relation in (3.10.9).

Starting with the first k raw moments of X, Theorems 3.10.1 and 3.10.2 will allow one to evaluate the first k raw moments of all order statistics for all sample sizes in a very simple recursive manner. Thus, for example, starting with $\alpha_{1:1} = E(X) = \ln 4$ and $\alpha_{1:1}^{(2)} = E(X^2) = \pi^2/3$, one can employ Theorems 3.10.1 and 3.10.2 to complete the means and variances of all order statistics for all sample sizes.

Theorem 3.10.3. *For the half logistic distribution with pdf as in* (3.10.1), *we have*

$$
\alpha_{i,i+1:n+1} = \alpha_{i:n+1}^{(2)} + \frac{2(n+1)}{n-i+1}\left\{\alpha_{i,i+1:n} - \alpha_{i:n}^{(2)} - \frac{1}{n-i}\,\alpha_{i:n}\right\}, \qquad 1 \le i \le n-1. \tag{3.10.14}
$$

Proof. From (2.3.5), we may write for $1 \le i \le n-1$ that

$$\alpha_{i:n} = E(X_{i:n}X^0_{i+1:n})$$
$$= \frac{n!}{(i-1)!(n-i-1)!} \int_0^\infty \int_x^\infty x\{F(x)\}^{i-1}\{1-F(y)\}^{n-i-1} f(x)f(y)\,dy\,dx$$
$$= \frac{n!}{(i-1)!(n-i-1)!} \int_0^\infty x\{F(x)\}^{i-1} f(x) I(x)\,dx, \qquad (3.10.15)$$

where

$$I(x) = \int_x^\infty \{1-F(y)\}^{n-i-1} f(y)\,dy. \qquad (3.10.16)$$

Upon using the relation in (3.10.4), we may write

$$I(x) = \int_x^\infty \{1-F(y)\}^{n-i}\,dy - \frac{1}{2}\int_x^\infty \{1-F(y)\}^{n-i+1}\,dy,$$

which, when integrated by parts, yields

$$I(x) = (n-i)\int_x^\infty y\{1-F(y)\}^{n-i-1} f(y)\,dy - x\{1-F(x)\}^{n-i}$$
$$+ \frac{1}{2}x\{1-F(x)\}^{n-i+1} - \frac{n-i+1}{2}\int_x^\infty y\{1-F(y)\}^{n-i} f(y)\,dy.$$

The recurrence relation in (3.10.14) follows immediately when one substitutes the above expression of $I(x)$ in (3.10.15) and simplifies the resulting equation.

Theorem 3.10.4. *For the half logistic distribution with pdf as in* (3.10.1), *we have*

$$\alpha_{2,3:n+1} = \alpha^{(2)}_{3:n+1} + (n+1)\left\{\alpha_{2:n} - \frac{n}{2}\alpha^{(2)}_{1:n-1}\right\}, \qquad n \ge 2, \quad (3.10.17)$$

and

$$\alpha_{i+1,i+2:n+1} = \alpha^{(2)}_{i+2:n+1} + \frac{(n+1)}{i(i+1)}[2\alpha_{i+1:n} + n\{\alpha_{i-1,i:n-1} - \alpha^{(2)}_{i:n-1}\}],$$
$$2 \le i \le n-1. \qquad (3.10.18)$$

Proof. From (2.3.5), let us write for $1 \le i \le n-1$

$$\alpha_{i+1:n} = E(X^0_{i:n} X_{i+1:n})$$
$$= \frac{n!}{(i-1)!(n-i-1)!} \int_0^\infty \int_0^y y\{F(x)\}^{i-1}\{1-F(y)\}^{n-i-1} f(x) f(y)\, dx\, dy$$
$$= \frac{n!}{(i-1)!(n-i-1)!} \int_0^\infty y\{1-F(y)\}^{n-i-1} f(y) J(y)\, dy, \qquad (3.10.19)$$

where

$$J(y) = \int_0^y \{F(x)\}^{i-1} f(x)\, dx. \qquad (3.10.20)$$

Upon using the relation in (3.10.5), we may write

$$J(y) = \frac{1}{2}\left[\int_0^y \{F(x)\}^{i-1}\, dx - \int_0^y \{F(x)\}^{i+1}\, dx\right],$$

which, when integrated by parts, yields for $i = 1$,

$$J(y) = \frac{1}{2}\left[y - y\{F(y)\}^2 + 2\int_0^y xF(x) f(x)\, dx\right],$$

and for $2 \le i \le n-1$,

$$J(y) = \frac{1}{2}\Bigg[y\{F(y)\}^{i-1} - (i-1)\int_0^y x\{F(x)\}^{i-2} f(x)\, dx$$
$$- y\{F(y)\}^{i+1} + (i+1)\int_0^y x\{F(x)\}^{i} f(x)\, dx\Bigg].$$

The recurrence relations in (3.10.17) and (3.10.18) follow immediately upon substitution of the above expressions of $J(y)$ in (3.10.19) and simplification of the resulting equations.

By setting $i = n-1$ in (3.10.18), in particular, we obtain the relation

$$\alpha_{n,n+1:n+1} = \alpha^{(2)}_{n+1:n+1} + \frac{(n+1)}{(n-1)n}[2\alpha_{n:n} + n\{\alpha_{n-2,n-1:n-1} - \alpha^{(2)}_{n-1:n-1}\}],$$
$$n \ge 3. \qquad (3.10.21)$$

It should be mentioned here that Theorems 3.10.3 and 3.10.4 will allow one to evaluate the product moments of order statistics for all sample sizes in a simple recursive manner. To see this, we first note from (3.3.8) that

$$\alpha_{1,2:2} = \alpha^2_{1:1} = (\ln 4)^2. \qquad (3.10.22)$$

The relations in (3.10.14), (3.10.17), and (3.10.18), when used along with (3.10.22), will enable one to calculate the product moments $\alpha_{i,i+1:n}$ ($1 \le i \le n-1$). This is sufficient for the evaluation of all the product moments $\alpha_{i,j:n}$, as the remaining product moments, viz., $\alpha_{i,j:n}$ for $1 \le i < j \le n$ and $j-i \ge 2$, can all be systematically computed by using Relation 3.3.4.

Proceeding similarly, Balakrishnan (1985) has established some more recurrence relations satisfied by the product moments $\alpha_{i,j:n}$ $(1 \le i < j \le n,$ $j-i \ge 2)$, which are presented in the following two theorems.

Theorem 3.10.5. *For the half logistic distribution with pdf as in* (3.10.1), *we have*

$$\alpha_{i,j:n+1} = \alpha_{i,j-1:n+1} + \frac{2(n+1)}{n-j+2}\left\{\alpha_{i,j:n} - \alpha_{i,j-1:n} - \frac{1}{n-j+1}\alpha_{i:n}\right\},$$

$$1 \le i \le n-2, j-i \ge 2. \tag{3.10.23}$$

Theorem 3.10.6. *For the half logistic distribution with pdf as in* (3.10.1), *we have*

$$\alpha_{2,j+1:n+1} = \alpha_{3,j+1:n+1} + (n+1)\left\{\alpha_{j:n} - \frac{n}{2}\alpha_{1,j-1:n-1}\right\}, \qquad 3 \le j \le n, \tag{3.10.24}$$

and

$$\alpha_{i+1,j+1:n+1} = \alpha_{i+2,j+1:n+1} + \frac{(n+1)}{i(i+1)}[2\alpha_{j:n} - n\{\alpha_{i,j-1:n-1} - \alpha_{i-1,j-1:n-1}\}],$$

$$2 \le i \le n-2, j-i \ge 2. \tag{3.10.25}$$

By making use of all these recurrence relations, Balakrishnan (1985) has tabulated the means, variances, and covariances of order statistics from the half logistic distribution with pdf as in (3.10.1) for sample sizes up to 15. Balakrishnan (1985) has also tabulated several percentage points of the smallest and the largest order statistics for sample sizes up to 15. He has further shown that the distribution of order statistics is unimodal and has also tabulated the modes of all order statistics for sample sizes 15 and less. Wong (1988) has recently extended the tables of means, variances, and covariances for sample sizes up to 20.

3.11. David and Johnson's Approximation

As explained already in Sections 2.4 and 3.4, the probability integral transformation $u = F(x)$ transforms the order statistic $X_{i:n}$ from a population with pdf $f(x)$ and cdf $F(x)$ into the uniform order statistic $U_{i:n}$ for $i = 1, 2, \ldots, n$. Hence, by inverting the above transformation we get for $1 \le i \le n$

$$X_{i:n} = F^{-1}(U_{i:n}) = G(U_{i:n}), \tag{3.11.1}$$

which, when expanded in a Taylor series around the point $E(U_{i:n}) = i/(n+1) = p_i$, gives

$$\begin{aligned} X_{i:n} = {} & G_i + G_i'(U_{i:n} - p_i) + \frac{1}{2} G_i''(U_{i:n} - p_i)^2 \\ & + \frac{1}{6} G_i'''(U_{i:n} - p_i)^3 + \frac{1}{24} G_i^{\text{iv}}(U_{i:n} - p_i)^4 + \cdots; \end{aligned} \tag{3.11.2}$$

here, G_i denotes $G(p_i)$, G_i' denotes $d/du\, G(u)|_{u=p_i}$, and similarly G_i'', G_i''', $G_i^{\text{iv}}, \ldots$, denote successive derivatives of $G(u)$ evaluated at $u = p_i$. Then, by taking expectation on both sides of (3.11.2) and by using the expressions of the central moments of uniform order statistics derived from (3.4.7) (written, however, in inverse powers of $n+2$ by David and Johnson (1954) for simplicity and computational ease), we obtain

$$\begin{aligned} \alpha_{i:n} \simeq {} & G_i + \frac{p_i q_i}{2(n+2)} G_i'' + \frac{p_i q_i}{(n+2)^2} \left[\frac{1}{3}(q_i - p_i) G_i''' + \frac{1}{8} p_i q_i G_i^{\text{iv}} \right] \\ & + \frac{p_i q_i}{(n+2)^3} \left[-\frac{1}{3}(q_i - p_i) G_i''' + \frac{1}{4}\{(q_i - p_i)^2 - p_i q_i\} G_i^{\text{iv}} \right. \\ & \left. + \frac{1}{6} p_i q_i (q_i - p_i) G_i^{\text{v}} + \frac{1}{48} p_i^2 q_i^2 G_i^{\text{vi}} \right], \end{aligned} \tag{3.11.3}$$

where $q_i = 1 - p_i = (n - i + 1)/(n+1)$. Similarly, by taking expectation on the series expansion for $X_{i:n}^2$ obtained from (3.11.2), and then subtracting from it the expression of $\alpha_{i:n}^2$ obtained from (3.11.3), we get an approximate formula for the variance of $X_{i:n}$:

$$\begin{aligned} \beta_{i,i:n} \simeq {} & \frac{p_i q_i}{n+2} (G_i')^2 + \frac{p_i q_i}{(n+2)^2} \left[2(q_i - p_i) G_i' G_i'' + p_i q_i \left\{ G_i' G_i''' + \frac{1}{2}(G_i'')^2 \right\} \right] \\ & + \frac{p_i q_i}{(n+2)^3} \left[-2(q_i - p_i) G_i' G_i'' + \{(q_i - p_i)^2 - p_i q_i\} \right. \\ & \times \left\{ 2 G_i' G_i''' + \frac{3}{2}(G_i'')^2 \right\} + p_i q_i (q_i - p_i) \left\{ \frac{5}{3} G_i' G_i^{\text{iv}} + 3 G_i'' G_i''' \right\} \\ & \left. + \frac{1}{4} p_i^2 q_i^2 \left\{ G_i' G_i^{\text{v}} + 2 G_i'' G_i^{\text{iv}} + \frac{5}{3}(G_i''')^2 \right\} \right]. \end{aligned} \tag{3.11.4}$$

Next, by taking expectation on the series expansion for $X_{i:n}X_{j:n}$ obtained from (3.11.2), and then subtracting from it the expression of $\alpha_{i:n}\alpha_{j:n}$ obtained from (3.11.3), we derive an approximate formula for the covariance of $X_{i:n}$ and $X_{j:n}$ $(1 \le i < j \le n)$:

$$\begin{aligned}
\beta_{i,j:n} \simeq{}& \frac{p_i q_j}{n+2} G_i' G_j' + \frac{p_i q_j}{(n+2)^2}\Big[(q_i - p_i)G_i'' G_j' + (q_j - p_j)G_i' G_j'' \\
&+ \frac{1}{2} p_i q_i G_i''' G_j' + \frac{1}{2} p_j q_j G_i' G_j''' + \frac{1}{2} p_i q_j G_i'' G_j''\Big] \\
&+ \frac{p_i q_j}{(n+2)^3}\Big[-(q_i - p_i)G_i'' G_j' - (q_j - p_j)G_i' G_j'' + \{(q_i - p_i)^2 - p_i q_i\} G_i''' G_j' \\
&+ \{(q_j - p_j)^2 - p_j q_j\} G_i' G_j''' + \Big\{\frac{3}{2}(q_i - p_i)(q_j - p_j) + \frac{1}{2} p_j q_i \\
&- 2 p_i q_j\Big\} G_i'' G_j'' + \frac{5}{6} p_i q_i (q_i - p_i) G_i^{\text{iv}} G_j' + \frac{5}{6} p_j q_j (q_j - p_j) G_i' G_j^{\text{iv}} \\
&+ \Big\{p_i q_j (q_i - p_i) + \frac{1}{2} p_i q_i (q_j - p_j)\Big\} G_i''' G_j'' + \Big\{p_i q_j (q_j - p_j) \\
&+ \frac{1}{2} p_j q_j (q_i - p_i)\Big\} G_i'' G_j''' + \frac{1}{8} p_i^2 q_i^2 G_i^{\text{v}} G_j' + \frac{1}{8} p_j^2 q_j^2 G_i' G_j^{\text{v}} \\
&+ \frac{1}{4} p_i^2 q_i q_j G_i^{\text{iv}} G_j'' + \frac{1}{4} p_i p_j q_j^2 G_i'' G_j^{\text{iv}} \\
&+ \frac{1}{12}(2 p_i^2 q_j^2 + 3 p_i p_j q_i q_j) G_i''' G_j'''\Big].
\end{aligned} \tag{3.11.5}$$

Similarly, series approximations for the first four cumulants and cross-cumulants of order statistics have been developed by David and Johnson (1954); also see K. Pearson and M. V. Pearson (1931). Clark and Williams (1958) have developed very similar series approximations by making use of the exact expressions of the central moments of uniform order statistics where the kth central moment is of order $\{(n+2)(n+3)\cdots(n+k)\}^{-1}$ instead of in inverse powers of $n+2$. A different kind of series approximation based on the logistic distribution rather than on the uniform distribution has been given by Plackett (1958). Saw (1960) has employed the Darboux form for the remainder in a Taylor series expansion in order to obtain bounds for the remainder term when the expansion for the first single moment $\alpha_{i:n}$ $(1 \le i \le n)$ in (3.11.3) is terminated after an even number of terms in the series. Details of these works may be had from Arnold and Balakrishnan (1989).

As pointed out by David and Johnson (1954), the evaluation of the derivatives of G_i is rather easy in most cases. First of all, we realize that

$$G_i' = \frac{d}{du} G(u)|_{u=p_i} = \frac{1}{f(F^{-1}(u))}\bigg|_{u=p_i} = \frac{1}{f(G_i)}, \qquad (3.11.6)$$

which is just the reciprocal of the pdf of the population evaluated at G_i. The above expression of G_i' in (3.11.6) allows us to write down the higher-order derivatives of G_i without great difficulty in most cases.

For example, for the standard normal distribution, by making use of the property $f'(x) = -xf(x)$ we get

$$\begin{aligned}
G_i' &= 1/f(G_i), \\
G_i'' &= G_i/\{f(G_i)\}^2, \\
G_i''' &= \{1+2G_i^2\}/\{f(G_i)\}^3, \\
G_i^{\text{iv}} &= G_i(7+6G_i^2)/\{f(G_i)\}^4, \\
G_i^{\text{v}} &= \{7+46G_i^2+24G_i^4\}/\{f(G_i)\}^5, \\
G_i^{\text{vi}} &= G_i\{127+326G_i^2+120G_i^4\}/\{f(G_i)\}^6,
\end{aligned}$$

etc.

Similarly, for the logistic population with pdf and cdf as

$$f(x) = \frac{e^{-x}}{(1+e^{-x})^2} \quad \text{and} \quad F(x) = \frac{1}{1+e^{-x}},$$

respectively, we obtain

$$\begin{aligned}
G_i &= F^{-1}(p_i) = \ln(p_i) - \ln(q_i), \\
G_i' &= \frac{1}{p_i} + \frac{1}{q_i}, \\
G_i'' &= -\frac{1}{p_i^2} + \frac{1}{q_i^2}, \\
G_i''' &= 2\left\{\frac{1}{p_i^3} + \frac{1}{q_i^3}\right\}, \\
G_i^{\text{iv}} &= 6\left\{-\frac{1}{p_i^4} + \frac{1}{q_i^4}\right\}, \\
G_i^{\text{v}} &= 24\left\{\frac{1}{p_i^5} + \frac{1}{q_i^5}\right\}, \\
G_i^{\text{vi}} &= 120\left\{-\frac{1}{p_i^6} + \frac{1}{q_i^6}\right\},
\end{aligned}$$

etc.

As illustrated by David and Johnson (1954), and also by several other authors, this simple approximation procedure works well in most cases. However, this procedure may not provide satisfactory results for the extreme order statistics. In this case, the convergence of this approximation to the exact value may be very slow, and even nonexistent in some cases.

Chapter 4 Linear Estimation Based on Order Statistics

4.1. Preliminary Remarks

In this chapter, we first describe the best linear unbiased estimation of the scale parameter of a population, in Section 2, and illustrate the method with an example from the Rayleigh population. In Section 3, we consider the one-parameter exponential distribution and use the formula in Section 2 to derive an explicit expression for the BLUE (best linear unbiased estimator) of the scale parameter and its variance. Next, in Section 4, by considering the location and scale parameter family of distributions, we describe the best linear unbiased estimation, and illustrate it with some real-life examples. In Section 5, we consider the two-parameter exponential distribution and make use of the formulas in Section 4 to derive explicit expressions for the BLUEs of the location and scale parameters and their variances. We describe in Section 6 the simplified linear estimators of the location and scale parameters given by Gupta (1952). In Section 7, we explain the unbiased nearly best linear estimators of the location and scale parameters derived by Blom (1958, 1962). We also present some numerical examples to illustrate the two simplified methods of estimation given in Sections 6 and 7. In Section 8, we describe the linear estimators with polynomial coefficients proposed by Downton (1966) and discuss their efficiency. Finally, in Section 9, we present details of all other work related to the problems discussed in this chapter.

4.2. BLUE of the Scale Parameter

Let $X_{r+1:n} \le X_{r+2:n} \le \cdots \le X_{n-s:n}$ be a Type-II censored sample from a scale-parameter distribution. Let us denote $Z_{i:n} = X_{i:n}/\sigma$ $(r+1 \le i \le n-s)$, $E(Z_{i:n}) = \alpha_{i:n}(r+1 \le i \le n-s)$, and $\mathrm{Cov}(Z_{i:n}, Z_{j:n}) = \beta_{i,j:n}(r+1 \le i \le j \le n-s)$. We shall use the notations

$$\mathbf{X} = (X_{r+1:n} X_{r+2:n} \cdots X_{n-s:n})', \tag{4.2.1}$$

$$\boldsymbol{\alpha} = (\alpha_{r+1:n} \alpha_{r+2:n} \cdots \alpha_{n-s:n})', \tag{4.2.2}$$

and

$$\boldsymbol{\beta} = \begin{bmatrix} \beta_{r+1,r+1:n} & \beta_{r+1,r+2:n} & \cdots & \beta_{r+1,n-s:n} \\ \beta_{r+1,r+2:n} & \beta_{r+2,r+2:n} & \cdots & \beta_{r+2,n-s:n} \\ \vdots & \vdots & \cdots & \vdots \\ \beta_{r+1,n-s:n} & \beta_{r+2,n-s:n} & \cdots & \beta_{n-s,n-s:n} \end{bmatrix}; \tag{4.2.3}$$

we may then write

$$\xi(\mathbf{X}) = \sigma\boldsymbol{\alpha} \quad \text{and} \quad \nu(\mathbf{X}) = \sigma^2\boldsymbol{\beta}. \tag{4.2.4}$$

Let us now consider the generalized variance given by

$$\begin{aligned} &(\mathbf{X} - \sigma\boldsymbol{\alpha})'\boldsymbol{\beta}^{-1}(\mathbf{X} - \sigma\boldsymbol{\alpha}) \\ &\quad = \mathbf{X}'\boldsymbol{\beta}^{-1}\mathbf{X} - \sigma\boldsymbol{\alpha}'\boldsymbol{\beta}^{-1}\mathbf{X} - \sigma\mathbf{X}'\boldsymbol{\beta}^{-1}\boldsymbol{\alpha} + \sigma^2\boldsymbol{\alpha}'\boldsymbol{\beta}^{-1}\boldsymbol{\alpha} \\ &\quad = \mathbf{X}'\boldsymbol{\beta}^{-1}\mathbf{X} - 2\sigma\boldsymbol{\alpha}'\boldsymbol{\beta}^{-1}\mathbf{X} + \sigma^2\boldsymbol{\alpha}'\boldsymbol{\beta}^{-1}\boldsymbol{\alpha}. \end{aligned} \tag{4.2.5}$$

By minimizing the expression of the generalized variance in (4.2.5) with respect to σ, we derive the BLUE of σ as

$$\sigma^* = \left(\frac{\boldsymbol{\alpha}'\boldsymbol{\beta}^{-1}}{\boldsymbol{\alpha}'\boldsymbol{\beta}^{-1}\boldsymbol{\alpha}}\right)\mathbf{X} = \sum_{i=r+1}^{n-s} a_i X_{i:n}, \tag{4.2.6}$$

and its variance is given by

$$\begin{aligned} \mathrm{Var}(\sigma^*) &= \frac{1}{(\boldsymbol{\alpha}'\boldsymbol{\beta}^{-1}\boldsymbol{\alpha})^2}\boldsymbol{\alpha}'\boldsymbol{\beta}^{-1}(\sigma^2\boldsymbol{\beta})\boldsymbol{\beta}^{-1}\boldsymbol{\alpha} \\ &= \sigma^2/(\boldsymbol{\alpha}'\boldsymbol{\beta}^{-1}\boldsymbol{\alpha}). \end{aligned} \tag{4.2.7}$$

We may note that if $\boldsymbol{\alpha}$ and $\boldsymbol{\beta}$ are known, the coefficients a_i needed for the calculation of the BLUE σ^* in (4.2.6) and the variance of σ^* in (4.2.7) can be tabulated once and for all.

Example 4.2.1. Let us consider the following censored sample of size seven (obtained from a random sample of size 12 by censoring the three smallest and the two largest observations) given by Dyer and Whisenand (1973a):

$$7.2113, 13.0397, 13.5380, 15.0124, 16.1683, 18.1613, 23.2829.$$

The above data has been simulated by them from a Rayleigh distribution with $\sigma = 12.38$.

For the case of the Rayleigh distribution, the coefficients a_i in (4.2.6) and the variance of σ^* in (4.2.7) have been tabulated by Dyer and Whisenand (1973a) for various choices of r and s and for sample sizes up to 15. From their tables, we obtain the BLUE of σ in this example to be

$$\begin{aligned}\sigma^* &= 0.09699(7.2113) + 0.05009(13.0397) + 0.05681(13.5380)\\ &\quad + 0.06372(15.0124) + 0.07105(16.1683) + 0.07911(18.1613)\\ &\quad + 0.28379(23.2829)\\ &= 12.27,\end{aligned}$$

and the standard errror of this estimate is obtained as

$$\text{S.E.}(\sigma^*) = 12.27(0.025)^{1/2} = 1.94.$$

Example 4.2.2. Let us consider the following data, which represent failure times, in minutes, for a specific type of electrical insulation that was subjected to a continuously increasing voltage stress as given by Lawless (1982, p. 138):

$$12.3, 21.8, 24.4, 28.6, 43.2, 46.9, 70.7, 75.3, 95.5, 98.1, 138.6, 151.9$$

Let us assume a one-parameter half logistic distribution with pdf

$$f(x; \sigma) = \frac{2}{\sigma} \frac{e^{-x/\sigma}}{(1 + e^{-x/\sigma})^2}, \qquad 0 \leq x \leq \infty, \sigma > 0,$$

for the above data. Chan (1989) has shown that this one-parameter half logistic distribution fits the above data extremely well. In this case, the coefficients a_i in (4.2.6) and the variance of σ^* in (4.2.7) have been tabulated by Chan (1989) for various choices of r and s and for sample sizes up to 20. From his tables, we obtain the BLUE of σ in this example to be

$$\begin{aligned}\sigma^* &= 0.01469(12.3) + 0.02238(21.8) + 0.03010(24.4) + 0.03770(28.6)\\ &\quad + 0.04501(43.2) + 0.05184(46.9) + 0.05796(70.7) + 0.06311(75.3)\\ &\quad + 0.06694(95.5) + 0.06897(98.1) + 0.06844(138.6) + 0.06356(151.9)\\ &= 48.01,\end{aligned}$$

with the standard error

$$\text{S.E.}(\sigma^*) = 48.01(0.05848)^{1/2} = 11.61.$$

Hence, an unbiased estimate of the expected failure time is given by

$$\sigma^* \ln 4 = 66.55 \text{ min},$$

with the standard error

$$11.61 \ln 4 = 16.09 \text{ min}.$$

Suppose the experiment had been stopped as soon as the eleventh failure occurred—that is, the largest observation had been censored. Then, from the tables prepared by Chan (1989) we obtain the BLUE of σ to be

$$\begin{aligned}\sigma^* &= 0.01582(12.3) + 0.02410(21.8) + 0.03242(24.4) + 0.04060(28.6)\\ &\quad + 0.04848(43.2) + 0.05582(46.9) + 0.06240(70.7) + 0.06791(75.3)\\ &\quad + 0.07199(95.5) + 0.07409(98.1) + 0.14034(138.6)\\ &= 50.50,\end{aligned}$$

and the standard error of this estimate to be

$$\text{S.E.}(\sigma^*) = 50.50(0.06302)^{1/2} = 12.68.$$

Thus, an unbiased estimate of the expected failure time is given by

$$\sigma^* \ln 4 = 70.01 \text{ min},$$

with the standard error

$$12.68 \ln 4 = 17.58 \text{ min}.$$

4.3. BLUE for the One-Parameter Exponential Distribution

In this section we assume that

$$X_{r+1:n} \le X_{r+2:n} \le \cdots \le X_{n-s:n} \tag{4.3.1}$$

is a Type-II censored sample available from the one-parameter exponential population with pdf

$$f(x;\sigma) = \frac{1}{\sigma} e^{-x/\sigma}, \qquad x \ge 0, \sigma > 0. \tag{4.3.2}$$

As we have already seen in Section 3.5, we have in this case

$$\alpha_{i:n} = \sum_{l=n-i+1}^{n} 1/l, \qquad 1 \le i \le n, \tag{4.3.3}$$

and

$$\beta_{i,i:n} = \beta_{i,j:n} = \sum_{l=n-i+1}^{n} 1/l^2, \qquad 1 \le i \le j \le n. \tag{4.3.4}$$

From the expressions of the variances and covariances in (4.3.4), we determine the inverse of the variance–covariance matrix $\boldsymbol{\beta}$ to be a symmetric matrix whose (i, j)th element is given by

$$(\boldsymbol{\beta}^{-1})_{ij} = \begin{cases} \left[\sum_{l=n-r}^{n} 1/l^2\right]^{-1} + (n-r-1)^2 & \text{for } i=j=r+1 \\ (n-i)^2 + (n-i+1)^2 & \text{for } r+2 \le i = j \le n-s-1 \\ (s+1)^2 & \text{for } i=j=n-s \\ -(n-i)^2 & \text{for } j=i+1,\ r+1 \le i \le n-s-1 \\ 0 & \text{otherwise.} \end{cases} \tag{4.3.5}$$

By using the expression of the expected value of exponential order statistics in (4.3.3) along with the expression of the inverse of the variance–covariance matrix given in (4.3.5), we obtain the vector $\mathbf{a}^{*\prime} = \boldsymbol{\alpha}'\boldsymbol{\beta}^{-1}$ to be

$$a_i^* = \begin{cases} \left\{\left[\sum_{l=n-r}^{n} 1/l\right] \Big/ \left[\sum_{l=n-r}^{n} 1/l^2\right]\right\} - (n-r-1) & \text{for } i=r+1 \\ 1 & \text{for } r+2 \le i \le n-s-1 \\ s+1 & \text{for } i=n-s. \end{cases} \tag{4.3.6}$$

We also similarly obtain

$$\begin{aligned} \boldsymbol{\alpha}'\boldsymbol{\beta}^{-1}\boldsymbol{\alpha} &= \mathbf{a}^{*\prime}\boldsymbol{\alpha} \\ &= \sum_{i=r+1}^{n-s} a_i^* \sum_{l=n-i+1}^{n} 1/l \\ &= \frac{\left\{\sum_{l=n-r}^{n} 1/l\right\}^2}{\sum_{l=n-r}^{n} 1/l^2} + (n-r-s-1) \\ &= K \quad \text{(say).} \end{aligned} \tag{4.3.7}$$

We then derive the BLUE of σ from (4.2.6) to be

$$\sigma^* = \sum_{i=r+1}^{n-s} a_i X_{i:n}, \tag{4.3.8}$$

where

$$a_i = a_i^* / K, \qquad r+1 \le i \le n-s, \tag{4.3.9}$$

and the variance of this estimator is given by

$$\text{Var}(\sigma^*) = \sigma^2 / K, \tag{4.3.10}$$

with K as defined in (4.3.7).

In the special case of the right-censored sample (that is, $r=0$), the estimator σ^* in (4.3.8) simplifies to

$$\sigma^* = \frac{1}{n-s}\left\{\sum_{i=1}^{n-s} X_{i:n} + sX_{n-s:n}\right\}, \tag{4.3.11}$$

and the variance in (4.3.10) reduces to

$$\text{Var}(\sigma^*) = \sigma^2/(n-s). \tag{4.3.12}$$

It is of interest to mention here that the BLUE σ^* in (4.3.11) is exactly the same as the maximum likelihood estimator of σ based on the right-censored sample.

By using the formulas in (4.3.6) through (4.3.10), Sarhan and Greenberg (1957) have tabulated the coefficients a_i and the variance of the BLUE for doubly censored samples in samples of size up to 10. Epstein (1956, 1962), Epstein and Sobel (1953, 1954), Herd (1956), and Balakrishnan *et al.* (1990a) have proposed some simplified estimators for σ in the case of doubly and multiply censored samples.

Example 4.3.1. The following data, given by Proschan (1963), are times between successive failures of air conditioning equipment in a Boeing 720 airplane, arranged in increasing order of magnitude:

12, 21, 26, 27, 29, 29, 48, 57, 59, 70, 74, 153, 326, 386, 502.

Let us assume an exponential distribution with pdf as in (4.3.2) for the time between successive failures of the air conditioning equipment.

Then, based on the above-given complete sample of size 15 we obtain the BLUE of σ to be

$$\sigma^* = \bar{X} = 121.2667,$$

and its standard error to be

$$\text{S.E.}(\sigma^*) = \sigma^*/\sqrt{15} = 31.3109.$$

Suppose in the above sample the largest observation is censored. We then have the BLUE of σ to be

$$\sigma^* = \frac{1}{14}\left\{\sum_{i=1}^{14} X_{i:15} + X_{14:15}\right\} = 121.6429,$$

and its standard error to be

$$\text{S.E.}(\sigma^*) = \sigma^*/\sqrt{14} = 32.5104.$$

Suppose the largest two observations in the sample are censored. Then the BLUE of σ is given by

$$\sigma^* = \frac{1}{13}\left\{\sum_{i=1}^{13} X_{i:15} + 2X_{13:15}\right\} = 121.7692,$$

and its standard error is given by

$$\text{S.E.}(\sigma^*) = \sigma^*/\sqrt{13} = 33.7727.$$

In addition to the largest two observations, suppose the smallest observation is also censored in this example. We get from Eqs. (4.3.6) and (4.3.7) that

$$a_i^* = \begin{cases} 1.46556 & \text{for } i = 2 \\ 1 & \text{for } 3 \le i \le 12 \\ 3 & \text{for } i = 13, \end{cases}$$

and

$$K = 12.99762.$$

Then the BLUE of σ is obtained in this case to be

$$\sigma^* = \frac{1}{K}\sum_{i=2}^{13} a_i^* X_{i:15} = 121.6205,$$

and the standard error of this estimate is given by

$$\text{S.E.}(\sigma^*) = \sigma^*/\sqrt{K} = 33.7345.$$

Example 4.3.2. The first eight observations in a random sample of 12 lifetimes from an assumed exponential distribution (in hours) are given by Lawless (1982) as

$$31, 58, 157, 185, 300, 470, 497, 673.$$

Here, we have $n = 12$, $r = 0$, and $s = 4$.

We then get the BLUE of σ from (4.3.11) to be

$$\sigma^* = \frac{1}{8}\left\{\sum_{i=1}^{8} X_{i:12} + 4X_{8:12}\right\} = 632.875,$$

and the standard error of this estimate to be

$$\text{S.E.}(\sigma^*) = \sigma^*/\sqrt{8} = 223.755.$$

As mentioned earlier, the above given estimate is also the maximum likelihood estimate of σ, because the censoring is only on the right in this case; see also Lawless (1982) and Balakrishnan *et al.* (1990a).

4.4. BLUEs of the Location and Scale Parameters

Let $X_{r+1:n} \leq X_{r+2:n} \leq \cdots \leq X_{n-s:n}$ be a Type-II censored sample from a location-scale parameter distribution. Let us now denote $Z_{i:n} = (X_{i:n} - \mu)/\sigma (r+1 \leq i \leq n-s)$ and, as before, $E(Z_{i:n}) = \alpha_{i:n}$ $(r+1 \leq i \leq n-s)$ and $\text{Cov}(Z_{i:n}, Z_{j:n}) = \beta_{i,j:n} (r+1 \leq i \leq j \leq n-s)$. Let us use $\mathbf{X}$, $\boldsymbol{\alpha}$, and $\boldsymbol{\beta}$ as given in (4.2.1), (4.2.2), and (4.2.3), respectively. In addition, let $\mathbf{1}$ denote a column vector of $(n-r-s)$ ones. With these notations, we may then write

$$\xi(\mathbf{X}) = \mu \mathbf{1} + \sigma \boldsymbol{\alpha} \quad \text{and} \quad \nu(\mathbf{X}) = \sigma^2 \boldsymbol{\beta}. \tag{4.4.1}$$

Let us consider the generalized variance given by

$$\begin{aligned}
&(\mathbf{X} - \mu\mathbf{1} - \sigma\boldsymbol{\alpha})'\boldsymbol{\beta}^{-1}(\mathbf{X} - \mu\mathbf{1} - \sigma\boldsymbol{\alpha}) \\
&\quad = \mathbf{X}'\boldsymbol{\beta}^{-1}\mathbf{X} - \mu\mathbf{1}'\boldsymbol{\beta}^{-1}\mathbf{X} - \sigma\boldsymbol{\alpha}'\boldsymbol{\beta}^{-1}\mathbf{X} - \mu\mathbf{X}'\boldsymbol{\beta}^{-1}\mathbf{1} + \mu^2\mathbf{1}'\boldsymbol{\beta}^{-1}\mathbf{1} \\
&\qquad + \mu\sigma\boldsymbol{\alpha}'\boldsymbol{\beta}^{-1}\mathbf{1} - \sigma\mathbf{X}'\boldsymbol{\beta}^{-1}\boldsymbol{\alpha} + \mu\sigma\mathbf{1}'\boldsymbol{\beta}^{-1}\boldsymbol{\alpha} + \sigma^2\boldsymbol{\alpha}'\boldsymbol{\beta}^{-1}\boldsymbol{\alpha} \\
&\quad = \mathbf{X}'\boldsymbol{\beta}^{-1}\mathbf{X} + \mu^2\mathbf{1}'\boldsymbol{\beta}^{-1}\mathbf{1} + \sigma^2\boldsymbol{\alpha}'\boldsymbol{\beta}^{-1}\boldsymbol{\alpha} - 2\mu\mathbf{1}'\boldsymbol{\beta}^{-1}\mathbf{X} \\
&\qquad - 2\sigma\boldsymbol{\alpha}'\boldsymbol{\beta}^{-1}\mathbf{X} + 2\mu\sigma\boldsymbol{\alpha}'\boldsymbol{\beta}^{-1}\mathbf{1}.
\end{aligned} \tag{4.4.2}$$

By minimizing the expression of the generalized variance in (4.4.2) with respect to μ and σ, we obtain the equations

$$\mu\mathbf{1}'\boldsymbol{\beta}^{-1}\mathbf{1} + \sigma\boldsymbol{\alpha}'\boldsymbol{\beta}^{-1}\mathbf{1} = \mathbf{1}'\boldsymbol{\beta}^{-1}\mathbf{X} \tag{4.4.3}$$

and

$$\mu\boldsymbol{\alpha}'\boldsymbol{\beta}^{-1}\mathbf{1} + \sigma\boldsymbol{\alpha}'\boldsymbol{\beta}^{-1}\boldsymbol{\alpha} = \boldsymbol{\alpha}'\boldsymbol{\beta}^{-1}\mathbf{X}. \tag{4.4.4}$$

Upon solving Eqs. (4.4.3) and (4.4.4), we derive the BLUEs of μ and σ as (Lloyd, 1952)

$$\begin{aligned}
\mu^* &= \left\{ \frac{\boldsymbol{\alpha}'\boldsymbol{\beta}^{-1}\boldsymbol{\alpha}\mathbf{1}'\boldsymbol{\beta}^{-1} - \boldsymbol{\alpha}'\boldsymbol{\beta}^{-1}\mathbf{1}\boldsymbol{\alpha}'\boldsymbol{\beta}^{-1}}{(\boldsymbol{\alpha}'\boldsymbol{\beta}^{-1}\boldsymbol{\alpha})(\mathbf{1}'\boldsymbol{\beta}^{-1}\mathbf{1}) - (\boldsymbol{\alpha}'\boldsymbol{\beta}^{-1}\mathbf{1})^2} \right\} \mathbf{X} \\
&= -\boldsymbol{\alpha}'\boldsymbol{\Delta}\mathbf{X} \\
&= \sum_{i=r+1}^{n-s} a_i X_{i:n}
\end{aligned} \tag{4.4.5}$$

and

$$\sigma^* = \left\{ \frac{\mathbf{1}'\boldsymbol{\beta}^{-1}\mathbf{1}\boldsymbol{\alpha}'\boldsymbol{\beta}^{-1} - \mathbf{1}'\boldsymbol{\beta}^{-1}\boldsymbol{\alpha}\mathbf{1}'\boldsymbol{\beta}^{-1}}{(\boldsymbol{\alpha}'\boldsymbol{\beta}^{-1}\boldsymbol{\alpha})(\mathbf{1}'\boldsymbol{\beta}^{-1}\mathbf{1}) - (\boldsymbol{\alpha}'\boldsymbol{\beta}^{-1}\mathbf{1})^2} \right\} \boldsymbol{X}$$

$$= \mathbf{1}'\boldsymbol{\Delta}\boldsymbol{X}$$

$$= \sum_{i=r+1}^{n-s} b_i X_{i:n}, \tag{4.4.6}$$

where $\boldsymbol{\Delta}$ is a skew-symmetric matrix of order $n-r-s$ given by

$$\boldsymbol{\Delta} = \left\{ \frac{\boldsymbol{\beta}^{-1}(\mathbf{1}\boldsymbol{\alpha}' - \boldsymbol{\alpha}\mathbf{1}')\boldsymbol{\beta}^{-1}}{(\boldsymbol{\alpha}'\boldsymbol{\beta}^{-1}\boldsymbol{\alpha})(\mathbf{1}'\boldsymbol{\beta}^{-1}\mathbf{1}) - (\boldsymbol{\alpha}'\boldsymbol{\beta}^{-1}\mathbf{1})^2} \right\}. \tag{4.4.7}$$

Furthermore, from Eqs. (4.4.5) and (4.4.6) we obtain the variances and covariance of the estimators to be

$$\operatorname{Var}(\mu^*) = \sigma^2 \left\{ \frac{\boldsymbol{\alpha}'\boldsymbol{\beta}^{-1}\boldsymbol{\alpha}}{(\boldsymbol{\alpha}'\boldsymbol{\beta}^{-1}\boldsymbol{\alpha})(\mathbf{1}'\boldsymbol{\beta}^{-1}\mathbf{1}) - (\boldsymbol{\alpha}'\boldsymbol{\beta}^{-1}\mathbf{1})^2} \right\}, \tag{4.4.8}$$

$$\operatorname{Var}(\sigma^*) = \sigma^2 \left\{ \frac{\mathbf{1}'\boldsymbol{\beta}^{-1}\mathbf{1}}{(\boldsymbol{\alpha}'\boldsymbol{\beta}^{-1}\boldsymbol{\alpha})(\mathbf{1}'\boldsymbol{\beta}^{-1}\mathbf{1}) - (\boldsymbol{\alpha}'\boldsymbol{\beta}^{-1}\mathbf{1})^2} \right\}, \tag{4.4.9}$$

and

$$\operatorname{Cov}(\mu^*, \sigma^*) = -\sigma^2 \left\{ \frac{\boldsymbol{\alpha}'\boldsymbol{\beta}^{-1}\mathbf{1}}{(\boldsymbol{\alpha}'\boldsymbol{\beta}^{-1}\boldsymbol{\alpha})(\mathbf{1}'\boldsymbol{\beta}^{-1}\mathbf{1}) - (\boldsymbol{\alpha}'\boldsymbol{\beta}^{-1}\mathbf{1})^2} \right\}. \tag{4.4.10}$$

For the case in which the population distribution is symmetric and the censoring is also symmetric (that is, $r = s$), some simplification in the formulas (4.4.5) through (4.4.10) is possible. In this case, since $\beta_{i,j:n} = \beta_{n-j+1,n-i+1:n}$, by defining the matrix

$$\mathbf{I}^* = \begin{bmatrix} 0 & 0 & 0 & \cdots & 0 & 1 \\ 0 & 0 & 0 & \cdots & 1 & 0 \\ \vdots & \vdots & \vdots & \cdots & \vdots & \vdots \\ 1 & 0 & 0 & \cdots & 0 & 0 \end{bmatrix}_{(n-r-s)\times(n-r-s)} \tag{4.4.11}$$

we see that

$$\boldsymbol{\beta} = \mathbf{I}^*\boldsymbol{\beta}\mathbf{I}^*; \tag{4.4.12}$$

also, since $\alpha_{i:n} = -\alpha_{n-i+1:n}$, we have

$$\boldsymbol{\alpha} = -\mathbf{I}^*\boldsymbol{\alpha}. \tag{4.4.13}$$

By noting that $\mathbf{I}^{*-1} = \mathbf{I}^*$, we obtain from (4.4.12) that

$$\boldsymbol{\beta}^{-1} = \mathbf{I}^*\boldsymbol{\beta}^{-1}\mathbf{I}^*. \tag{4.4.14}$$

Upon using (4.4.13) and (4.4.14), we have

$$\begin{aligned}\boldsymbol{\alpha}'\boldsymbol{\beta}^{-1}\mathbf{1} &= (-\mathbf{I}^*\boldsymbol{\alpha})'(\mathbf{I}^*\boldsymbol{\beta}^{-1}\mathbf{I}^*)\mathbf{1}\\ &= -\boldsymbol{\alpha}'\mathbf{I}^{*2}\boldsymbol{\beta}^{-1}\mathbf{I}^*\mathbf{1}\\ &= -\boldsymbol{\alpha}'\mathbf{I}\boldsymbol{\beta}^{-1}1 \quad (\text{since } \mathbf{I}^{*2} = \mathbf{I} \text{ and } \mathbf{I}^*\mathbf{1} = \mathbf{1})\\ &= -\boldsymbol{\alpha}'\boldsymbol{\beta}^{-1}\mathbf{1},\end{aligned}$$

so that

$$\boldsymbol{\alpha}'\boldsymbol{\beta}^{-1}\mathbf{1} = 0. \tag{4.4.15}$$

Hence, in this case, we note from (4.4.10) that the estimators μ^* and σ^* are uncorrelated. Also, the estimators simplify to

$$\mu^* = \frac{\mathbf{1}'\boldsymbol{\beta}^{-1}}{\mathbf{1}'\boldsymbol{\beta}^{-1}\mathbf{1}}\boldsymbol{X} \tag{4.4.16}$$

and

$$\sigma^* = \frac{\boldsymbol{\alpha}'\boldsymbol{\beta}^{-1}}{\boldsymbol{\alpha}'\boldsymbol{\beta}^{-1}\boldsymbol{\alpha}}\boldsymbol{X}, \tag{4.4.17}$$

and their variances simplify to

$$\operatorname{Var}(\mu^*) = \frac{\sigma^2}{\mathbf{1}'\boldsymbol{\beta}^{-1}\mathbf{1}} \tag{4.4.18}$$

and

$$\operatorname{Var}(\sigma^*) = \frac{\sigma^2}{\boldsymbol{\alpha}'\boldsymbol{\beta}^{-1}\boldsymbol{\alpha}}. \tag{4.4.19}$$

For the complete sample case ($r = s = 0$), we note that the BLUE μ^* in (4.4.16) becomes the sample mean if

$$\mathbf{1}'\boldsymbol{\beta}^{-1} = \mathbf{1}' \quad \text{or} \quad \boldsymbol{\beta}\mathbf{1} = \mathbf{1}; \tag{4.4.20}$$

that is, if

$$\sum_{j=1}^{n} \beta_{i,j:n} = 1 \quad \text{for } i = 1, 2, \ldots, n,$$

which is precisely the case for the standard normal distribution (see Corollary 3.9.2). It can be shown that the variance of the estimator μ^* given in

(4.4.18) is strictly less than σ^2/n, except when the condition in (4.4.20) is satisfied. For a proof of this result and some similar results for the case in which the population distribution is skewed, interested readers may refer to Lloyd (1952), Downton (1953), Govindarajulu (1968), and Bondesson (1976). In particular, the last author has shown that the sample mean is the BLUE for the population mean when and only when the population distribution is either normal or gamma.

Example 4.4.1. Let us consider the following data of Gupta (1952), showing the number X' of days to death of the first seven in a sample of 10 mice after inoculation with a uniform culture of human tuberculosis:

X':	41	44	46	54	55	58	60
$X = \log X'$:	1.613	1.644	1.663	1.732	1.740	1.763	1.778

Gupta (1952) has assumed $X = \log X'$ to be distributed as normal $N(\mu, \sigma^2)$.

For the normal distribution, the coefficients a_i and b_i in (4.4.5) and (4.4.6), respectively, and the variances and covariance of μ^* and σ^* in (4.4.8)–(4.4.10) have been tabulated by Sarhan and Greenberg (1962) for various choices of r and s and for sample sizes up to 20. These tables have been extended recently by Balakrishnan (1990e) for sample sizes up to 40. From these tables, we obtain the BLUEs of μ and σ to be

$$\begin{aligned}\mu^* &= 0.0244(1.613) + 0.0636(1.644) + 0.0818(1.663)\\ &\quad + 0.0962(1.732) + 0.1089(1.740) + 0.1207(1.763)\\ &\quad + 0.5045(1.778)\\ &= 1.746\end{aligned}$$

and

$$\begin{aligned}\sigma^* &= -0.3252(1.613) - 0.1758(1.644) - 0.1058(1.663)\\ &\quad - 0.0502(1.732) - 0.0006(1.740) + 0.0469(1.763)\\ &\quad + 0.6107(1.778)\\ &= 0.091,\end{aligned}$$

and the standard errors of the above estimates are obtained to be

$$\text{S.E.}(\mu^*) = 0.091(0.1167)^{1/2} = 0.031$$

and

$$\text{S.E.}(\sigma^*) = 0.091(0.0989)^{1/2} = 0.029.$$

Example 4.4.2. Let us consider the following data of Darwin (Fisher, 1966), which represent the differences in heights between cross- and self-fertilized plants of the same pair grown together in one pot:

$$49, -67, 8, 16, 6, 23, 28, 41, 14, 29, 56, 24, 75, 60, -48.$$

Based on robustness considerations, let us censor the two smallest and the two largest observations, as suggested by Tiku *et al.* (1986). The remaining 11 observations, arranged in increasing order of magnitude,

$$6, 8, 14, 16, 23, 24, 28, 29, 41, 49, 56,$$

constitute a symmetrically Type-II censored sample. By using the tables of Sarhan and Greenberg (1962), we obtain the BLUEs of μ and σ to be

$$\begin{aligned}\mu^* &= 0.1835(6)+0.0701(8)+0.0703(14)+0.0704(16)\\ &\quad +0.0705(23)+0.0705(24)+0.0705(28)+0.0704(29)\\ &\quad +0.0703(41)+0.0701(49)+0.1835(56)\\ &= 27.695\end{aligned}$$

and

$$\begin{aligned}\sigma^* &= -0.4139(6)-0.0838(8)-0.0607(14)-0.0395(16)\\ &\quad -0.0195(23)+0.0(24)+0.0195(28)+0.0395(29)\\ &\quad +0.0607(41)+0.0838(49)+0.4139(56)\\ &= 26.381,\end{aligned}$$

and the standard errors of the above estimates are obtained to be

$$\text{S.E.}(\mu^*) = 26.381(0.0706)^{1/2} = 7.009$$

and

$$\text{S.E.}(\sigma^*) = 26.381(0.0594)^{1/2} = 6.430.$$

Example 4.4.3. Next, let us consider the following arranged sample of size 30 given by D'agostino and Stephens (1986, p. 529) (simulated from a normal distribution with mean $\mu = 100$ and standard deviation $\sigma = 10$):

79.43, 83.53, 83.67, 84.27, 85.29, 89.90,
90.03, 90.87, 92.45, 92.55, 95.45, 96.13,
96.20, 98.70, 98.98, 100.42, 101.58, 102.56,
103.61, 105.48, 106.82, 108.52, 109.26, 111.56,
112.69, 112.97, 113.75, 115.92, 116.87, 118.52.

By making use of the tables prepared by Balakrishnan (1990e), we have computed the BLUEs of μ and σ based on complete and symmetrically

TABLE 4.4.1
The BLUEs of μ and σ and Their Standard Errors for the Data of D'agostino and Stephens

r	μ^*	σ^*	S.E.(μ^*)	S.E.(σ^*)
0	99.92269	11.33213	2.06885	1.49824
1	100.01080	11.71832	2.14992	1.63679
2	99.96206	12.46629	2.30137	1.84316
3	99.82979	12.89126	2.39791	2.02438
4	99.87046	13.51498	2.53563	2.26310
5	100.50230	13.02747	2.46870	2.33770
6	100.31435	13.87997	2.65974	2.68569
7	100.01015	13.79553	2.67683	2.90034
8	100.20452	14.12685	2.77980	3.25929
9	99.77651	14.93689	2.98551	3.83066
10	100.23788	13.12885	2.67068	3.81008
11	99.83684	12.74440	2.64366	4.29847
12	99.47579	15.19162	3.22191	6.22245
13	100.09911	11.55664	2.51261	6.28080
14	99.70000	17.35657	3.88182	16.82209

Type-II censored samples and have also computed the standard errors of these estimates. These values are presented in Table 4.4.1 for $r = s = 0(1)14$.

We note from Table 4.4.1 that the standard errors of the estimates of both μ and σ increase in general with an increased amount of censoring. In particular, we note that this increase in the standard error is quite pronounced for the BLUE of σ.

Example 4.4.4. Let us consider the following data (see Example 4.2.2), which represent failure times, in minutes, for a specific type of electrical insulation in an experiment in which the insulation was subjected to a continuously increasing voltage stress (Lawless, 1982, p. 138):

$$12.3, 21.8, 24.4, 28.6, 43.2, 46.9, 70.7, 75.3, 95.5, 98.1, 138.6, —.$$

Here, the largest observation is censored because the experiment was stopped as soon as the eleventh failure occurred.

Balakrishnan and Puthenpura (1986) and Balakrishnan and Wong (1990c) have assumed for the above data a two-parameter half logistic distribution with pdf

$$f(x; \mu, \sigma) = \frac{2\, e^{-(x-\mu)/\sigma}}{\sigma\{1+e^{-(x-\mu)/\sigma}\}^2}, \qquad \mu \le x < \infty,\ \sigma > 0. \tag{4.4.21}$$

In this case, Balakrishnan and Puthenpura (1986) have tabulated the coefficients required for the BLUEs of μ and σ, and also the variances and covariance of the BLUEs when samples are complete, for sample sizes up to 15. These tables have been extended by Balakrishnan and Wong (1990c) to the case in which the samples are doubly censored for sample sizes up to 15. More extensive tables have been prepared by Wong (1988). By using the tables of Balakrishnan and Wong (1990c), we compute the BLUEs of μ and σ to be

$$\begin{aligned}\mu^* &= 1.07759(12.3)+0.00412(21.8)+0.00194(24.4)-0.00038(28.6)\\ &\quad -0.00279(43.2)-0.00525(46.9)-0.00770(70.7)-0.01007(75.3)\\ &\quad -0.01225(95.5)-0.01413(98.1)-0.03108(138.6)\\ &= 4.847511\end{aligned}$$

and

$$\begin{aligned}\sigma^* &= -0.66733(12.3)+0.02149(21.8)+0.03119(24.4)+0.04085(28.6)\\ &\quad +0.05025(43.2)+0.05915(46.9)+0.06728(70.7)+0.07429(75.3)\\ &\quad +0.07976(95.5)+0.08305(98.1)+0.16005(138.6)\\ &= 47.432552;\end{aligned}$$

using the above estimates, we obtain the BLUE of the expected failure time as

$$\begin{aligned}\mu^*+\sigma^* \ln 4 &= 4.847511+47.432552 \ln 4\\ &= 70.60299 \text{ min},\end{aligned}$$

and the standard error of this estimate is obtained from the tables of Balakrishnan and Wong (1990c) to be

$$\begin{aligned}&\sigma^*\{0.02440+0.07282(\ln 4)^2-2(0.01547)\ln 4\}^{1/2}\\ &\quad = 0.348503\sigma^* = 16.53039 \text{ min}.\end{aligned}$$

These values may be compared with the best linear unbiased estimate of 70.01 min and the standard error of 17.58 min based on the one-parameter half logistic distribution (see Section 4.2), and with the best linear unbiased estimate of 71.55333 min and the standard error of 19.4965 min based on the two-parameter exponential distribution (see Section 4.5). We may note here that we get a smaller standard error in estimating the expected failure time when we assume a half logistic distribution for the failure time data instead of an exponential distribution. A reason for this, as rightly pointed out by Balakrishnan and Puthenpura (1986), is the fact that a half logistic distribution fits the data better than an exponential distribution.

Example 4.4.5. Sarhan and Greenberg (1962, p. 212) have given the measurements resulting from an experiment to measure the strontium-90 concentrations in samples of milk. The test substance was supposed to contain 9.22 picocuries per liter. Ten measurements were taken, but because of relatively large measurement error known to exist at the extremes, the two smallest and the three largest observations were censored. The remaining five observations, arranged in increasing order, are presented below:

$$8.2, 8.4, 9.1, 9.8, 9.9.$$

In this case, we have $n = 10$, $r = 2$, and $s = 3$.

Let us now assume that the above censored sample has come from a logistic population with unknown mean μ and variance σ^2. For the logistic distribution, the coefficients a_i and b_i in (4.4.5) and (4.4.6), respectively, and the variances and covariance of μ^* and σ^* in (4.4.8) through (4.4.10) have been tabulated by Gupta *et al.* (1967) for sample sizes $n = 2(1)5(5)25$ and some selected choices of r and s. Recently, Balakrishnan (1990f) has prepared more exhaustive tables covering sample sizes $n = 2(1)25(5)40$ and all possible choices of r and s. From these tables, we obtain the BLUEs of μ and σ to be

$$\begin{aligned}\mu^* &= 0.15237(8.2) + 0.13241(8.4) + 0.15377(9.1) + 0.16153(9.8)\\ &\quad + 0.39992(9.9)\\ &= 9.3032\end{aligned}$$

and

$$\begin{aligned}\sigma^* &= -0.93064(8.2) - 0.17440(8.4) - 0.04984(9.1) + 0.07908(9.8)\\ &\quad + 1.07581(9.9)\\ &= 1.8758,\end{aligned}$$

and the standard errors of the above estimates are obtained to be

$$\text{S.E.}(\mu^*) = 1.8758(0.09921)^{1/2} = 0.5908$$

and

$$\text{S.E.}(\sigma^*) = 1.8758(0.18338)^{1/2} = 0.8033.$$

4.5. BLUEs for the Two-Parameter Exponential Distribution

In this section we shall assume that

$$X_{r+1:n} \leq X_{r+2:n} \leq \cdots \leq X_{n-s:n} \tag{4.5.1}$$

is a Type-II censored sample available from the two-parameter exponential distribution with pdf

$$f(x; \mu, \sigma) = \frac{1}{\sigma} e^{-(x-\mu)/\sigma}, \qquad x \geq \mu, \sigma > 0. \tag{4.5.2}$$

Here, μ and σ are the location and scale parameters, respectively; in a life-testing context, μ may represent the guarantee or warranty period and σ the mean life of the component, so that $\mu + \sigma$ represents the expected lifetime of the component.

By using the expressions of the expected value of the standard exponential order statistics in (4.3.3) and that of the inverse of the variance-covariance matrix in (4.3.5), we simply find

$$\boldsymbol{\alpha}'\boldsymbol{\beta}^{-1}\boldsymbol{\alpha} = \frac{\left(\sum_{l=n-r}^{n} 1/l\right)^2}{\sum_{l=n-r}^{n} 1/l^2} + (n-r-s-1), \tag{4.5.3}$$

$$\mathbf{1}'\boldsymbol{\beta}^{-1}\mathbf{1} = 1 \Big/ \left\{ \sum_{l=n-r}^{n} 1/l^2 \right\} \tag{4.5.4}$$

and

$$\boldsymbol{\alpha}'\boldsymbol{\beta}^{-1}\mathbf{1} = \frac{\sum_{l=n-r}^{n} 1/l}{\sum_{l=n-r}^{n} 1/l^2}, \tag{4.5.5}$$

so that

$$(\boldsymbol{\alpha}'\boldsymbol{\beta}^{-1}\boldsymbol{\alpha})(\mathbf{1}'\boldsymbol{\beta}^{-1}\mathbf{1}) - (\boldsymbol{\alpha}'\boldsymbol{\beta}^{-1}\mathbf{1})^2 = (n-r-s-1) \Big/ \left\{ \sum_{l=n-r}^{n} 1/l^2 \right\}. \tag{4.5.6}$$

We also find that

$$(\boldsymbol{\alpha}'\boldsymbol{\beta}^{-1}\boldsymbol{\alpha}\mathbf{1}'\boldsymbol{\beta}^{-1})_i = \begin{cases} \dfrac{\left(\sum_{l=n-r}^{n} 1/l\right)^2}{\left(\sum_{l=n-r}^{n} 1/l^2\right)^2} + \dfrac{(n-r-s-1)}{\sum_{l=n-r}^{n} 1/l^2}, & \text{for } i = r+1 \\ 0, & \text{for } r+2 \leq i \leq n-s \end{cases} \tag{4.5.7}$$

and

$$(\boldsymbol{\alpha}'\boldsymbol{\beta}^{-1}\mathbf{1}\boldsymbol{\alpha}'\boldsymbol{\beta}^{-1})_i = \begin{cases} \dfrac{\left(\sum_{l=n-r}^{n} 1/l\right)^2}{\left(\sum_{l=n-r}^{n} 1/l^2\right)^2} - (n-r-1)\dfrac{\sum_{l=n-r}^{n} 1/l}{\sum_{l=n-r}^{n} 1/l^2}, & \text{for } i=r+1 \\ \left(\sum_{l=n-r}^{n} 1/l\right)\Big/\left(\sum_{l=n-r}^{n} 1/l^2\right), & \text{for } r+2 \le i \\ & \le n-s-1. \\ (s+1)\left(\sum_{l=n-r}^{n} 1/l\right)\Big/\left(\sum_{l=n-r}^{n} 1/l^2\right), & \text{for } i=n-s \end{cases} \tag{4.5.8}$$

so that the coefficients for the BLUE of μ are obtained from (4.4.5) to be

$$a_i = \begin{cases} 1+\dfrac{(n-r-1)}{(n-r-s-1)}\sum_{l=n-r}^{n} 1/l, & \text{for } i=r+1 \\ -\dfrac{1}{(n-r-s-1)}\sum_{l=n-r}^{n} 1/l, & \text{for } r+2 \le i \le n-s-1. \\ -\dfrac{(s+1)}{(n-r-s-1)}\sum_{l=n-r}^{n} 1/l, & \text{for } i=n-s \end{cases} \tag{4.5.9}$$

Similarly, we find

$$(\mathbf{1}'\boldsymbol{\beta}^{-1}\mathbf{1}\boldsymbol{\alpha}'\boldsymbol{\beta}^{-1})_i = \begin{cases} \dfrac{\sum_{l=n-r}^{n} 1/l}{\left(\sum_{l=n-r}^{n} 1/l^2\right)^2} - \dfrac{(n-r-1)}{\sum_{l=n-r}^{n} 1/l^2}, & \text{for } i=r+1 \\ 1\Big/\left(\sum_{l=n-r}^{n} 1/l^2\right), & \text{for } r+2 \le i \le n-s-1 \\ (s+1)\Big/\left(\sum_{l=n-r}^{n} 1/l^2\right), & \text{for } i=n-s \end{cases} \tag{4.5.10}$$

and

$$(\mathbf{1}'\boldsymbol{\beta}^{-1}\boldsymbol{\alpha}\mathbf{1}'\boldsymbol{\beta}^{-1})_i = \begin{cases} \left(\sum_{l=n-r}^{n} 1/l\right)\Big/\left(\sum_{l=n-r}^{n} 1/l^2\right)^2, & \text{for } i=r+1 \\ 0, & \text{for } r+2 \le i \le n-s \end{cases} \tag{4.5.11}$$

so that the coefficients for the BLUE of σ are obtained from (4.4.6) to be

$$b_i = \begin{cases} -(n-r-1)/(n-r-s-1), & \text{for } i=r+1 \\ 1/(n-r-s-1), & \text{for } r+2 \le i \le n-s-1. \\ (s+1)/(n-r-s-1), & \text{for } i=n-s \end{cases} \tag{4.5.12}$$

By making use of the expressions of $\boldsymbol{\alpha}'\boldsymbol{\beta}^{-1}\boldsymbol{\alpha}$, $\mathbf{1}'\boldsymbol{\beta}^{-1}\mathbf{1}$ and $\boldsymbol{\alpha}'\boldsymbol{\beta}^{-1}\mathbf{1}$ derived in (4.5.3) through (4.5.5), respectively, in the formulas for the variances and covariance of the BLUEs in (4.4.8) through (4.4.10), we obtain

$$\operatorname{Var}(\mu^*) = \sigma^2 \left\{ \frac{1}{n-r-s-1} \left(\sum_{l=n-r}^{n} 1/l \right)^2 + \sum_{l=n-r}^{n} 1/l^2 \right\}, \tag{4.5.13}$$

$$\operatorname{Var}(\sigma^*) = \sigma^2/(n-r-s-1), \tag{4.5.14}$$

and

$$\operatorname{Cov}(\mu^*, \sigma^*) = -\sigma^2 \left(\sum_{l=n-r}^{n} 1/l \right) \Big/ (n-r-s-1). \tag{4.5.15}$$

By using the expressions of the coefficients a_i and b_i in (4.5.9) and (4.5.12), respectively, we derive the BLUE of the expected life time, viz., $\theta = \mu + \sigma$, to be

$$\theta^* = \mu^* + \sigma^* = \sum_{i=r+1}^{n-s} c_i X_{i:n}, \tag{4.5.16}$$

where

$$c_i = \begin{cases} 1 - \dfrac{(n-r-1)}{(n-r-s-1)} \left\{ 1 - \displaystyle\sum_{l=n-r}^{n} 1/l \right\}, & \text{for } i=r+1 \\ \dfrac{1}{(n-r-s-1)} \left\{ 1 - \displaystyle\sum_{l=n-r}^{n} 1/l \right\}, & \text{for } r+2 \le i \le n-s-1. \\ \dfrac{(s+1)}{(n-r-s-1)} \left\{ 1 - \displaystyle\sum_{l=n-r}^{n} 1/l \right\}, & \text{for } i=n-s \end{cases} \tag{4.5.17}$$

From (4.5.13)–(4.5.15) we also find the variance of the BLUE θ^* of θ to be

$$\begin{aligned} \operatorname{Var}(\theta^*) &= \operatorname{Var}(\mu^*) + \operatorname{Var}(\sigma^*) + 2\operatorname{Cov}(\mu^*, \sigma^*) \\ &= \sigma^2 \left[\sum_{l=n-r}^{n} 1/l^2 + \frac{1}{(n-r-s-1)} \left\{ 1 + \left(\sum_{l=n-r}^{n} 1/l \right)^2 - 2 \sum_{l=n-r}^{n} 1/l \right\} \right] \\ &= \sigma^2 \left[\sum_{l=n-r}^{n} 1/l^2 + \frac{1}{(n-r-s-1)} \left\{ 1 - \sum_{l=n-r}^{n} 1/l \right\}^2 \right]. \end{aligned} \tag{4.5.18}$$

In particular, when the available sample is complete (that is, $r = s = 0$), we obtain from (4.5.9) and (4.5.12) the BLUEs of μ and σ, which are

$$\mu^* = \left(1 + \frac{1}{n}\right) X_{1:n} - \frac{1}{n(n-1)} \sum_{i=2}^{n} X_{i:n}$$

$$= (nX_{1:n} - \bar{X})/(n-1) \tag{4.5.19}$$

and

$$\sigma^* = -X_{1:n} + \frac{1}{n-1} \sum_{i=2}^{n} X_{i:n}$$

$$= n(\bar{X} - X_{1:n})/(n-1), \tag{4.5.20}$$

where $\bar{X}$ is the sample mean. From (4.5.13) through (4.5.15) we also obtain the variances and covariance of the above estimators:

$$\text{Var}(\mu^*) = \frac{\sigma^2}{n(n-1)}, \tag{4.5.21}$$

$$\text{Var}(\sigma^*) = \sigma^2/(n-1), \tag{4.5.22}$$

and

$$\text{Cov}(\mu^*, \sigma^*) = -\frac{\sigma^2}{n(n-1)}. \tag{4.5.23}$$

For the complete sample case, we also derive from (4.5.16)-(4.5.18) that the BLUE of θ is

$$\theta^* = \bar{X}, \tag{4.5.24}$$

and that

$$\text{Var}(\theta^*) = \sigma^2/n. \tag{4.5.25}$$

Furthermore, by comparing the variances of the BLUEs of μ, σ, and θ based on doubly Type-II censored samples to those based on complete samples, we get the relative efficiencies of the three estimators to be

$$\text{R.E.}(\mu^*) = \frac{1}{n(n-1)\left\{\frac{1}{n-r-s-1}\left(\sum_{l=n-r}^{n} 1/l\right)^2 + \sum_{l=n-r}^{n} 1/l^2\right\}}, \tag{4.5.26}$$

$$\text{R.E.}(\sigma^*) = (n-r-s-1)/(n-1), \tag{4.5.27}$$

and

$$\text{R.E.}(\theta^*) = \frac{1}{n\left[\sum_{l=n-r}^{n} 1/l^2 + \frac{1}{n-r-s-1}\left\{1 - \sum_{l=n-r}^{n} 1/l\right\}^2\right]}. \quad (4.5.28)$$

The coefficients necessary for the BLUEs μ^*, σ^*, and θ^*, the variances and covariance of the estimators, and relative efficiencies have all been tabulated by Sarhan and Greenberg (1957) for sample sizes up to 10 for all choices of censoring (viz., r and s). Epstein (1956, 1962), Epstein and Sobel (1953, 1954), and Balakrishnan (1990d) have proposed some simplified estimators for μ and σ in the case of doubly and multiply Type-II censored samples. The construction of confidence intervals for μ and σ has also been explained by Epstein (1962).

Example 4.5.1. Let us consider once again the following data (see Example 4.4.4), which represent failure times, in minutes, for a specific type of electrical insulation in an experiment in which the insulation was subjected to a continuously increasing voltage stress (Lawless, 1982, p. 138):

12.3, 21.8, 24.4, 28.6, 43.2, 46.9, 70.7, 75.3, 95.5, 98.1, 138.6, —

The largest observation is censored here, as the experiment was stopped as soon as the eleventh failure occurred.

By assuming a two-parameter exponential distribution with pdf as in (4.5.2) for the above given failure time data, we obtain the BLUEs of μ and σ as (see Lawless, 1982, p. 127)

$$\mu^* = 6.91333 \quad \text{and} \quad \sigma^* = 64.64,$$

so that the BLUE of the expected failure time is given by

$$\theta^* = \mu^* + \sigma^* = 71.55333 \text{ min},$$

and the standard error of this estimate is calculated from (4.5.18) to be

$$\text{S.E.}(\theta^*) = 19.4965 \text{ min}.$$

Example 4.5.2. The following data, presented by Sarhan and Greenberg (1957), is part of an experiment in which 10 rabbits were inoculated with 0.2 ml of graded inoculum containing varying numbers of *Treponema pallidum.* Each rabbit received six injections from solutions containing 10^1, 10^2, 10^3, 10^4, 10^5, and 10^6 spirochetes per milliliter, and each was then observed for a period of 90 days to observe whether a syphilitic lesion developed at the site of injection.

The incubation time required for a lesion to appear is an index of the amount and potency of the inoculation, as well as the susceptibility of the individual rabbit. The distribution of incubation periods is assumed to be a two-parameter exponential distribution.

Knowledge of the reaction mechanism in rabbits has indicated that censoring the observations at 90 days after inoculation is desirable because only an infinitesimal proportion of rabbits will have an incubation period beyond that. In fact, the present data has been considered at one-half that period, viz., 45 days.

During the experiment, the rabbits were examined about twice a week for lesions. Those lesions that developed in the interim between examinations were undetected until the next period. Lesions that were one or two days old at time of first observation could be distinguished by their greater size. The first examination was performed approximately one week following inoculation. This meant that certain observations could be considered censored from the left if the size of the lesion was large at the first examination period.

TABLE 4.5.1
Days of Incubation among 10 Rabbits Following Inoculation with Graded Amounts of *Treponema pallidum*; BLUEs

Rabbit number	Inoculum					
	10^6	10^5	10^4	10^3	10^2	10^1
7	<7	<11	18	<18	<25	>45
8	11	11	18	18	40	>45
9	<7	<11	14	>45	>45	>45
10	7	11	18	<18	<25	<25
11	11	14	18	25	25	25
12	14	14	18	21	29	25
13	7	11	18	18	29	32
14	>45	35	40	25	>45	>45
15	7	14	18	25	29	40
16	11	14	18	21	29	>45
μ^*	5.544	9.271	13.356	16.096	20.899	18.174
σ^*	4.333	5.143	6.444	5.667	12.200	32.333
θ^*	9.877	14.414	19.800	21.763	33.099	50.507
S.E.(θ^*)	1.446	1.634	2.038	1.892	4.333	15.500

The data from the above-described experiment are presented in Table 4.5.1, along with the BLUEs of μ, σ and θ calculated from the censored samples and the standard error of the estimates of θ. There is, however, a slight deviation from the original data in one respect. For illustrative purposes of censoring from the left, Sarhan and Greenberg (1957) have assumed that some of the rabbits had lesions large enough at the time of first examination to allow the presumption that the true incubation period ended a few days earlier. For example, we may note that when 10^6 inoculum was used, rabbits 7 and 9 were considered to have lesions large enough that it could be presumed that the incubation was less than seven days.

4.6. Gupta's Simplified Linear Estimators

The best linear estimators derived in Section 4.4 require the exact values of the means, variances, and covariances of order statistics. The covariances may sometimes be difficult to evaluate. Moreover, even if the variances and covariances are available, the inversion of the variance–covariance matrix involved in the BLU estimation may be difficult to perform particularly when the sample size gets large. For these reasons, Gupta (1952) has proposed the simplified linear estimators for μ and σ obtained from the BLUEs of μ and σ simply by replacing the variance–covariance matrix $\boldsymbol{\beta}$ by an identity matrix $\mathbf{I}$. Even though this method appears to be crude, it gives surprisingly good results in the case of the normal distribution, as displayed by Chernoff and Lieberman (1954), Sarhan and Greenberg (1956), and Ali and Chan (1964).

Now, with $\boldsymbol{\beta}$ replaced by the identity matrix $\mathbf{I}$ of order $n-r-s$, we have $\boldsymbol{\beta}^{-1}$ also to be $\mathbf{I}$. We then have

$$\mathbf{1}'\boldsymbol{\beta}^{-1}\mathbf{1} = \mathbf{1}'\mathbf{I}\mathbf{1} = n-r-s,$$

$$\boldsymbol{\alpha}'\boldsymbol{\beta}^{-1}\boldsymbol{\alpha} = \boldsymbol{\alpha}'\mathbf{I}\boldsymbol{\alpha} = \sum_{i=r+1}^{n-s} \alpha_{i:n}^2,$$

$$\boldsymbol{\alpha}'\boldsymbol{\beta}^{-1}\mathbf{1} = \boldsymbol{\alpha}'\mathbf{I}\mathbf{1} = \sum_{i=r+1}^{n-s} \alpha_{i:n},$$

and

$$\begin{aligned}(\boldsymbol{\alpha}'\boldsymbol{\beta}^{-1}\boldsymbol{\alpha})(\mathbf{1}'\boldsymbol{\beta}^{-1}\mathbf{1}) - (\boldsymbol{\alpha}'\boldsymbol{\beta}^{-1}\mathbf{1})^2 &= (n-r-s)\sum_{i=r+1}^{n-s} \alpha_{i:n}^2 - \left(\sum_{i=r+1}^{n-s} \alpha_{i:n}\right)^2 \\ &= (n-r-s)\sum_{i=r+1}^{n-s} (\alpha_{i:n} - \bar{\alpha})^2,\end{aligned}$$

where

$$\bar{\alpha} = \frac{1}{n-r-s} \sum_{i=r+1}^{n-s} \alpha_{i:n}. \tag{4.6.1}$$

Also, the BLUE of μ in (4.4.5) reduces in this case to

$$\mu^{**} = \sum_{j=r+1}^{n-s} a_j^* X_{j:n}, \tag{4.6.2}$$

where

$$a_j^* = \frac{\sum_{i=r+1}^{n-s} \alpha_{i:n}^2 - \alpha_{j:n} \sum_{i=r+1}^{n-s} \alpha_{i:n}}{(n-r-s) \sum_{i=r+1}^{n-s} (\alpha_{i:n} - \bar{\alpha})^2}$$

$$= \frac{\sum_{i=r+1}^{n-s} (\alpha_{i:n} - \bar{\alpha})^2 + \bar{\alpha} \sum_{i=r+1}^{n-s} \alpha_{i:n} - \alpha_{j:n} \sum_{i=r+1}^{n-s} \alpha_{i:n}}{(n-r-s) \sum_{i=r+1}^{n-s} (\alpha_{i:n} - \bar{\alpha})^2}$$

$$= \frac{1}{n-r-s} - \frac{\bar{\alpha}(\alpha_{j:n} - \bar{\alpha})}{\sum_{i=r+1}^{n-s} (\alpha_{i:n} - \bar{\alpha})^2}, \qquad r+1 \le j \le n-s. \tag{4.6.3}$$

Similarly, the BLUE of σ in (4.4.6) simplifies to

$$\sigma^{**} = \sum_{j=r+1}^{n-s} b_j^* X_{j:n}, \tag{4.6.4}$$

where

$$b_j^* = \frac{(n-r-s)\alpha_{j:n} - \sum_{i=r+1}^{n-s} \alpha_{i:n}}{(n-r-s) \sum_{i=r+1}^{n-s} (\alpha_{i:n} - \bar{\alpha})^2}$$

$$= \frac{\alpha_{j:n} - \bar{\alpha}}{\sum_{i=r+1}^{n-s} (\alpha_{i:n} - \bar{\alpha})^2}, \qquad r+1 \le j \le n-s, \tag{4.6.5}$$

with $\bar{\alpha}$ as defined in (4.6.1).

The variances and covariance of these simplified linear estimators may be obtained as

$$\operatorname{Var}(\mu^{**}) = \sigma^2 \sum_{i=r+1}^{n-s} \sum_{j=r+1}^{n-s} a_i^* a_j^* \beta_{i,j:n}, \tag{4.6.6}$$

$$\operatorname{Var}(\sigma^{**}) = \sigma^2 \sum_{i=r+1}^{n-s} \sum_{j=r+1}^{n-s} b_i^* b_j^* \beta_{i,j:n}, \tag{4.6.7}$$

and

$$\operatorname{Cov}(\mu^{**}, \sigma^{**}) = \sigma^2 \sum_{i=r+1}^{n-s} \sum_{j=r+1}^{n-s} a_i^* b_j^* \beta_{i,j:n}, \tag{4.6.8}$$

where a_i^* and b_i^* are as given in (4.6.3) and (4.6.5), respectively.

From (4.6.3) and (4.6.5) we easily observe that

$$\sum_{j=r+1}^{n-s} a_j^* = 1, \qquad \sum_{j=r+1}^{n-s} a_j^* \alpha_{j:n} = 0,$$

$$\sum_{j=r+1}^{n-s} b_j^* = 0 \quad \text{and} \quad \sum_{j=r+1}^{n-s} b_j^* \alpha_{j:n} = 1,$$

so that the simplified linear estimators μ^{**} and σ^{**} given in (4.6.2) and (4.6.4) are unbiased for μ and σ, respectively. Moreover, when the population distribution is symmetric and the censoring in the available sample is also symmetric (i.e., $r = s$), we note from (4.6.8) that the estimators μ^{**} and σ^{**} are uncorrelated.

Ali and Chan (1964) have shown that the estimator σ^{**} is asymptotically normal (which also follows from the result of Stigler (1974), with σ^{**} being a linear function of order statistics) and fully efficient in the normal case; in addition, the loss in the efficiency of σ^{**} relative to the BLUE σ^* is negligible even for small sample sizes. For example, in the normal case, when the available sample is complete we know that the maximum likelihood estimator of σ (corrected for its bias) is given by

$$\tilde{\sigma} = \frac{\Gamma\left(\frac{1}{2}(n-1)\right)}{\sqrt{2}\,\Gamma(n/2)} \left\{ \sum_{i=1}^{n} (X_i - \bar{X})^2 \right\}^{1/2}, \tag{4.6.9}$$

where $\Gamma(\cdot)$ denotes the complete gamma function. By realizing that

$$\sum_{i=1}^{n} (X_i - \bar{X})^2 / \sigma^2 \to \chi^2_{n-1},$$

we immediately have

$$\operatorname{Var}(\tilde{\sigma}) = \sigma^2 \left[\frac{(n-1)}{2} \left\{ \frac{\Gamma(\frac{1}{2}(n-1))}{\Gamma(n/2)} \right\}^2 - 1 \right]. \tag{4.6.10}$$

TABLE 4.6.1
Comparison of the Estimators $\tilde{\sigma}$, σ^*, and σ^{**} for the Normal Population

n	$\text{Var}(\tilde{\sigma})/\sigma^2$	$\text{Var}(\sigma^*)/\sigma^2$	$\text{Var}(\sigma^{**})/\sigma^2$	R.E.(σ^*)	R.E.(σ^{**})
2	0.57080	0.57080	0.57080	100.000	100.000
3	0.27324	0.27548	0.27548	99.187	99.187
4	0.17810	0.18005	0.18013	99.917	99.873
5	0.13177	0.13332	0.13342	98.837	98.763
6	0.10447	0.10571	0.10580	98.827	98.743
7	0.08650	0.08750	0.08759	98.857	98.756
8	0.07379	0.07461	0.07469	98.901	98.795
9	0.06432	0.06502	0.06509	98.923	98.817
10	0.05701	0.05760	0.05766	98.976	98.873
11	0.05118	0.0517	0.05175	98.994	98.899
12	0.04644	0.0469	0.04693	99.019	98.956
13	0.04250	0.0429	0.04293	99.068	98.998
14	0.03917	0.0395	0.03956	99.165	99.014
15	0.03633	0.0366	0.03667	99.262	99.073
16	0.03387	0.0341	0.03418	99.326	99.093
17	0.03172	0.0320	0.03200	99.125	99.125
18	0.02983	0.0301	0.03008	99.103	99.169
19	0.02815	0.0284	0.02838	99.120	99.190
20	0.02665	0.0268	0.02686	99.440	99.218

From (4.6.10) we have computed the values of $\text{Var}(\tilde{\sigma})/\sigma^2$ for $n = 2(1)20$, and we present them in Table 4.6.1; also presented are the values of $\text{Var}(\sigma^*)/\sigma^2$ taken from the tables of Sarhan and Greenberg (1962), and the values of $\text{Var}(\sigma^{**})/\sigma^2$ computed from Eq. (4.6.7). Also given in Table 4.6.1 are the efficiencies of σ^* and σ^{**} relative to the estimator $\tilde{\sigma}$.

We note from Table 4.6.1 that the BLUE of σ is very efficient as compared to the MLE of σ, and that the smallest relative efficiency observed is for sample size six, viz., 98.827%. We also note that the simplified linear estimator σ^{**} has negligible loss in efficiency relative to the BLUE σ^*. Moreover, we observe that the estimator σ^{**} is very efficient as compared to the MLE of σ, and that the smallest relative efficiency observed is for sample size six once again, viz., 98.743%.

For the purpose of illustrating this simplified linear estimation method due to Gupta (1952), we shall revisit Examples 4.4.1 and 4.4.2 here.

Example 4.6.1. Let us consider once again the following data due to Gupta (1952), giving the number X' of days to death of the first seven in a sample of 10 mice after inoculation with a uniform culture of human tuberculosis:

X':	41	44	46	54	55	58	60
$X = \log X'$:	1.613	1.644	1.663	1.732	1.740	1.763	1.778

Gupta (1952) has assumed $X = \log X'$ to be distributed as $N(\mu, \sigma^2)$.

From the above Type-II censored sample, we obtain the simplified linear estimates of μ and σ as

$$\begin{aligned}\mu^{**} &= -0.043316(1.613) + 0.049137(1.644) + 0.108542(1.663) \\ &\quad + 0.156763(1.732) + 0.200306(1.740) + 0.242513(1.763) \\ &\quad + 0.286055(1.778) \\ &= 1.7476\end{aligned}$$

and

$$\begin{aligned}\sigma^{**} &= -0.407742(1.613) - 0.205259(1.644) - 0.075155(1.663) \\ &\quad + 0.030456(1.732) + 0.125819(1.740) + 0.218258(1.763) \\ &\quad + 0.313622(1.778) \\ &= 0.0940.\end{aligned}$$

The standard errors of the above estimates are computed to be

$$\text{S.E.}(\mu^{**}) = 0.094(0.1215)^{1/2} = 0.0328$$

and

$$\text{S.E.}(\sigma^{**}) = 0.094(0.1075)^{1/2} = 0.0308.$$

We note here that the above estimates are quite close to those obtained in Section 4.4 by the BLU estimation method.

Example 4.6.2. Let us consider the well-known Darwin data (Fisher, 1966) that represent the differences in heights between cross- and self-fertilized plants of the same pair grown together in one pot:

$$49, -67, 8, 16, 6, 23, 28, 41, 14, 29, 56, 24, 75, 60, -48.$$

Due to robustness reasoning (see Tiku *et al.*, 1986), let us censor the two smallest and the two largest observations and consider the resulting symmetrically Type-II censored sample

$$6, 8, 14, 16, 23, 24, 28, 29, 41, 49, 56.$$

From this censored sample, we obtain the simplified linear estimates of μ and σ as

$$\mu^{**} = \tfrac{1}{11}(6+8+14+16+23+24+28+29+41+49+56)$$
$$= 26.7273$$

and

$$\sigma^{**} = -0.261092(6) - 0.196951(8) - 0.142077(14) - 0.092375(16)$$
$$- 0.045540(23) + 0(24) + 0.045540(28) + 0.092375(29)$$
$$+ 0.142077(41) + 0.196951(49) + 0.261092(56)$$
$$= 26.3942.$$

The standard errors of the above estimates are computed to be

$$\text{S.E.}(\mu^{**}) = 26.3942(0.0719)^{1/2} = 7.0774$$

and

$$\text{S.E.}(\sigma^{**}) = 26.3942(0.0630)^{1/2} = 6.6249.$$

We may note here that the above estimates are close to those computed in Section 4.4 by the BLU estimation method. It is also of interest to observe that the simplified linear estimate of μ in this case is exactly the same as the 10% trimmed mean (see Dixon, 1957; Dixon and Tukey, 1968; Tukey and McLaughlin, 1963), since the population distribution (assumed to be normal) is symmetric and the censoring in the sample is also symmetric so that $\bar{\alpha} = 0$ in (4.6.3).

The efficiencies of the simplified linear estimators μ^{**} and σ^{**} relative to the BLUEs μ^* and σ^*, respectively, have been tabulated by Sarhan and Greenberg (1962) for sample sizes 10, 12, and 15 and for all choices of censoring. The relative efficiencies in most cases are over 90%, the lowest for μ^{**} being 84.66% when $n = 15$ and r or $s = 10$, and the lowest for σ^{**} being 86.75% when $n = 15$ and r or $s = 9$. The table of coefficients b_j^* in (4.6.5) required for the computation of the simplified linear estimator σ^{**} has been extended by Prescott (1970), who has also given a simple approximation for the variance of σ^{**} in (4.6.7). For the normal case, Stephens (1975) has related the estimator σ^{**} to the second eigenvector of the stochastic matrix $((\beta_{i,j:n}))$. Some simplified unbiased estimators of μ and σ based on the averages of the k_1 smallest and k_2 largest observations have been proposed and studied in detail by Abe (1971).

4.7. Blom's Unbiased Nearly Best Linear Estimators

Another simplified method of estimation of the location and scale parameters of an arbitrary distribution has been proposed by Blom (1958, 1962). These estimators, referred to as "unbiased nearly best linear estimators" in the literature, require only the exact values of the means of order statistics, just as Gupta's (1952) estimators do. However, Blom's estimators make use of the asymptotic approximations for the variances and covariances of order statistics. If the exact values of the means of order statistics are not available, we may use the asymptotic approximation for the means in the following derivation of the estimators and obtain what are referred to as "nearly unbiased, nearly best linear estimators."

Let $Z_{i:n} = (X_{i:n} - \mu)/\sigma$, $1 \le i \le n$, be the standardized order statistics from a population with pdf $f(z)$ and cdf $F(z)$. Then, we may approximate the variances and covariances of these standardized order statistics by the first term of (3.10.4) and (3.10.5), respectively, and write in general

$$\mathrm{Cov}(Z_{i:n}, Z_{j:n}) \simeq \frac{p_i q_j}{(n+2)P_i P_j}, \qquad 1 \le i \le j \le n, \tag{4.7.1}$$

where $p_i = i/(n+1)$, $q_i = 1 - p_i$ and $P_i = f(F^{-1}(p_i))$ for $i = 1, 2, \ldots, n$. Then from (4.7.1) we immediately have

$$\mathrm{Cov}(P_i Z_{i:n}, P_j Z_{j:n}) \simeq \frac{p_i q_j}{n+2}, \qquad 1 \le i \le j \le n. \tag{4.7.2}$$

Now with $P_0 = P_{n+1} = 0$, if we define

$$Z^*_{i:n} = P_{i+1} Z_{i+1:n} - P_i Z_{i:n}, \qquad 0 \le i \le n, \tag{4.7.3}$$

we find

$$E(Z^*_{i:n}) = P_{i+1}\alpha_{i+1:n} - P_i \alpha_{i:n}, \qquad 0 \le i \le n, \tag{4.7.4}$$

$$\mathrm{Var}(Z^*_{i:n}) \simeq \frac{n}{(n+1)^2(n+2)}, \qquad 0 \le i \le n, \tag{4.7.5}$$

and

$$\operatorname{Cov}(Z_{i:n}^*, Z_{j:n}^*) \simeq -\frac{1}{(n+1)^2(n+2)}, \qquad 0 \le i < j \le n, \tag{4.7.6}$$

independently of the parent distribution F.

In order to estimate the unknown parameter $\delta = l_1\mu + l_2\sigma$, for specified constants l_1 and l_2, let us consider the linear estimator

$$\delta_* = \sum_{i=1}^{n} Q_i X_{i:n} = \sum_{i=1}^{n} Q_i(\mu + \sigma Z_{i:n}). \tag{4.7.7}$$

From (4.7.3) we can write

$$P_i Z_{i:n} = \sum_{j=0}^{i-1} Z_{i:n}^*, \qquad 1 \le i \le n; \tag{4.7.8}$$

by writing the coefficients Q_i in (4.7.7) as

$$Q_i = P_i(R_i - R_{i-1}), \qquad 1 \le i \le n, \tag{4.7.9}$$

and by using (4.7.8), we get

$$\begin{aligned}
\delta_* &= \sum_{i=1}^{n} P_i(R_i - R_{i-1})(\mu + \sigma Z_{i:n}) \\
&= \mu \sum_{i=1}^{n} P_i(R_i - R_{i-1}) + \sigma \sum_{i=1}^{n} (R_i - R_{i-1}) \sum_{j=0}^{i-1} Z_{j:n}^* \\
&= \sum_{i=0}^{n} R_i\{\mu(P_i - P_{i+1}) - \sigma Z_{i:n}^*\}.
\end{aligned} \tag{4.7.10}$$

From (4.7.10) we get the expected value of δ_* by using (4.7.4) to be

$$\begin{aligned}
E(\delta_*) &= \mu \sum_{i=0}^{n} R_i(P_i - P_{i+1}) - \sigma \sum_{i=0}^{n} R_i(P_{i+1}\alpha_{i+1:n} - P_i\alpha_{i:n}) \\
&= \mu \sum_{i=0}^{n} R_i S_{1i} + \sigma \sum_{i=0}^{n} R_i S_{2i},
\end{aligned} \tag{4.7.11}$$

where

$$S_{1i} = P_i - P_{i+1}, \qquad 0 \le i \le n, \tag{4.7.12}$$

and

$$S_{2i} = P_i \alpha_{i:n} - P_{i+1} \alpha_{i+1:n}, \qquad 0 \leq i \leq n. \tag{4.7.13}$$

Furthermore, from (4.7.10) we obtain the variance of δ_* by using (4.7.5) and (4.7.6) to be

$$\begin{aligned}
\mathrm{Var}(\delta_*) &= \sigma^2 \left[\sum_{i=0}^{n} R_i^2 \,\mathrm{Var}(Z_{i:n}^*) + \sum_{\substack{i=0 \\ i \neq j}}^{n} \sum_{j=0}^{n} R_i R_j \,\mathrm{Cov}(Z_{i:n}^*, Z_{j:n}^*) \right] \\
&\simeq \frac{\sigma^2}{(n+1)^2(n+2)} \left[n \sum_{i=0}^{n} R_i^2 - \sum_{\substack{i=0 \\ i \neq j}}^{n} \sum_{j=0}^{n} R_i R_j \right] \\
&= \frac{\sigma^2}{(n+1)(n+2)} \sum_{i=0}^{n} (R_i - \bar{R})^2,
\end{aligned} \tag{4.7.14}$$

where $\bar{R} = \sum_{i=0}^{n} R_i/(n+1)$. Hence, the problem of determining the unbiased nearly best linear estimator of δ has been reduced to minimizing $\mathrm{Var}(\delta_*)$ in (4.7.14) subject to the unbiasedness conditions

$$\sum_{i=0}^{n} R_i S_{1i} = l_1 \quad \text{and} \quad \sum_{i=0}^{n} R_i S_{2i} = l_2. \tag{4.7.15}$$

By the standard Lagrangian method, we find the optimal solutions to be

$$R_i = \bar{R} + \lambda_1 S_{1i} + \lambda_2 S_{2i}, \tag{4.7.16}$$

where λ_1 and λ_2 are the Lagrangian multipliers given by

$$\lambda_1 = \frac{1}{T_{11} T_{22} - T_{12}^2} (T_{22} l_1 - T_{12} l_2) \tag{4.7.17}$$

and

$$\lambda_2 = \frac{1}{T_{11} T_{22} - T_{12}^2} (T_{11} l_2 - T_{12} l_1); \tag{4.7.18}$$

here,

$$T_{ij} = \sum_{k=0}^{n} S_{ik} S_{jk}, \qquad \text{for } 1 \leq i \leq j \leq 2. \tag{4.7.19}$$

With R_is as determined in (4.7.16), δ_* in (4.7.10) is the unbiased nearly best linear estimator of $\delta = l_1\mu + l_2\sigma$. From (4.7.7) we then have

$$\delta_* = \sum_{i=1}^{n} P_i(R_i - R_{i-1})X_{i:n}$$
$$= \sum_{i=1}^{n} P_i\{\lambda_1(S_{1i} - S_{1,i-1}) + \lambda_2(S_{2i} - S_{2,i-1})\}X_{i:n}, \quad (4.7.20)$$

where λ_1 and λ_2 are as given in (4.7.17) and (4.7.18), respectively.

In particular, by setting $l_1 = 1$, $l_2 = 0$ and $l_1 = 0$, $l_2 = 1$, we obtain the unbiased nearly best linear estimators of μ and σ as

$$\mu_* = \sum_{i=1}^{n} Q_{1i}X_{i:n} \quad (4.7.21)$$

and

$$\sigma_* = \sum_{i=1}^{n} Q_{2i}X_{i:n}, \quad (4.7.22)$$

where the coefficients Q_{1i} and Q_{2i} are given by

$$Q_{1i} = \frac{P_i}{T_{11}T_{22} - T_{12}^2}\{T_{22}(S_{1i} - S_{1,i-1}) - T_{12}(S_{2i} - S_{2,i-1})\} \quad (4.7.23)$$

and

$$Q_{2i} = \frac{P_i}{T_{11}T_{22} - T_{12}^2}\{T_{11}(S_{2i} - S_{2,i-1}) - T_{12}(S_{1i} - S_{1,i-1})\}. \quad (4.7.24)$$

Moreover, we obtain the variances and covariance of the estimators μ_* and σ_* from (4.7.14) to be

$$\operatorname{Var}(\mu_*) \simeq \frac{\sigma^2}{(n+1)(n+2)} \frac{1}{(T_{11}T_{22} - T_{12}^2)^2} \sum_{i=0}^{n} (T_{22}S_{1i} - T_{12}S_{2i})^2$$
$$= \frac{\sigma^2}{(n+1)(n+2)} \frac{1}{(T_{11}T_{22} - T_{12}^2)^2} (T_{22}^2T_{11} - T_{12}^2T_{22})$$
$$= \frac{\sigma^2}{(n+1)(n+2)} \frac{T_{22}}{(T_{11}T_{22} - T_{12}^2)}, \quad (4.7.25)$$

$$\operatorname{Var}(\sigma_*) \simeq \frac{\sigma^2}{(n+1)(n+2)} \frac{1}{(T_{11}T_{22} - T_{12}^2)^2} \sum_{i=0}^{n} (T_{11}S_{2i} - T_{12}S_{1i})^2$$
$$= \frac{\sigma^2}{(n+1)(n+2)} \frac{1}{(T_{11}T_{22} - T_{12}^2)^2} (T_{11}^2T_{22} - T_{12}^2T_{11})$$
$$= \frac{\sigma^2}{(n+1)(n+2)} \frac{T_{11}}{(T_{11}T_{22} - T_{12}^2)}, \quad (4.7.26)$$

and

$$\text{Cov}(\mu_*, \sigma_*) \simeq -\frac{\sigma^2}{(n+1)(n+2)} \frac{T_{12}}{(T_{11}T_{22} - T_{12}^2)}. \tag{4.7.27}$$

We may note that these estimators, unlike the BLUEs, do not need the inversion of the variance–covariance matrix of order n. However, since the estimators μ_* and σ_* are based on the asymptotic approximations of the variances and covariances of order statistics, they may not be good for small sample sizes. But, for large sample sizes the estimators μ_* and σ_* have been noted to be highly efficient in many situations.

When the sample is Type-II censored with r smallest and s largest observations missing, the formulas of μ_* and σ_* in (4.7.21) and (4.7.22) and their variances and covariance in (4.7.25) through (4.7.27) continue to hold with S_{1i} and S_{2i} replaced by S_{1i}^* and S_{2i}^*, respectively, where

$$S_{1i}^* = \begin{cases} -\dfrac{1}{(r+1)} P_{r+1}, & 0 \le i \le r \\ S_{1i} = P_i - P_{i+1}, & r+1 \le i \le n-s-1 \\ \dfrac{1}{(s+1)} P_{n-s}, & n-s \le i \le n \end{cases} \tag{4.7.28}$$

and

$$S_{2i}^* = \begin{cases} -\dfrac{1}{(r+1)} P_{r+1}\alpha_{r+1:n}, & 0 \le i \le r \\ S_{2i} = P_i\alpha_{i:n} - P_{i+1}\alpha_{i+1:n}, & r+1 \le i \le n-s-1. \\ \dfrac{1}{(s+1)} P_{n-s}\alpha_{n-s:n}, & n-s \le i \le n \end{cases} \tag{4.7.29}$$

We may note, in this case, that the unbiased nearly best linear estimator δ_* in (4.7.20) becomes

$$\delta_* = \sum_{i=r+1}^{n-s} P_i\{\lambda_1(S_{1i}^* - S_{1,i-1}^*) + \lambda_2(S_{2i}^* - S_{2,i-1}^*)\}X_{i:n}, \tag{4.7.30}$$

where λ_1 and λ_2 are as given in (4.7.17) and (4.7.18), respectively. Furthermore, the estimators μ_* and σ_* in (4.7.21) and (4.7.22) become

$$\mu_* = \sum_{i=r+1}^{n-s} Q_{1i}X_{i:n} \tag{4.7.31}$$

and

$$\sigma_* = \sum_{i=r+1}^{n-s} Q_{2i} X_{i:n}, \tag{4.7.32}$$

where

$$Q_{1i} = \frac{P_i}{T_{11}T_{22} - T_{12}^2} \{T_{22}(S_{1i}^* - S_{1,i-1}^*) - T_{12}(S_{2i}^* - S_{2,i-1}^*)\} \tag{4.7.33}$$

and

$$Q_{2i} = \frac{P_i}{T_{11}T_{22} - T_{12}^2} \{T_{11}(S_{2i}^* - S_{2,i-1}^*) - T_{12}(S_{1i}^* - S_{1,i-1}^*)\}. \tag{4.7.34}$$

By considering the data given by Gupta (1952) and Fisher (1966), we shall demonstrate the above method of estimation of μ and σ.

Example 4.7.1. Let us consider once again the data of Gupta (1952), which gives the log-number of days to death of the first seven in a sample of 10 mice after inoculation with a uniform culture of human tuberculosis:

1.613, 1.644, 1.663, 1.732, 1.740, 1.763, 1.778.

Here, we have $n = 10$, $r = 0$, $s = 3$, and

i	P_i	$\alpha_{i:n}$	S_{1i}^*	S_{2i}^*
0	0.00000	—	−0.16361	0.25175
1	0.16361	−1.53875	−0.10045	0.01267
2	0.26406	−1.00136	−0.06825	−0.04640
3	0.33231	−0.65606	−0.04310	−0.07695
4	0.37540	−0.37576	−0.02095	−0.09244
5	0.39635	−0.12267	0.00000	−0.09724
6	0.39635	0.12267	0.02095	−0.09244
7	0.37540	0.37576	0.09385	0.03527
8	0.33231	0.65606	0.09385	0.03527
9	0.26406	1.00136	0.09385	0.03527
10	0.16361	1.53875	0.09385	0.03527

Then,

$$T_{11}=0.07948,\qquad T_{12}=-0.02274,\qquad T_{22}=0.10313,$$

$$T_{22}/(T_{11}T_{22}-T_{12}^2)=13.4284,$$

$$T_{12}/(T_{11}T_{22}-T_{12}^2)=-2.9606,$$

and

$$T_{11}/(T_{11}T_{22}-T_{12}^2)=10.3488.$$

Further, we have:

i	Q_{1i}	Q_{2i}
1	0.02296	−0.37421
2	0.06800	−0.13625
3	0.08217	−0.08032
4	0.09444	−0.03556
5	0.10587	0.00490
6	0.11714	0.04427
7	0.50943	0.57717

We then obtain the unbiased nearly best linear estimates of μ and σ as

$$\begin{aligned}\mu_* &= 0.02296(1.613)+0.06800(1.644)+0.08217(1.663)+0.09444(1.732)\\ &\quad +0.10587(1.740)+0.11714(1.763)+0.50943(1.778)\\ &= 1.7455\end{aligned}$$

and

$$\begin{aligned}\sigma_* &= -0.37421(1.613)-0.13625(1.644)-0.08032(1.663)-0.03556(1.732)\\ &\quad +0.00490(1.740)+0.04427(1.763)+0.57717(1.778)\\ &= 0.0900.\end{aligned}$$

The standard errors of the above estimates are computed from (4.7.25) and (4.7.26) to be

$$\text{S.E.}(\mu_*)=0.09\left\{\frac{13.4284}{11\times 12}\right\}^{1/2}=0.0287$$

and

$$\text{S.E.}(\sigma_*) = 0.09\left\{\frac{10.3488}{11\times 12}\right\}^{1/2} = 0.0252.$$

Note that these values are very close to those obtained by the BLU estimation method in Section 4.4.

Example 4.7.2. Let us consider the data given by Fisher (1966), which represent the differences in heights between cross- and self-fertilized plants of the same pair grown together in one pot, with the smallest two and the largest two censored:

$$6, 8, 14, 16, 23, 24, 28, 41, 59, 56.$$

Here, we have $n = 15$, $r = 2$, $s = 2$, and

i	P_i	$\alpha_{i:n}$	S^*_{1i}	S^*_{2i}
0	0.00000	—	−0.08972	0.08503
1	0.12298	−1.73591	−0.08972	0.08503
2	0.20585	−1.24794	−0.08972	0.08503
3	0.26916	−0.94769	−0.04862	−0.02791
4	0.31778	−0.71488	−0.03625	−0.04460
5	0.35402	−0.51570	−0.02517	−0.05543
6	0.37920	−0.33530	−0.01484	−0.06201
7	0.39404	−0.16530	−0.00491	−0.06513
8	0.39894	0.00000	0.00491	−0.06513
9	0.39404	0.16530	0.01484	−0.06201
10	0.37920	0.33530	0.02517	−0.05543
11	0.35402	0.51570	0.03625	−0.04460
12	0.31778	0.71488	0.04862	−0.02791
13	0.26916	0.94769	0.08972	0.08503
14	0.20585	1.24794	0.08972	0.08503
15	0.12298	1.73591	0.08972	0.08503

Then,

$$T_{11} = 0.05741, \qquad T_{12} = 0, \qquad T_{22} = 0.07124,$$

$$T_{22}/(T_{11}T_{22} - T_{12}^2) = 1/T_{11} = 17.4189,$$

$$T_{12}/(T_{11}T_{22} - T_{12}^2) = 0,$$

and

$$T_{11}/(T_{11}T_{22}-T_{12}^2)=1/T_{22}=14.0385.$$

Further, we have:

i	Q_{1i}	Q_{2i}
3	0.19270	−0.42676
4	0.06847	−0.07446
5	0.06833	−0.05382
6	0.06823	−0.03503
7	0.06816	−0.01726
8	0.06824	0.00000
9	0.06816	0.01726
10	0.06823	0.03503
11	0.06833	0.05382
12	0.06847	0.07446
13	0.19270	0.42676

We then obtain the unbiased nearly best linear estimates of μ and σ as

$$\begin{aligned}\mu_* &= 0.19270(6)+0.06847(8)+0.06833(14)+0.06823(16)\\ &\quad +0.06816(23)+0.06824(24)+0.06816(28)+0.06823(29)\\ &\quad +0.06833(41)+0.06847(49)+0.19270(56)\\ &= 27.79\end{aligned}$$

and

$$\begin{aligned}\sigma_* &= -0.42676(6)-0.07446(8)-0.05382(14)-0.03503(16)\\ &\quad -0.01726(23)+0(24)+0.01726(28)+0.03503(29)\\ &\quad +0.05382(41)+0.07446(49)+0.42676(56)\\ &= 26.38.\end{aligned}$$

The standard errors of the above estimates are computed from (4.7.25) and (4.7.26) to be

$$\text{S.E.}(\mu_*) = 26.38\left\{\frac{17.4189}{16\times 17}\right\}^{1/2} = 6.676$$

and

$$\text{S.E.}(\sigma_*) = 26.38\left\{\frac{14.0385}{16\times 17}\right\}^{1/2} = 5.993.$$

4.8. Downton's Linear Estimators with Polynomial Coefficients

Downton (1966a) has proposed linear estimators for the location and scale parameters in which the general structure of the coefficients has been chosen for mathematical tractability, both in the determination of these coefficients and in the computation of the standard error of the estimates so obtained. He has shown that these estimators have variances and covariance that, in general, behave asymptotically as $1/n$, and he has discussed the class of distributions for which they may be suitable. By considering the normal and extreme value distributions, Downton (1966a) has demonstrated that these estimators are quite highly efficient.

In order to describe this method of estimation, let us consider an ordered random sample $X_{1:n} \leq X_{2:n} \leq \cdots \leq X_{n:n}$ from a location-scale parameter distribution. Based on this complete ordered sample, let us consider estimators of the form

$$\mu_{**} = \sum_{k=0}^{p} (k+1)\theta_k \sum_{i=1}^{n} (i-1)^{(k)} X_{i:n} / n^{(k+1)} \tag{4.8.1}$$

and

$$\sigma_{**} = \sum_{k=0}^{p} (k+1)\phi_k \sum_{i=1}^{n} (i-1)^{(k)} X_{i:n} / n^{(k+1)}, \tag{4.8.2}$$

where $m^{(r)}$, with m and r integers, denotes the rth factorial power of m, that is,

$$m^{(r)} = m(m-1)\dots(m-r+1). \tag{4.8.3}$$

As before, let us denote $Z_{i:n} = (X_{i:n} - \mu)/\sigma (1 \leq i \leq n)$, $E(Z_{i:n})$ by $\alpha_{i:n}$ $(1 \leq i \leq n)$, and $E(Z_{i:n} Z_{j:n})$ by $\alpha_{i,j:n} (1 \leq i \leq j \leq n)$.

Now by making use of the identities

$$\sum_{i=1}^{n} (i-1)^{(k)} = n^{(k+1)}/(k+1) \tag{4.8.4}$$

and

$$\sum_{i=1}^{n} (i-1)^{(k)} \alpha_{i:n} = n^{(k+1)} \alpha_{k+1:k+1}/(k+1) \tag{4.8.5}$$

(see Downton, 1966a; Arnold and Balakrishnan, 1989), we get from (4.8.1) and (4.8.2) that

$$E(\mu_{**}) = \mu \sum_{k=0}^{p} \theta_k + \sigma \sum_{k=0}^{p} \theta_k \alpha_{k+1:k+1} \tag{4.8.6}$$

and

$$E(\sigma_{**}) = \mu \sum_{k=0}^{p} \phi_k + \sigma \sum_{k=0}^{p} \phi_k \alpha_{k+1:k+1}. \tag{4.8.7}$$

Now let us denote

$$\boldsymbol{\theta} = (\theta_0 \quad \theta_1 \quad \cdots \quad \theta_p)',$$

$$\boldsymbol{\phi} = (\phi_0 \quad \phi_1 \quad \cdots \quad \phi_p)',$$

$$\boldsymbol{\alpha} = (\alpha_{1:1} \quad \alpha_{2:2} \quad \cdots \quad \alpha_{p+1:p+1})',$$

and $\mathbf{1}$ for the column vector with $(p+1)$ ones. Then, the estimators μ_{**} and σ_{**} in (4.8.6) and (4.8.7), respectively, are unbiased if

$$\begin{bmatrix} \boldsymbol{\theta}' \\ \hdashline \boldsymbol{\phi}' \end{bmatrix} [\mathbf{1} \quad \boldsymbol{\alpha}] = \begin{bmatrix} 1 & \vdots & 0 \\ \hdashline 0 & \vdots & 1 \end{bmatrix} \tag{4.8.8}$$

If we now let ψ_k be the random variable defined by

$$\psi_k = (k+1) \sum_{i=1}^{n} (i-1)^{(k)} Z_{i:n} / n^{(k+1)} \tag{4.8.9}$$

and let $\boldsymbol{\Omega}$ be the symmetric matrix with elements

$$\Omega_{ij} = \Omega_{ji} = \mathrm{Cov}(\psi_i, \psi_j), \qquad i \leq j, \tag{4.8.10}$$

the variance matrix of the estimators μ_{**} and σ_{**} is given by

$$\mathrm{Var} \begin{bmatrix} \mu_{**} \\ \hdashline \sigma_{**} \end{bmatrix} = \sigma^2 \begin{bmatrix} \boldsymbol{\theta}' \\ \hdashline \boldsymbol{\phi}' \end{bmatrix} \boldsymbol{\Omega} [\boldsymbol{\theta} \quad \boldsymbol{\phi}]. \tag{4.8.11}$$

The best unbiased estimators (in the least-squares sense) of μ and σ having the form (4.8.1) and (4.8.2) are obtained by choosing $\boldsymbol{\theta}$ and $\boldsymbol{\phi}$ such that

$$\begin{bmatrix} \boldsymbol{\theta}' \\ \hdashline \boldsymbol{\phi}' \end{bmatrix} \boldsymbol{\Omega} [\boldsymbol{\theta} \vdots \boldsymbol{\phi}] + 2 \begin{bmatrix} \boldsymbol{\theta}' \\ \hdashline \boldsymbol{\phi}' \end{bmatrix} [\mathbf{1} \vdots \boldsymbol{\alpha}] \begin{bmatrix} \lambda_1 & \vdots & \lambda_3 \\ \hdashline \lambda_2 & \vdots & \lambda_4 \end{bmatrix} \tag{4.8.12}$$

is a minimum. The values of $\boldsymbol{\theta}$ and $\boldsymbol{\phi}$ corresponding to this minimum are obtained by solving the equation

$$\begin{bmatrix} \boldsymbol{\Omega} & \vdots & \mathbf{1} & \vdots & \boldsymbol{\alpha} \\ \hdashline \mathbf{1}' & \vdots & 0 & \vdots & 0 \\ \hdashline \boldsymbol{\alpha}' & \vdots & 0 & \vdots & 0 \end{bmatrix} \begin{bmatrix} \boldsymbol{\theta} & \vdots & \boldsymbol{\phi} \\ \hdashline \lambda_1 & \vdots & \lambda_3 \\ \hdashline \lambda_2 & \vdots & \lambda_4 \end{bmatrix} = \begin{bmatrix} \mathbf{0} & \vdots & \mathbf{0} \\ \hdashline 1 & \vdots & 0 \\ \hdashline 0 & \vdots & 1 \end{bmatrix} \tag{4.8.13}$$

that is,

$$[\boldsymbol{\theta} \vdots \boldsymbol{\phi}] = \boldsymbol{\Omega}^{-1} [\mathbf{1} \vdots \boldsymbol{\alpha}] \left[\begin{bmatrix} \mathbf{1}' \\ \hdashline \boldsymbol{\alpha}' \end{bmatrix} \boldsymbol{\Omega}^{-1} [\mathbf{1} \vdots \boldsymbol{\alpha}] \right]^{-1}. \tag{4.8.14}$$

Thus the computation of the estimators μ_{**} and σ_{**} requires the evaluation of the elements Ω_{ij} of $\mathbf{\Omega}$ followed by the inversion of two matrices, one of order $p+1$ and the other of order 2. It may also be noted that in this process the variances and covariance of the estimators μ_{**} and σ_{**} are also computed, since using (4.8.11) and (4.8.14) we have

$$\begin{bmatrix} \mathrm{Var}(\mu_{**}) & \mathrm{Cov}(\mu_{**}, \sigma_{**}) \\ & \mathrm{Var}(\sigma_{**}) \end{bmatrix} = \sigma^2 \left[\begin{bmatrix} \mathbf{1}' \\ \boldsymbol{\alpha}' \end{bmatrix} \mathbf{\Omega}^{-1} [\mathbf{1} \quad \boldsymbol{\alpha}] \right]^{-1}. \tag{4.8.15}$$

By making use of the identity in (4.8.5), we obtain from (4.8.9) that

$$E(\psi_k) = \alpha_{k+1:k+1} \tag{4.8.16}$$

and, hence, for $i \le j$ we have

$$\begin{aligned} \Omega_{ij} &= \Omega_{ji} \\ &= (i+1)(j+1)E\left\{ \sum_{r=1}^{n} \sum_{s=1}^{n} (r-1)^{(i)}(s-1)^{(j)} Z_{r:n} Z_{s:n} \right\} \Big/ \{n^{(i+1)} n^{(j+1)}\} \\ &\quad - \alpha_{i+1:i+1}\alpha_{j+1:j+1} \\ &= (i+1)(j+1) \sum_{r=1}^{n} \sum_{s=1}^{n} (r-1)^{(i)}(s-1)^{(j)} \alpha_{r,s:n} / \{n^{(i+1)} n^{(j+1)}\} \\ &\quad - \alpha_{i+1:i+1}\alpha_{j+1:j+1}. \end{aligned} \tag{4.8.17}$$

Now by making use of the identities

$$(a+b)^{(m)} = \sum_{r=0}^{m} \binom{m}{r} a^{(r)} b^{(m-r)} \tag{4.8.18}$$

and

$$(a-b)^{(m)} = \sum_{r=0}^{m} (-1)^r \binom{m}{r} (a-r)^{(m-r)} b^{(r)}, \tag{4.8.19}$$

and the relations

$$\sum_{i=1}^{n} (i-1)^{(k)}(n-i)^{(l)} \alpha_{i,i:n} = k!\,l! \binom{n}{k+l+1} \alpha_{k+1,k+1:k+l+1} \tag{4.8.20}$$

and

$$\sum\sum_{i<j} (i-1)^{(k)}(n-j)^{(l)} \alpha_{i,j:n} = k!\,l! \binom{n}{k+l+2} \alpha_{k+1,k+2:k+l+2} \tag{4.8.21}$$

(see Downton, 1966a; Arnold and Balakrishnan, 1989), we have

$$\sum_{r=1}^{n}\sum_{s=1}^{n}(r-1)^{(i)}(s-1)^{(j)}\alpha_{r,s:n}$$

$$=\sum_{r=1}^{n}(r-1)^{(i)}(r-1)^{(j)}\alpha_{r,r:n}$$

$$+\sum\sum_{r<s}\{(r-1)^{(i)}(s-1)^{(j)}+(s-1)^{(i)}(r-1)^{(j)}\}\alpha_{i,j:n}$$

$$=\sum_{t=0}^{i}\binom{i}{t}j^{(i-t)}(j+t)!\binom{n}{j+t+1}\alpha_{j+t+1,j+t+1:j+t+1}$$

$$+\sum_{t=0}^{i}(-1)^{t}\binom{i}{t}(n-t-1)^{(i-t)}j!t!\binom{n}{j+t+2}\alpha_{j+1,j+2:j+t+2}$$

$$+\sum_{t=0}^{j}(-1)^{t}\binom{j}{t}(n-t-1)^{(j-t)}i!t!\binom{n}{i+t+2}\alpha_{i+1,i+2:i+t+2}. \qquad (4.8.22)$$

By using the identities (4.8.18) and (4.8.19), we can express (4.8.22) as a factorial power series of the form

$$\sum_{r=1}^{n}\sum_{s=1}^{n}(r-1)^{(i)}(s-1)^{(j)}\alpha_{r,s:n}=\sum_{t=j+1}^{i+j+2}a_{ij}^{(t)}n^{(t)}, \qquad (4.8.23)$$

where

$$a_{ij}^{(i+j+2)}=\alpha_{i+1:i+1}\alpha_{j+1:j+1}/\{(i+1)(j+1)\} \qquad (4.8.24)$$

and, for $j+1\leq s\leq i+j+1$,

$$a_{ij}^{(s)}=i!j!\alpha_{s,s:s}/\{(i+j+1-s)!\,(s-i-1)!\,(s-j-1)!\,s\}$$

$$+i!j!(j+1)^{(i+j+2-s)}\sum_{r=0}^{s-j-2}(-1)^{r}\frac{\alpha_{j+1,j+2:j+2+r}}{(i+j+2-s)!\,(j+r+2)!\,(s-j-2-r)!}$$

$$-i!j!(i+1)^{(i+j+2-s)}\sum_{r=0}^{i}(-1)^{r}\frac{\alpha_{s-i-1,s-i:s-i+r}}{(i+j+2-s)!\,(i-r)!\,(s-i+r)!}$$

$$+i!j!\alpha_{i+1:i+1}\alpha_{s-i-1:s-i-1}/\{(i+j+2-s)!\,(s-i-1)!\,(s-j-1)!\}, \qquad (4.8.25)$$

where the second term vanishes when $s=j+1$, while the third and fourth terms vanish when $i=j$ and $s=j+1$.

Because of (4.8.24), we see from the expression of Ω_{ij} in (4.8.17) that the coefficient of $n^{(i+j+2)}$ becomes zero leaving, for $i \leq j$,

$$\Omega_{ij} = \Omega_{ji} = (i+1)(j+1) \sum_{s=0}^{i} b_{ij}^{(s)} (n-j-1)^{(s)} / n^{(i+1)}, \tag{4.8.26}$$

where

$$\begin{aligned} b_{ij}^{(s)} &= i!j!\alpha_{s+j+1,s+j+1:s+j+1} / \{(i-s)!\,(s+j-i)!\,s!\,(s+j+1)\} \\ &\quad + i!j!(j+1)^{(i+1-s)} \sum_{r=0}^{s-1} (-1)^r \frac{\alpha_{j+1,j+2:j+2+r}}{(i+1-s)!\,(j+2+r)!\,(s-1-r)!} \\ &\quad - i!j!(i+1)^{(i+1-s)} \sum_{r=0}^{i} (-1)^r \frac{\alpha_{s+j-i,s+j-i+1:s+j-i+1+r}}{(i+1-s)!\,(i-r)!\,(s+j-i+1+r)!} \\ &\quad + i!j!\alpha_{i+1:i+1}(\alpha_{s+j-i:s+j-i} - \alpha_{j+1:j+1}) / \{(i+1-s)!\,(s+j-i)!\,s!\}. \end{aligned} \tag{4.8.27}$$

We want to mention that the first term on the RHS of (4.8.25) and (4.8.27) as reported by Downton (1966a) contains an error. It should be noted here that these coefficients depend only upon diagonal and immediate upper-diagonal product moments of relatively small variance matrices of ordered observations and upon the expected values of the largest observations. The amount of computation required to obtain estimates of the form (4.8.1) and (4.8.2) is considerably less than what is required to obtain the BLUEs.

By using the expected values and product moments of normal order statistics prepared by Teichroew (1956) for sample sizes up to 20, Downton (1966a) has computed the values of the coefficients $b_{ij}^{(s)}$ that are needed to

TABLE 4.8.1
Coefficients $b_{ij}^{(s)}$ for Computing Elements of $\mathbf{\Omega}$ in the Normal Distribution

s	0	1	2	3
$i=0, j=0$	1.000000000	—	—	—
$i=0, j=1$	0.500000000	—	—	—
$i=0, j=2$	0.333333333	—	—	—
$i=0, j=3$	0.250000000	—	—	—
$i=1, j=1$	0.340845057	0.290687893	—	—
$i=1, j=2$	0.464866288	0.207354561	—	—
$i=1, j=3$	0.531084792	0.161889237	—	—
$i=2, j=2$	0.372978136	0.670263303	0.153448650	—
$i=2, j=3$	0.860369202	0.791066611	0.122697037	—
$i=3, j=3$	0.737572855	2.042440665	0.953422919	0.099725871

Produced with kind permission of Biometrika Trustees.

evaluate the variance matrix $\mathbf{\Omega}$. These values are presented in Table 4.8.1 for $0 \le i \le j \le 3$.

The case $p = 1$, which is an estimator consisting simply of two terms having constant and linear coefficients, respectively, produces a particularly simple estimator. In this case minimization is not necessary, as only two coefficients are required for each estimator and these are determined by the two conditions due to unbiasedness. For the case of the normal distribution, the estimator of μ will always be the sample mean $\bar{X}$. Also, since $\alpha_{1:1} = 0$ the estimator of σ is derived as

$$\sigma_{**} = C \sum_{i=1}^{n} (2i - n - 1) X_{i:n} / \{n(n-1)\}, \tag{4.8.28}$$

where $C = 1/\alpha_{2:2} = 1.77245385$. It should be mentioned that the estimator σ_{**} in (4.8.28) is the weighted sum of the quasi-ranges used by Dixon (1957) in order to obtain simple estimators of σ. The variance of this estimator, unlike those of Dixon's estimators, may be explicitly written down as

$$\operatorname{Var}(\sigma_{**}) = \frac{\sigma^2}{n} \{0.511299138 + 0.630293516/(n-1)\}. \tag{4.8.29}$$

Asymptotically the efficiency of σ_{**} relative to the most efficient estimator of σ is 97.8%, and it is considerably more efficient than the sample range for moderate sample sizes. The efficiencies of the sample range relative to σ_{**} in (4.8.28) have been computed by Downton (1966a) and are presented in Table 4.8.2. For $n = 2$ and 3, the estimator σ_{**} is identical to the sample range, and the table of variance of the range given by Harley and Pearson (1957) was used to compute the efficiencies for $n = 30$, 45, and 200. Also, the relative efficiency tends to zero as $n \to \infty$.

Because of the symmetry of the normal distribution, there is no advantage in using a three-term (quadratic) approximation rather than a two-term

TABLE 4.8.2
Efficiency of Range as an Estimate of σ in the Normal Distribution Relative to $\sigma_{**} = C \sum_{i=1}^{n} (2i - n - 1) X_{i:n} / \{n(n-1)\}$ $[C = 1.77245385]$

n	Efficiency	n	Efficiency	n	Efficiency
4	98.75	8	90.66	20	71.52
5	96.93	9	88.62	30	61.81
6	94.88	10	86.67	45	52.27
7	92.76	15	78.17	200	23.01

(linear) one. If, on the other hand, a four-term (cubic) approximation is used, the asymptotic efficiency of σ_{**} so obtained increases to 99.6%, and from the efficiency point of view the estimators σ_{**} so obtained are almost identical to the BLUEs. In order to illustrate the calculation of the estimators, we present below the computation of four-term (cubic) coefficients for a sample of size 10.

By using the coefficients in Table 4.8.1 we obtain for $n=10$ that

$$\mathbf{\Omega} = 10^{-1}\begin{bmatrix} 1.00000000 & 1.00000000 & 1.00000000 & 1.00000000 \\ & 1.18504363 & 1.27756548 & 1.33548464 \\ & & 1.43870807 & 1.54794666 \\ & & & 1.70038124 \end{bmatrix},$$

$$\mathbf{\Omega}^{-1} = 10\begin{bmatrix} 30.49 & -120.90 & 160.46 & -69.05 \\ & 626.27 & -943.68 & 438.31 \\ & & 1521.74 & -738.52 \\ & & & 369.26 \end{bmatrix},$$

and

$$\begin{bmatrix} \mathbf{1}' \\ \cdots \\ \boldsymbol{\alpha}' \end{bmatrix} = \begin{bmatrix} 1 & 1 & 1 & 1 \\ 0 & 0.56418958 & 0.84628438 & 1.02937537 \end{bmatrix},$$

which yield the estimators of μ and σ to be

$$\mu_{**} = \sum_{i=1}^{10} X_{i:10}/10^{(1)}$$

and

$$\sigma_{**} = -2.016\sum_{i=1}^{10} X_{i:10}/10^{(1)} + 3.404\sum_{i=1}^{10}(i-1)X_{i:10}/10^{(2)}$$
$$-2.772\sum_{i=1}^{10}(i-1)^{(2)}X_{i:10}/10^{(3)} + 1.384\sum_{i=1}^{10}(i-1)^{(3)}X_{i:10}/10^{(4)},$$

with variances and covariance given by

$$\begin{bmatrix} \mathrm{Var}(\mu_{**}) & \mathrm{Cov}(\mu_{**}, \sigma_{**}) \\ & \mathrm{Var}(\sigma_{**}) \end{bmatrix} = \sigma^2\begin{bmatrix} 0.1000 & 0 \\ & 0.0577 \end{bmatrix}.$$

This may be compared with the variance–covariance matrix of the BLUEs of μ and σ, taken from the tables of Sarhan and Greenberg (1956), where

$$\begin{bmatrix} \mathrm{Var}(\mu^*) & \mathrm{Cov}(\mu^*, \sigma^*) \\ & \mathrm{Var}(\sigma^*) \end{bmatrix} = \sigma^2\begin{bmatrix} 0.1000 & 0 \\ & 0.0576 \end{bmatrix}.$$

We thus note that the estimator σ_{**} is remarkably efficient.

TABLE 4.8.3
Coefficients $b_{ij}^{(s)}$ for Computing Elements of $\mathbf{\Omega}$ in the Logistic Distribution

s	0	1	2	3
$i=0, j=0$	1.000000000	—	—	—
$i=0, j=1$	0.500000000	—	—	—
$i=0, j=2$	0.348018224	—	—	—
$i=0, j=3$	0.272027336	—	—	—
$i=1, j=1$	0.348018224	0.303963551	—	—
$i=1, j=2$	0.500000000	0.227972662	—	—
$i=1, j=3$	0.598018224	0.185755503	—	—
$i=2, j=2$	0.413363707	0.759908875	0.177312070	—
$i=2, j=3$	1.000000000	0.937220948	0.147760060	—
$i=3, j=3$	0.879407749	2.482369003	1.177858760	0.124962794

By using the means, variances and covariances of logistic order statistics prepared by Balakrishnan and Malik (1990), Balakrishnan (1990a) has computed the values of the coefficients $b_{ij}^{(s)}$ that are needed to evaluate the variance matrix $\mathbf{\Omega}$. These values are presented in Table 4.8.3 for $0 \le i \le j \le 3$.

By making use of the values of $b_{ij}^{(s)}$ presented in Table 4.8.3, the values of Ω_{ij} $(=\Omega_{ji})$ for $0 \le i \le j \le 3$ have been computed for sample size $n = 20$ and are given in Table 4.8.4. These values were used in Eqs. (4.8.14) and (4.8.15) to compute the coefficients $\boldsymbol{\theta}$ and $\boldsymbol{\phi}$ and the variances and covariance of the estimators μ_{**} and σ_{**} for $n = 20$ that are presented in Table 4.8.5. Because of the symmetry of the logistic distribution, we have the covariance of the estimators μ_{**} and σ_{**} to be zero. Also given in the last column of this table are the efficiencies of these estimators relative to the BLUEs of μ and σ based on the entire sample.

From Table 4.8.5, we note that even a two-term (linear) estimator of μ and σ is highly efficient as compared to the BLUEs. While the three-term (quadratic) estimator of μ is observed to be as efficient as the BLUE of μ,

TABLE 4.8.4
Values of Ω_{ij} for the Logistic Distribution When $n = 20$

i j	0	1	2	3
0	0.050000000	0.050000000	0.052202733	0.054405467
1	—	0.061256443	0.069087398	0.075160132
2	—	—	0.081000917	0.090277104
3	—	—	—	0.102257778

TABLE 4.8.5
Coefficients, Variances, and Relative Efficiencies of the Estimators μ_{**} and σ_{**} When $n = 20$

p		$k=0$	$k=1$	$k=2$	$k=3$	$(\text{Var})/\sigma^2$	Rel. Eff.
1	$\boldsymbol{\theta}^T$	1.00000	0.00000	—	—	0.05000	92.34%
	$\boldsymbol{\phi}^T$	−1.81380	1.81380	—	—	0.03703	98.84%
2	$\boldsymbol{\theta}^T$	0.13171	2.60487	−1.73658	—	0.04618	99.98%
	$\boldsymbol{\phi}^T$	−1.81380	1.81380	0.00000	—	0.03703	98.84%
3	$\boldsymbol{\theta}^T$	0.13189	2.60496	−1.73747	0.00062	0.04617	100.00%
	$\boldsymbol{\phi}^T$	−1.47986	−0.52309	4.00554	−2.00259	0.03660	100.00%

no improvement is achieved in using a three-term (quadratic) estimator of σ instead of a two-term (linear) one, because of the symmetry of the logistic distribution. However, a four-term (cubic) estimator of σ turns out to be just as efficient as the BLUE of σ.

Example 4.8.1. Davis (1952) has given lifetimes in hours of 417 40-watt incandescent lamps taken from 42 weekly forced-life test samples. Davis has indicated that the normal distribution provides a good model for these data, although the frequency distribution of the data appears to be somewhat more peaked in the center and flatter in the flanks than the normal distribution. This is indeed how the logistic distribution compares with the normal, as pointed out by Chew (1968), and so the assumption of the logistic distribution will be quite appropriate for this data. The first 20 observations from the data are given below, after having been arranged in increasing order of magnitude:

$$785, 855, 905, 918, 919, 920, 929, 936, 948, 950,$$

$$972, 1035, 1045, 1067, 1092, 1126, 1156, 1162, 1170, 1196.$$

From the above data, we find that

$$\sum_{i=1}^{20} (i-1)^{(0)} X_{i:20}/20^{(1)} = 1004.3,$$

$$\sum_{i=1}^{20} (i-1)^{(1)} X_{i:20}/20^{(2)} = 536.0078947,$$

$$\sum_{i=1}^{20} (i-1)^{(2)} X_{i:20}/20^{(3)} = 369.4295322,$$

and

$$\sum_{i=1}^{20} (i-1)^{(3)} X_{i:20}/20^{(4)} = 282.8421569.$$

Now by using Table 4.8.5, we obtain the best unbiased estimates with polynomial coefficients of μ and σ (with $p=3$) to be

$$\begin{aligned}\mu_{**} &= 0.13189(1004.3) + 2.60496(536.0078947) - 1.73747(369.4295322) \\ &\quad + 0.00062(282.8421569) \\ &= 1000.098638 \text{ hours}\end{aligned}$$

and

$$\begin{aligned}\sigma_{**} &= -1.47986(1004.3) - 0.52309(536.0078947) + 4.00554(369.4295322) \\ &\quad - 2.00259(282.8421569) \\ &= 126.642668 \text{ hours},\end{aligned}$$

and their standard errors to be

$$\text{S.E.}(\mu_{**}) = 126.642668(0.04617)^{1/2} = 27.212 \text{ hours}$$

and

$$\text{S.E.}(\sigma_{**}) = 126.642668(0.03660)^{1/2} = 24.228 \text{ hours}.$$

On some occasions it may be more convenient to consider estimators of the form

$$\mu_{***} = \sum_{k=0}^{p} (k+1)\bar{\theta}_k \sum_{i=1}^{n} (n-i)^{(k)} X_{i:n}/n^{(k+1)} \tag{4.8.30}$$

and

$$\sigma_{***} = \sum_{k=0}^{p} (k+1)\bar{\phi}_k \sum_{i=1}^{n} (n-i)^{(k)} X_{i:n}/n^{(k+1)}. \tag{4.8.31}$$

As before, let us denote

$$\bar{\boldsymbol{\theta}} = (\bar{\theta}_0 \quad \bar{\theta}_1 \quad \cdots \quad \bar{\theta}_p)',$$

$$\bar{\boldsymbol{\phi}} = (\bar{\phi}_0 \quad \bar{\phi}_1 \quad \cdots \quad \bar{\phi}_p)',$$

and

$$\bar{\boldsymbol{\alpha}} = (\alpha_{1:1} \quad \alpha_{1:2} \quad \cdots \quad \alpha_{1:p+1})'.$$

Then the choices of the vectors $\bar{\boldsymbol{\theta}}$ and $\bar{\boldsymbol{\phi}}$ giving a best unbiased estimator are given by

$$[\bar{\boldsymbol{\theta}} \vdots \bar{\boldsymbol{\phi}}] = \bar{\boldsymbol{\Omega}}^{-1}[\mathbf{1} \vdots \bar{\boldsymbol{\alpha}}]\left[\begin{bmatrix}\mathbf{1}' \\ \cdots \\ \bar{\boldsymbol{\alpha}}'\end{bmatrix}\bar{\boldsymbol{\Omega}}^{-1}[\mathbf{1} \quad \bar{\boldsymbol{\alpha}}]\right]^{-1}, \tag{4.8.32}$$

where the elements of $\bar{\boldsymbol{\Omega}}$ are $\bar{\Omega}_{ij}$, given by $(i \leq j)$

$$\bar{\Omega}_{ij} = \bar{\Omega}_{ji} = (i+1)(j+1)\sum_{s=0}^{i} \bar{b}_{ij}^{(s)}(n-j-1)^{(s)}/n^{(i+1)}, \tag{4.8.33}$$

with

$$\begin{aligned}
\bar{b}_{ij}^{(s)} = {} & i!j!\alpha_{1,1:s+j+1}/\{(i-s)!\,(s+j-i)!\,s!\,(s+j+1)\} \\
& + i!j!(j+1)^{(i+1-s)}\sum_{r=0}^{s-1}(-1)^r \frac{\alpha_{r+1,r+2:r+j+2}}{(i+1-s)!\,(r+j+2)!\,(s-1-r)!} \\
& - i!j!(i+1)^{(i+1-s)}\sum_{r=0}^{i}(-1)^r \frac{\alpha_{r+1,r+2:s+j-i+r+1}}{(i+1-s)!\,(i-r)!\,(s+j-i+r+1)!} \\
& + i!j!\alpha_{1:i+1}\{\alpha_{1:s+j-i} - \alpha_{1:j+1}\}/\{(i+1-s)!\,(s+j-i)!\,s!\}.
\end{aligned} \tag{4.8.34}$$

This method of estimation has, however, a major difficulty; in some cases the matrix $\boldsymbol{\Omega}$ or $\bar{\boldsymbol{\Omega}}$ is so ill-conditioned that it may be difficult to achieve sufficient accuracy in the final results, and it is also impracticable for some distributions to go beyond three-term (quadratic) coefficients. However, such coefficients may often yield highly efficient estimators.

4.9. Details of Other Related Work

Mann (1969c) has shown that the minimum-mean squared error invariant estimators of μ and σ can be obtained from the BLUEs μ^* and σ^*. By noting that the tables for the BLUEs are usually available only for small samples sizes, McCool (1965) and Chu and Ya'coub (1968) have considered the estimation for large sample sizes based on the average BLUEs obtained for subsamples small enough to allow the use of tables. Interested readers may refer to Grundy (1952), Hammersley and Morton (1954), Swamy (1962), and David and Mishriky (1968). Maxwell (1973) has considered the estimation problem with both Type I and Type II censoring based on Fraser's (1968) structural inference. Lwin (1976) has derived the linear Bayes estimator of $\boldsymbol{\theta} = \binom{\mu}{\sigma}$ by minimizing the expected risk function after assuming that $\boldsymbol{\theta}$ has been drawn from a population with cdf $F(\boldsymbol{\theta})$ whose first two moments are known.

In the following we list a number of articles dealing with the estimation of the location and scale parameters from complete and censored samples for various populations. The BLUEs have been given by Sarhan and Greenberg (1962) for several populations for sample size five and less.

Logistic: Gupta *et al.* (1967), Gajjar and Khatri (1969), Raghunandanan and Srinivasan (1970), Hall (1975), D'agostino and Lee (1976), Balakrishnan (1990f).

Double exponential: Govindarajulu (1966), Raghunandanan and Srinivasan (1971).

Power function: Likeš (1967), Kabir and Ahsanullah (1974).

Pareto: Malik (1970), Kulldorff and Vännman (1973), Vännman (1976), Arnold (1983).

Raleigh: Dyer and Whisenand (1973a), D'agostino and Lee (1975).

Student's t: Resek (1976).

Symmetric power: Tiao and Lund (1970).

Half logistic: Balakrishnan and Puthenpura (1986), Wong (1988), Balakrishnan and Wong (1990c), Chan (1989).

Log-logistic: Ragab and Green (1984, 1987), Balakrishnan *et al.* (1987).

Double quadratic: Gravel and van Eeden (1981).

Double Weibull: Balakrishnan and Kocherlakota (1985).

Cauchy: Barnett (1966).

Chi (1 d.f.): Govindarajulu and Eisenstat (1965).

Log normal: Gajjar and Khatri (1969), Munro and Wixley (1970), Gibbons and McDonald (1975), Cohen (1976), Nelson and Schmee (1979).

Generalized logistic: Balakrishnan and Leung (1988b).

Gamma: Mehrotra and Nanda (1974), Wilk *et al.* (1962, 1963, 1966).

Extreme value and Weibull: Lieblein (1954), Lieblein and Zelen (1956), Gumbel (1958), Winer (1963), White (1964), Harter and Moore (1965a, b), Downton (1966b), Bain and Antle (1967), Mann (1967a, b, 1968, 1969a, b, 1970, 1971), Govindarajulu and Joshi (1968), Danziger (1970), D'agostino (1971), Bain (1972), Billman *et al.* (1972), Mann and Fertig (1973, 1975a, b), Engelhardt and Bain (1973, 1974), Engelhardt (1975), Murthy and Swartz (1975), Balakrishnan *et al.* (1990b).

Chapter 5 Maximum Likelihood Estimation

5.1. Preliminary Remarks

Readers might wish to study Chapter 8 along with or before reading this chapter in order to compare modified estimators that make use of order statistics with the maximum likelihood estimators (MLEs) presented here. In most instances, the MLEs are optimal estimators, but in some situations they are rendered invalid because of regularity problems. In general, the modified estimators are free from regularity problems, and in most applications, they are at least nearly optimal.

The likelihood function of a random sample that consists of observations $\{x_i\}$, $i = 1, 2, \ldots, n$, from a distribution with pdf $f(x; \theta_1, \theta_2)$, where θ_1 and θ_2 are parameters, may be expressed as

$$L(x_1, x_2, \ldots, x_n; \theta_1, \theta_2) = \prod_{i=1}^{n} f(x_i; \theta_1, \theta_2). \tag{5.1.1}$$

The loglikelihood function becomes

$$\ln L(x_1, x_2, \ldots, x_n; \theta_1, \theta_2) = \sum_{i=1}^{n} \ln f(x_i; \theta_1, \theta_2). \tag{5.1.2}$$

In the interest of keeping notations simple, only two parameters are considered in (5.1.1) and (5.1.2). However, more than two parameters are included in many distributions.

For distributions that are unimodal (i.e., bell-shaped as opposed to reverse J-shaped), maximum likelihood estimating equations are obtained by equating to zero the partial derivatives of $\ln L$ with respect to the parameters. Parameter estimates are then obtained as the simultaneous solution of these equations. Asymptotic variances and covariances are obtained by inverting the Fisher information matrix in which elements are negative of expected values of second partial derivatives of $\ln L$ with respect to the parameters.

The MLEs are well behaved and easy to calculate from sample data from the normal distributions and from various skewed distributions with discernible modes. Unfortunately, regularity problems are encountered in the reverse J-shaped distributions. Therefore, in those cases, the estimators based on order statistics, which are presented in Chapter 8, become more appealing.

5.2. The Weibull Distribution

The Weibull is a skewed distribution that is often employed as a model for distributions of life spans and reaction times. The pdf of the three-parameter Weibull is

$$\begin{aligned} f(x; \gamma, \delta, \beta) &= (\delta/\beta^{\delta})(x-\gamma)^{\delta-1} \exp -[(x-\gamma)/\beta]^{\delta}, \qquad \gamma < x < \infty, \\ &= 0, \qquad \text{elsewhere}, \end{aligned} \tag{5.2.1}$$

where γ is the threshold parameter, δ is the shape parameter, and β is the scale parameter. The cdf is

$$F(x; \gamma, \delta, \beta) = 1 - \exp\{-[(x-\gamma)/\beta]^{\delta}\}. \tag{5.2.2}$$

The Weibull distribution is bell-shaped for $\delta > 1$, with a discernible mode given by

$$M_0(X) = \gamma + \beta[(\delta-1)/\delta]^{1/\delta}. \tag{5.2.3}$$

It is reverse J-shaped when $\delta \leq 1$, and for the special case $\delta = 1$, it becomes the exponential distribution.

The likelihood function of a random sample $\{x_i\}$, $i = 1, 2, \ldots, n$ from this distribution is

$$L(x_1, x_2, \ldots, x_n; \gamma, \delta, \beta) = (\delta/\beta^{\delta})^n \prod_{i=1}^{n} (x_i-\gamma)^{\delta-1}\{\exp -[(x_i-\gamma)/\beta]^{\delta}\}. \tag{5.2.4}$$

In the bell-shaped case, with $\delta > 1$, the estimating equations follow as

$$\frac{\partial \ln L}{\partial \gamma} = \frac{\delta}{\theta}\sum_1^n (x_i - \gamma)^{\delta-1} - (\delta - 1)\sum_1^n (x_i - \gamma)^{-1} = 0,$$

$$\frac{\partial \ln L}{\partial \delta} = \frac{n}{\delta} + \sum_1^n \ln(x_i - \gamma) - \frac{1}{\theta}\sum_1^n (x_i - \gamma)^{\delta} \ln(x_i - \gamma) = 0, \tag{5.2.5}$$

$$\frac{\partial \ln L}{\partial \theta} = -\frac{n}{\theta} + \frac{1}{\theta^2}\sum_1^n (x_i - \gamma)^{\delta} = 0,$$

where, in order to simplify the derivatives, the scale parameter β has been replaced by

$$\theta = \beta^{\delta}. \tag{5.2.6}$$

The three equations of (5.2.5) do not yield explicit solutions for the estimates. However, as shown by Cohen (1965), θ can be eliminated from the last two equations to give

$$\left[\frac{\sum_1^n (x_i - \hat{\gamma})^{\hat{\delta}} \ln(x_i - \hat{\gamma})}{\sum_1^n (x_i - \hat{\gamma})^{\hat{\delta}}} - \frac{1}{\hat{\delta}}\right] - \frac{1}{n}\sum_1^n \ln(x_i - \hat{\gamma}) = 0. \tag{5.2.7}$$

From the last equation of (5.2.5), we have

$$\hat{\theta} = \frac{1}{n}\sum_{i=1}^n (x_i - \hat{\gamma})^{\hat{\delta}}. \tag{5.2.8}$$

When γ is known, it is quite easy to solve (5.2.7) iteratively for $\hat{\delta}$. When γ is unknown, and must therefore be estimated from the sample data, we select a first approximation $\gamma_1 < x_1$ and employ a trial-and-error procedure to calculate the required estimates. Any one of various standard iterative procedures could be employed, but the trial-and-error technique is simple in concept and easy to apply in practice. With γ_1 fixed, (5.2.7) is solved for δ_1, and θ_1 follows from (5.2.8). We then calculate $(\partial \ln L/\partial \gamma)_1$ by substituting γ_1, δ_1, and θ_1 into the first equation of (5.2.5). If $(\partial \ln L/\partial \gamma)_1 = 0$, then $\hat{\gamma} = \gamma_1$, $\hat{\delta} = \delta_1$, $\hat{\theta} = \theta_1$, and our task is completed. Otherwise, we repeat the cycle of computations with a new approximation γ_2 and continue until we find a pair of values (γ_i, γ_j) such that $|\gamma_i - \gamma_j|$ is sufficiently small and such that $(\partial \ln L/\partial \gamma)_i \gtreqless 0 \gtreqless (\partial \ln L/\partial \gamma)_j$. We then interpolate linearly for the required estimates. When desired, we calculate $\hat{\beta} = \hat{\theta}^{1/\hat{\delta}}$. A more complete account of the Weibull distribution and of maximum likelihood estimators for its parameters is given by Cohen and Whitten (1988).

A. Errors of Estimates

The asymptotic variance-covariance matrix for the MLE is obtained by inverting the Fisher information matrix in which elements are negatives of expected values of the second partial derivatives of the loglikelihood function. When this is accomplished, it follows that

$$\begin{aligned} &\mathrm{V}(\hat{\gamma}) = \frac{\beta^2}{n}\phi_{11}, && \mathrm{V}(\hat{\delta}) = \frac{\delta^2}{n}\phi_{22}, \\ &\mathrm{V}(\hat{\beta}) = \frac{\beta^2}{n}\phi_{33}, && \mathrm{Cov}(\hat{\gamma}, \hat{\delta}) = \frac{\beta}{n}\phi_{12}, \\ &\mathrm{Cov}(\hat{\gamma}, \hat{\beta}) = \frac{\beta^2}{n}\phi_{13}, && \mathrm{Cov}(\hat{\delta}, \hat{\beta}) = \frac{\beta}{n}\phi_{23}, \end{aligned} \tag{5.2.9}$$

where

$$\begin{aligned} &\phi_{11} = \frac{\psi'(1)}{\delta^2 M}, && \phi_{22} = \frac{C - \Gamma^2(2 - 1/\delta)}{M}, \\ &\phi_{33} = \frac{KC - J^2}{\delta^2 M}, && \phi_{12} = \frac{J + \psi(2)\Gamma(2 - 1/\delta)}{M}, \\ &\phi_{13} = -\frac{J\psi(2) + K\Gamma(2 - 1/\delta)}{\delta^2 M}, && \phi_{23} = \frac{C\psi(2) - J\Gamma(2 - 1/\delta)}{M}, \end{aligned} \tag{5.2.10}$$

where

$$M = KC - 2J\psi(2)\Gamma\left(2 - \frac{1}{\delta}\right) - C\psi^2(2) - K\Gamma^2\left(2 - \frac{1}{\delta}\right) - J^2, \tag{5.2.11}$$

and where

$$\begin{aligned} &A = 1 + \psi\left(2 - \frac{1}{\delta}\right), \\ &C = \left[\Gamma\left(1 - \frac{2}{\delta}\right) + \delta\Gamma\left(2 - \frac{2}{\delta}\right)\right]\frac{(\delta - 1)}{\delta^2}, \\ &J = \Gamma\left(1 - \frac{1}{\delta}\right) - A\Gamma\left(2 - \frac{1}{\delta}\right), \\ &K = \psi'(1) + \psi^2(2). \end{aligned} \tag{5.2.12}$$

TABLE 5.2.1
Variance-Covariance Factors for Maximum Likelihood Estimates of Weibull Parameters[a]

α_3	δ	ϕ_{11}	ϕ_{22}	ϕ_{33}	ϕ_{12}	ϕ_{13}	ϕ_{23}
0.05	3.40325	1.81053	3.11756	2.11502	-7.25440	1.91206	3.64075
0.06	3.36564	1.73452	3.02575	2.03759	-6.89240	1.83425	3.36154
0.07	3.32873	1.66133	2.93695	1.96301	-6.54777	1.75926	3.10390
0.08	3.29249	1.59085	2.85105	1.89116	-6.21964	1.68696	2.86620
0.09	3.25691	1.52298	2.76792	1.82193	-5.90715	1.61727	2.64695
0.10	3.22197	1.45760	2.68748	1.75523	-5.60953	1.55009	2.44475
0.11	3.18766	1.39464	2.60962	1.69097	-5.32601	1.48531	2.25832
0.12	3.15397	1.33400	2.53424	1.62905	-5.05590	1.42284	2.08648
0.13	3.12087	1.27559	2.46125	1.56939	-4.79851	1.36261	1.92813
0.14	3.08836	1.21932	2.39057	1.51190	-4.55323	1.30452	1.78226
0.15	3.05642	1.16513	2.32212	1.45651	-4.31945	1.24849	1.64794
0.16	3.02503	1.11292	2.25580	1.40313	-4.09661	1.19446	1.52430
0.17	2.99419	1.06263	2.19156	1.35170	-3.88417	1.14235	1.41054
0.18	2.96388	1.01419	2.12930	1.30215	-3.68162	1.09208	1.30591
0.19	2.93409	0.96753	2.06897	1.25440	-3.48850	1.04359	1.20974
0.20	2.90481	0.92259	2.01049	1.20840	-3.30433	0.99681	1.12138
0.21	2.87603	0.87930	1.95381	1.16409	-3.12869	0.95169	1.04024
0.22	2.84773	0.83760	1.89884	1.12140	-2.96118	0.90817	0.96579
0.23	2.81990	0.79744	1.84555	1.08027	-2.80141	0.86618	0.89752
0.24	2.79254	0.75876	1.79387	1.04066	-2.64900	0.82567	0.83495
0.25	2.76563	0.72150	1.74374	1.00251	-2.50361	0.78660	0.77766
0.26	2.73917	0.68563	1.69511	0.96578	-2.36491	0.74890	0.72525
0.27	2.71314	0.65109	1.64794	0.93040	-2.23260	0.71254	0.67734
0.28	2.68753	0.61783	1.60216	0.89635	-2.10636	0.67746	0.63359
0.29	2.66234	0.58581	1.55774	0.86357	-1.98591	0.64363	0.59367
0.30	2.63756	0.55498	1.51463	0.83201	-1.87100	0.61099	0.55730
0.31	2.61317	0.52530	1.47279	0.80165	-1.76136	0.57950	0.52420
0.32	2.58918	0.49674	1.43217	0.77244	-1.65675	0.54914	0.49412
0.33	2.56557	0.46926	1.39273	0.74434	-1.55694	0.51985	0.46681
0.34	2.54233	0.44281	1.35444	0.71732	-1.46171	0.49161	0.44206
0.35	2.51946	0.41736	1.31726	0.69133	-1.37085	0.46437	0.41967
0.36	2.49694	0.39289	1.28115	0.66636	-1.28417	0.43811	0.39946
0.37	2.47478	0.36934	1.24608	0.64236	-1.20147	0.41279	0.38123
0.38	2.45296	0.34670	1.21201	0.61930	-1.12258	0.38837	0.36485
0.39	2.43148	0.32494	1.17891	0.59716	-1.04732	0.36483	0.35014
0.40	2.41032	0.30401	1.14675	0.57590	-0.97554	0.34215	0.33699
0.41	2.38950	0.28390	1.11551	0.55550	-0.90707	0.32028	0.32525
0.42	2.36899	0.26458	1.08514	0.53593	-0.84178	0.29921	0.31482
0.43	2.34879	0.24601	1.05564	0.51716	-0.77951	0.27891	0.30557
0.44	2.32889	0.22819	1.02695	0.49917	-0.72014	0.25935	0.29741
0.45	2.30929	0.21107	0.99907	0.48193	-0.66353	0.24050	0.29025
0.50	2.21560	0.13533	0.87087	0.40630	-0.41794	0.15627	0.26648
0.55	2.12856	0.07410	0.75933	0.34639	-0.22546	0.08681	0.25688
0.60	2.04757	0.02514	0.66209	0.30002	-0.07555	0.02991	0.25552
0.63	2.00166	0.00083	0.60978	0.27791	-0.00249	0.00100	0.25695

[a] Valid only if $\delta > 2$.

Reproduced from Cohen and Whitten (1988), Table 3.3, page 48, by courtesy of Marcel Dekker, Inc.

In the notation employed here, $\Gamma(\ )$ is the gamma function, $\psi(\ )$ is the digamma function, $\psi'(\ )$ is the trigamma function, and $\Gamma^2(\ )=[\Gamma(\ )]^2$, $\psi^2(\ )=[\psi(\ )]^2$, etc.

The ϕ_{ij} are thus functions of the shape parameter δ alone, and since α_3 is a function of δ alone, the ϕ_{ij} can be considered as functions of α_3 rather than of δ. In order to facilitate the calculation of estimate variances and covariances, Table 5.2.1, in which entries of the ϕ_{ij} are given as functions of α_3, has been included.

Unfortunately, variances and covariances given here are valid only if $\delta > 2$, and they are strictly applicable only for the MLE. Computational difficulties might be encountered unless δ is greater than approximately 2.2. Although these results do not strictly apply to the estimators of Chapter 8, simulation investigations by Cohen and Whitten (1982b) and by Cohen *et al.* (1984) indicate that the variances and covariances of (5.2.9) also provide close approximations to those of the MME of Chapter 8. *In the two-parameter case with γ known*, asymptotic variances for the MLE $\hat{\delta}$ and $\hat{\beta}$ are valid for $\delta > 1$. These results, as given by Cohen and Whitten (1982b), are

$$
\begin{aligned}
\mathrm{V}(\hat{\delta}) &= 0.607927\left(\frac{\delta^2}{n}\right),\\
\mathrm{Cov}(\hat{\delta}, \hat{\beta}) &= \frac{0.257022\beta}{n},\\
\mathrm{V}(\hat{\beta}) &= 1.108665\left(\frac{\beta^2}{n\delta^2}\right).
\end{aligned}
\tag{5.2.13}
$$

The variances of (5.2.13) also give close approximations to corresponding variances of the MME.

5.3. The Lognormal Distribution

The name of this distribution is derived from the relation that exists between random variables X and $Y=\ln(X-\gamma)$. If Y is distributed normally (μ, σ^2), then X is lognormal (γ, μ, σ^2). When γ is known, it is a simple matter to make the transformation from X to Y. Subsequent analyses, including parameter estimation, can then be made by employing the well-known theory of normal distributions. When the threshold parameter γ is unknown, estimation procedures become more complex. The probability density function (pdf) of the three-parameter lognormal distribution follows from the definition as

$$f(x;\gamma,\mu,\sigma^2)=\frac{1}{\sigma\sqrt{2\pi}(x-\gamma)}$$

$$\times\exp\left\{-\frac{[\ln(x-\gamma)-\mu]^2}{2\sigma^2}\right\},\qquad \gamma<x<\infty,\sigma^2>0,$$

$$=0,\qquad \text{otherwise.} \tag{5.3.1}$$

We seek estimates that will maximize the likelihood function of a random sample consisting of observations $\{x_i\}$, $i=1,\ldots,n$. The likelihood function may be written as

$$L=\left(\frac{1}{\sigma\sqrt{2\pi}}\right)^n\left[\prod_{i=1}^n(x_i-\gamma)^{-1}\right]\exp\left\{-\frac{1}{2\sigma^2}\sum_{i=1}^n[\ln(x_i-\gamma)-\mu]^2\right\}. \tag{5.3.2}$$

It is immediately obvious that $L(\ )$ approaches infinity as $\gamma\to x_1$. It would thus appear that we should take $\hat{\gamma}=x_1$ as our estimate. However, Hill (1963) demonstrated the existence of paths along which the likelihood function of any ordered sample $x_1,\ldots,x_n$ tends to ∞ as (γ,μ,σ^2) approach $(x_1,-\infty,\infty)$. On some occasions, when no misunderstanding is likely to occur, the notation $x_{1:n}$ or $x_{1:N}$ for the first order statistic will be abbreviated to x_1. This global maximum thereby leads to the inadmissible estimates $\hat{\mu}=-\infty$ and $\hat{\sigma}^2=\infty$ regardless of the sample.

As an alternative, Cohen (1951), Cohen and Whitten (1980), and Harter and Moore (1966a) equated partial derivatives of the loglikelihood function to zero and solved the resulting equations to obtain *local maximum likelihood estimators* (*LMLE*), which in most cases would be considered reasonable in comparison with corresponding moment estimators. Harter and Moore, and later Calitz (1973), noted that these LMLE appear to possess most of the desirable properties ordinarily associated with MLE.

On differentiating the logarithm of the loglikelihood function and equating to zero, we obtain the LML estimating equations

$$\begin{aligned}
\frac{\partial\ln L}{\partial\mu}&=\frac{1}{\sigma^2}\sum_1^n[\ln(x_i-\gamma)-\mu]=0,\\
\frac{\partial\ln L}{\partial\sigma}&=-\frac{n}{\sigma}+\frac{1}{\sigma^3}\sum_1^n[\ln(x_i-\gamma)-\mu]^2=0,\\
\frac{\partial\ln L}{\partial\gamma}&=\frac{1}{\sigma^2}\sum_1^n\frac{\ln(x_i-\gamma)-\mu}{x_i-\gamma}+\sum_1^n(x_i-\gamma)^{-1}=0.
\end{aligned} \tag{5.3.3}$$

When σ^2 and μ are eliminated from these equations, as was done by Cohen (1951), the resulting equation in γ becomes

$$\lambda(\hat{\gamma}) = \sum_1^n (x_i - \hat{\gamma})^{-1} \left[\sum_1^n \ln(x_i - \hat{\gamma}) - \sum_1^n \ln^2(x_i - \hat{\gamma}) + \frac{1}{n} \left\{ \sum_1^n \ln(x_i - \hat{\gamma}) \right\}^2 \right]$$

$$- n \sum_1^n \frac{\ln(x_i - \hat{\gamma})}{x_i - \hat{\gamma}} = 0. \tag{5.3.4}$$

Equation (5.3.4) may be solved iteratively for $\hat{\gamma}$. It then follows from the first two equations of (5.3.3) that

$$\begin{aligned} \hat{\mu} &= \frac{1}{n} \sum_1^n \ln(x_i - \hat{\gamma}), \\ \hat{\sigma}^2 &= \frac{1}{n} \sum_1^n \ln^2(x_i - \hat{\gamma}) - \left[\frac{1}{n} \sum_1^n \ln(x_i - \hat{\gamma}) \right]^2. \end{aligned} \tag{5.3.5}$$

In solving (5.3.4) for $\hat{\gamma}$, we accept only admissible roots for which $\gamma < x_1$. Usually, only a single admissible root will be found. In the event that multiple admissible roots occur, we choose as our estimate the root that results in the closest agreement between $\bar{x}$ and $\hat{E}(X)$.

Standard iterative procedures such as the Newton–Raphson method are satisfactory for solving equation (5.3.4), but in many instances it is more convenient to use the trial-and-error technique with linear interpolation. We begin with a first approximation $\gamma_1 < x_1$ and evaluate $\lambda(\gamma_1)$. If this value is zero, then no further calculations are required. Otherwise we continue until we find a pair of values γ_i and γ_j in a sufficiently narrow interval such that $\lambda(\gamma_i) \gtrless 0 \gtrless \lambda(\gamma_j)$ and interpolate for the final estimate, $\hat{\gamma}$.

Wilson and Worcester (1945) and Lambert (1964) attempted to solve the three equations of (5.3.3) simultaneously without simplifying to the form given in (5.3.4), but they encountered convergence problems. Calitz (1973) examined the convergence problem further and concluded that the simplification of Cohen (1951) using equation (5.3.4) led to fewer convergence problems and was therefore to be recommended.

When $\gamma = 0$, the regularity problems described above are not encountered. It is necessary only that we make the transformation $y_i = \ln x_i$ and the usual normal estimators are applicable.

A. Asymptotic Variances and Covariances

In the absence of regularity restrictions, asymptotic variances and covariances of MLE of distribution parameters can be obtained by inverting the Fisher information matrix in which elements are negatives of expected values of second partial derivatives of the likelihood function with respect to the parameters. Unfortunately, the lognormal distribution is subject to regularity problems as previously mentioned, and this raises questions about the validity of asymptotic variances and covariances thus obtained as they might apply to the local MLE. Nevertheless, they have been found to be in reasonably close agreement with estimate variances, obtained in simulation studies by Cohen and Whitten (1980), by Harter and Moore (1966a), and by Cohen *et al.* (1984). They are therefore offered here as possible useful approximations to the applicable variances and covariances. Based on the information matrix for $\hat{\gamma}$, $\hat{\beta}$, $\hat{\sigma}$ given by Cohen (1951) or that for $\hat{\gamma}$, $\hat{\mu}$, $\hat{\sigma}^2$ given by Hill (1963), we have

$$\begin{aligned} V(\hat{\gamma}) &= \frac{\sigma^2}{n}\frac{\beta^2}{\omega}H, & \mathrm{Cov}(\hat{\gamma}, \hat{\beta}) &= \frac{-\sigma^3}{n}\frac{\beta^2}{\sqrt{\omega}}H, \\ V(\hat{\beta}) &= \frac{\sigma^2}{n}\beta^2[1+H], & \mathrm{Cov}(\hat{\gamma}, \hat{\sigma}) &= \frac{-\sigma^3}{n}\frac{\beta^2}{\sqrt{\omega}}H, \\ V(\hat{\sigma}) &= \frac{\sigma^2}{2n}[1+2\sigma^2 H], & \mathrm{Cov}(\hat{\beta}, \hat{\sigma}) &= \frac{-\sigma^3}{n}\beta^2 H, \\ V(\hat{\mu}) &= \frac{\sigma^2}{n}[1+H], & V(\hat{\sigma}^2) &= \frac{2\sigma^4}{n}[1+2\sigma^2 H], \end{aligned} \tag{5.3.6}$$

where

$$H = [\omega(1+\sigma^2) - (1+2\sigma^2)]^{-1}.$$

For large samples, the central limit theorem can, of course, be employed to approximate the variance of the estimate $\hat{m}$ of the distribution mean $[m = E(X)]$ as

$$V(\hat{m}) = \frac{V(X)}{n} \quad \text{where } V(X) = \beta^2\omega(\omega-1) \quad \text{and} \quad \omega = \exp(\sigma^2).$$

The variances and covariances of (5.3.6) can be expressed in a simpler format as

$$\begin{aligned} V(\hat{\gamma}) &= \frac{\sigma^2}{n}\beta^2\phi_{11}, & V(\hat{\sigma}^2) &= \frac{2\sigma^4}{n}\phi_{33}, \\ V(\hat{\beta}) &= \frac{\sigma^2}{n}\beta^2\phi_{22}, & \mathrm{Cov}(\hat{\gamma}, \hat{\beta}) &= \frac{\sigma^2}{n}\beta^2\phi_{12}, \\ V(\hat{\sigma}) &= \frac{\sigma^2}{2n}\phi_{33}, & \mathrm{Cov}(\hat{\gamma}, \hat{\sigma}) &= \frac{\sigma^2}{n}\beta^2\phi_{13}, \\ V(\hat{\mu}) &= \frac{\sigma^2}{n}\phi_{22}, & \mathrm{Cov}(\hat{\beta}, \hat{\sigma}) &= \frac{\sigma^2}{n}\beta^2\phi_{23}, \end{aligned} \tag{5.3.7}$$

TABLE 5.3.1
Variance-Covariance Factors for Maximum Likelihood Estimates of Lognormal Parameters

α_3	ω	ϕ_{11}	ϕ_{22}	ϕ_{33}	ϕ_{12}	ϕ_{23}
0.50	1.02728	885.23263	910.38125	49.95010	-147.19413	-149.18831
0.55	1.03289	607.65422	628.63763	41.61692	-111.08790	-112.89975
0.60	1.03898	431.38171	449.19715	35.27801	-85.98520	-87.64504
0.65	1.04555	315.03444	330.38447	30.34403	-67.98669	-69.51786
0.70	1.05258	235.69195	249.08577	26.42824	-54.74136	-56.16220
0.75	1.06007	180.05004	191.86570	23.26838	-44.77401	-46.09919
0.80	1.06799	140.07179	150.59584	20.68148	-37.12694	-38.36839
0.85	1.07634	110.73199	120.18559	18.53674	-31.15990	-32.32744
0.90	1.08510	88.79275	97.34930	16.73867	-26.43369	-27.53552
0.95	1.09426	72.11193	79.90929	15.21622	-22.64022	-23.68324
1.00	1.10380	59.23866	66.38784	13.91565	-19.55896	-20.54904
1.05	1.11372	49.16952	55.76087	12.79571	-17.02923	-17.97141
1.10	1.12398	41.19733	47.30516	11.82430	-14.93205	-15.83068
1.15	1.13460	34.81515	40.50113	10.97615	-13.17803	-14.03690
1.20	1.14554	29.65392	34.96965	10.23114	-11.69914	-12.52156
1.25	1.15679	25.44114	30.43014	9.57311	-10.44293	-11.23182
1.30	1.16835	21.97300	26.67221	8.98892	-9.36858	-10.12653
1.35	1.18020	19.09523	23.53622	8.46783	-8.44393	-9.17324
1.40	1.19233	16.68980	20.89971	8.00100	-7.64347	-8.34620
1.45	1.20472	14.66545	18.66773	7.58107	-6.94674	-7.62472
1.50	1.21736	12.95095	16.76599	7.20190	-6.33721	-6.99210
1.55	1.23025	11.49026	15.13585	6.85830	-5.80144	-6.43475
1.60	1.24336	10.23888	13.73062	6.54590	-5.32841	-5.94150
1.65	1.25669	9.16122	12.51284	6.26099	-4.90903	-5.50313
1.70	1.27023	8.22860	11.45225	6.00038	-4.53575	-5.11201
1.75	1.28397	7.41778	10.52422	5.76133	-4.20229	-4.76172
1.80	1.29790	6.70977	9.70861	5.54149	-3.90336	-4.44691
1.85	1.31200	6.08899	8.98879	5.33881	-3.63449	-4.16304
1.90	1.32628	5.54257	8.35100	5.15152	-3.39191	-3.90627
1.95	1.34071	5.05983	7.78379	4.97805	-3.17240	-3.67330
2.00	1.35530	4.63185	7.27756	4.81705	-2.97322	-3.46134
2.25	1.43024	3.08881	5.41775	4.16174	-2.20975	-2.64270
2.50	1.50791	2.16869	4.27019	3.68628	-1.70671	-2.09578
2.75	1.58758	1.58683	3.51921	3.32880	-1.35930	-1.71271
3.00	1.66869	1.20085	3.00385	3.05208	-1.11001	-1.43389
3.25	1.75079	0.93447	2.63607	2.83263	-.92535	-1.22440
3.50	1.83355	0.74441	2.36492	2.65498	-.78485	-1.06276
3.75	1.91669	0.60491	2.15943	2.50865	-.67550	-.93519
4.00	2.00000	0.50000	2.00000	2.38629	-.58871	-.83255
4.25	2.08331	0.41942	1.87378	2.28263	-.51863	-.74858
4.50	2.16650	0.35637	1.77207	2.19380	-.46121	-.67886
5.00	2.33211	0.26578	1.61983	2.04972	-.37349	-.57037
6.00	2.65871	0.16333	1.43425	1.84925	-.26335	-.42941
7.00	2.97764	0.11031	1.32847	1.71682	-.19884	-.34311
8.00	3.28840	0.07956	1.26164	1.62290	-.15742	-.28546

$\phi_{13} = -\phi_{12}$.

Reproduced from Cohen and Whitten (1988), Table 4.1, page 65, by courtesy of Marcel Dekker, Inc.

where

$$\phi_{11}=\frac{H}{\omega},\qquad \phi_{22}=1+H,\qquad \phi_{33}=1+2\sigma^2 H,$$
$$\phi_{12}=-\left(\frac{\sigma}{\sqrt{\omega}}\right)H,\qquad \phi_{13}=-\phi_{12},\qquad \phi_{23}=-\sigma H. \tag{5.3.8}$$

The ϕ_{ij} of (5.3.8) are functions of σ alone and thus of α_3. A table of these factors, which will facilitate the calculation of variances and covariances, is included as Table 5.3.1 with α_3 as the argument, where $\alpha_3=(\omega+2)\sqrt{\omega-1}$, and $\omega=\exp(\sigma^2)$.

As a consequence of the regularity problems encountered with the MLE, the modified estimators presented in Chapter 8 become more attractive. Although the variances and covariances given above are not strictly applicable to the MME of Chapter 8, the simulation studies, previously cited, disclose that they provide close approximations to actual observed values of these quantities.

5.4. The Inverse Gaussian Distribution

In the parametrization of Chan *et al.* (1983), the pdf of the inverse Gaussian (IG) distribution is

$$f(x;\,\gamma,\mu,\sigma)=\frac{1}{\sigma\sqrt{2\pi}}\left(\frac{\mu}{x-\gamma}\right)^{3/2}\exp\left\{-\frac{1}{2}\left(\frac{\mu}{x-\gamma}\right)\left[\frac{(x-\gamma)-\mu}{\sigma}\right]^2\right\},$$
$$\gamma<x<\infty,\ \mu>0,\ \sigma>0,$$
$$=0\quad\text{elsewhere,} \tag{5.4.1}$$

where γ is the threshold parameter, μ is a location parameter, σ^2 is the variance, $(\gamma+\mu)$ is the mean, and $\alpha_3=3\sigma/\mu$ is the shape parameter. The likelihood function of a random sample $\{x_i\}$, $i=1,2,\ldots,n$ from this distribution is

$$L(x_1,\ldots,x_n;\,\gamma,\mu,\sigma)=\left(\frac{1}{\sigma\sqrt{2\pi}}\right)^n\left[\prod_{i=1}^{n}\left(\frac{\mu}{x_i-\gamma}\right)^{3/2}\right]$$
$$\times\exp\left[-\frac{\mu}{2\sigma^2}\sum_{i=1}^{n}\frac{(x_i-\gamma-\mu)^2}{x_i-\gamma}\right]. \tag{5.4.2}$$

Maximum likelihood estimating equations follow by taking logarithms of (5.4.2), differentiating with respect to γ, μ, and σ in turn, and equating to zero. We thus obtain

$$\begin{aligned}
\frac{\partial \ln L}{\partial \gamma} &= \frac{3}{2}\sum_1^n (x_i-\gamma)^{-1} + \frac{n\mu}{2\sigma^2} - \frac{\mu^3}{2\sigma^2}\sum_1^n (x_i-\gamma)^{-2} = 0, \\
\frac{\partial \ln L}{\partial \mu} &= \frac{3n}{2\mu} - \frac{1}{2\sigma^2}\sum_1^n \frac{(x_i-\gamma-\mu)^2}{x_i-\gamma} + \frac{n\mu}{2\sigma} - \frac{\mu^2}{\sigma^2}\sum_1^n (x_i-\gamma)^{-1} = 0, \\
\frac{\partial \ln L}{\partial \sigma} &= \frac{-n}{\sigma} + \frac{\mu}{\sigma^3}\sum_1^n \frac{(x_i-\gamma-\mu)^2}{x_i-\gamma} = 0.
\end{aligned} \tag{5.4.3}$$

The three equations of (5.4.3) must now be solved simultaneously for estimates $\hat{\gamma}$, $\hat{\mu}$, and $\hat{\sigma}$. To accomplish this, we first eliminate μ and σ from these three equations to obtain the following equation in γ only:

$$\begin{aligned}
n + \frac{3(\bar{x}-\hat{\gamma})^2}{n}\left[\sum_1^n (x_i-\hat{\gamma})^{-1}\right]^2 \\
- 3(\bar{x}-\hat{\gamma})\sum_1^n (x_i-\hat{\gamma})^{-1} - (\bar{x}-\hat{\gamma})^2\sum_1^n (x_i-\hat{\gamma})^{-2} = 0.
\end{aligned} \tag{5.4.4}$$

On eliminating σ from the last two equations of (5.4.3), we obtain

$$\hat{\mu} = \bar{x} - \hat{\gamma}, \tag{5.4.5}$$

and thus demonstrate that when $\partial \ln L/\partial\mu = 0$ and $\partial \ln L/\partial\sigma = 0$, then $E(X) = \bar{x}$. Accordingly, the MLE of the mean $(\gamma+\mu)$ is unbiased.

From the last equation of (5.4.3), $\hat{\sigma}$ follows as

$$\hat{\sigma} = (\bar{x}-\hat{\gamma})\left[\frac{(\bar{x}-\hat{\gamma})}{n}\sum_1^n (x_i-\hat{\gamma})^{-1} - 1\right]^{1/2}. \tag{5.4.6}$$

Equation (5.4.4), in which $\hat{\gamma}$ is the only unknown and $\bar{x}$ is the sample mean, can be solved by using standard iterative procedures or even by employing simple "trial-and-error" techniques coupled with linear interpolation. With $\hat{\gamma}$ thus determined, $\hat{\mu}$ follows from (5.4.5) and $\hat{\sigma}$ follows from (5.4.6).

An alternate form of (5.4.4) that might be easier to visualize is

$$n - (\bar{x}-\hat{\gamma})^2 S_1 + \frac{3S_2S_3}{n} = 0, \tag{5.4.7}$$

where

$$\begin{aligned}
S_1 &= \sum_1^n (x_i-\gamma)^{-2}, \qquad S_2 = \sum_1^n (x_i-\gamma)^{-1}, \\
S_3 &= \sum_1^n [(x_i-\bar{x})^2/(x_i-\gamma)] = (\bar{x}-\gamma)^2 S_2 - n(\bar{x}-\gamma).
\end{aligned} \tag{5.4.8}$$

Padgett and Wei (1979) derived equivalent maximum likelihood estimators for μ and for $\lambda = \mu^3/\sigma^2$, and proved their existence provided $a_3 > 0$. Their results therefore guarantee that a solution exists for equation (5.4.4) and further guarantee the existence of $\hat{\mu}$ and $\hat{\sigma}$ as given by (5.4.5) and (5.4.6) provided that $a_3 > 0$. It is of course possible to obtain samples in which $a_3 < 0$ even though $\alpha_3 > 0$ when α_3 and/or n are small. The primary usefulness of the IG distribution, however, is in situations where α_3 is large. Consequently, existence or nonexistence problems are unlikely to prove troublesome in most practical applications. The occurrence of a sample with negative skewness might suggest that the IG distribution is an inappropriate model.

A. *Asymptotic Variances and Covariances*

By inverting the Fisher information matrix, we obtain the following estimate variances and covariances:

$$\begin{aligned} &V(\hat{\gamma}) = \frac{\sigma^2}{n}\phi_{11}, \quad V(\hat{\mu}) = \frac{\sigma^2}{n}\phi_{22}, \quad V(\hat{\sigma}) = \frac{\sigma^2}{n}\phi_{33}, \\ &\mathrm{Cov}(\hat{\gamma}, \hat{\mu}) = \frac{\sigma^2}{n}\phi_{12}, \quad \mathrm{Cov}(\hat{\gamma}, \hat{\sigma}) = \frac{\sigma^2}{n}\phi_{13}, \quad \mathrm{Cov}(\hat{\mu}, \hat{\sigma}) = \frac{\sigma^2}{n}\phi_{23}, \end{aligned} \tag{5.4.9}$$

where

$$\begin{aligned} &\phi_{11} = \frac{2}{D}, \quad \phi_{22} = \phi_{11} + 1, \quad \phi_{33} = \frac{(BC - E^2)}{D}, \\ &\phi_{12} = -\phi_{11}, \quad \phi_{13} = \frac{\alpha_3^3}{9D}, \quad \phi_{23} = -\alpha_3 \frac{(C - AE)}{D}, \end{aligned} \tag{5.4.10}$$

and where

$$\begin{aligned} &A = \frac{\alpha_3^2}{9} + 1, \quad B = \frac{\alpha_3^2}{2} + 1, \quad C = \frac{7}{54}\alpha_3^4 A + B, \\ &E = \frac{\alpha_3^2}{2} A + 1, \quad D = 2(C - 1) - \alpha_3^2 A^2. \end{aligned} \tag{5.4.11}$$

These results were originally given by Cohen and Whitten (1985). Additional details concerning their derivation can be found in Cohen and Whitten (1988). Note that the ϕ_{ij} are functions of α_3 alone. In order to facilitate calculation of estimate variances and covariances, these functions are entered in Table 5.4.1 for selected values of α_3. It must be remembered

TABLE 5.4.1
Variance-Covariance Factors for Maximum Likelihood Estimates of Inverse Gaussian Parameters

α_3	ϕ_{11}	ϕ_{33}	ϕ_{13}	ϕ_{23}
0.50	777.60000	0.60000	5.40000	5.15000
0.55	520.18733	0.62007	4.80812	4.53312
0.60	359.19540	0.64172	4.31034	4.01034
0.65	254.68604	0.66491	3.88573	3.56073
0.70	184.68582	0.68956	3.51929	3.16929
0.75	136.53333	0.71563	3.20000	2.82500
0.80	102.64044	0.74304	2.91955	2.51955
0.85	78.30306	0.77177	2.67155	2.24655
0.90	60.51803	0.80176	2.45098	2.00098
0.95	47.31804	0.83298	2.25385	1.77885
1.00	37.38462	0.86538	2.07692	1.57692
1.05	29.81606	0.89895	1.91755	1.39255
1.10	23.98443	0.93364	1.77352	1.22352
1.15	19.44521	0.96945	1.64299	1.06799
1.20	15.87907	1.00634	1.52439	0.92439
1.25	13.05348	1.04431	1.41639	0.79139
1.30	10.79709	1.08335	1.31784	0.66784
1.35	8.98215	1.12344	1.22775	0.55275
1.40	7.51246	1.16458	1.14523	0.44523
1.45	6.31489	1.20677	1.06954	0.34454
1.50	5.33333	1.25000	1.00000	0.25000
1.55	4.52442	1.29427	0.93602	0.16102
1.60	3.85435	1.33958	0.87708	0.07708
1.65	3.29660	1.38594	0.82271	-0.00229
1.70	2.83020	1.43335	0.77249	-0.07751
1.75	2.43851	1.48180	0.72605	-0.14895
1.80	2.10821	1.53131	0.68306	-0.21694
1.85	1.82858	1.58188	0.64322	-0.28178
1.90	1.59098	1.63352	0.60625	-0.34375
1.95	1.38836	1.68622	0.57192	-0.40308
2.00	1.21500	1.74000	0.54000	-0.46000
2.25	0.64831	2.02524	0.41026	-0.71474
2.50	0.36593	2.33824	0.31765	-0.93235
2.75	0.21650	2.67964	0.25014	-1.12486
3.00	0.13333	3.05000	0.20000	-1.30000
3.25	0.08500	3.44977	0.16210	-1.46290
3.50	0.05584	3.87931	0.13300	-1.61700
3.75	0.03766	4.33890	0.11034	-1.76466
4.00	0.02601	4.82877	0.09247	-1.90753
4.25	0.01833	5.34909	0.07819	-2.04681
4.50	0.01317	5.90000	0.06667	-2.18333
5.00	0.00713	7.09404	0.04954	-2.45046
6.00	0.00245	9.85294	0.02941	-2.97059
7.00	0.00099	13.10854	0.01882	-3.48118
8.00	0.00045	16.86226	0.01274	-3.98726

$\phi_{22} = \phi_{11} + 1; \quad \phi_{12} = -\phi_{11}.$

Reproduced from Cohen and Whitten (1988), Table 5.1, page 82, by courtesy of Marcel Dekker, Inc.

that they are strictly applicable only for the MLE. However, simulation results presented by Chan *et al.* (1984) and by Cohen and Whitten (1985) indicate that they closely approximate corresponding variances and covariances of the MME, which are presented in Chapter 8.

In the special case of the two-parameter IG distribution with $\gamma = 0$, the variances and covariance of $\hat{\mu}$ and $\hat{\sigma}$ become

$$V(\hat{\mu}) = \frac{\sigma^2}{n}, \qquad V(\hat{\sigma}) = \frac{\sigma^2}{2n}\left(\frac{\alpha_3^2}{2} + 1\right), \qquad \mathrm{Cov}(\hat{\mu}, \hat{\sigma}) = \frac{\sigma^2}{2n}\alpha_3. \tag{5.4.12}$$

When $\alpha_3 = 0$, the well-known normal distribution estimate variances and covariance follow from (5.4.12) as

$$V(\hat{\mu}) = \frac{\sigma^2}{n}, \qquad V(\hat{\sigma}) = \frac{\sigma^2}{2n}, \qquad \mathrm{Cov}(\hat{\mu}, \hat{\sigma}) = 0. \tag{5.4.13}$$

5.5. The Gamma Distribution

The pdf of the three-parameter gamma distribution is

$$\begin{aligned} f(x; \gamma, \rho, \beta) &= 1/(\beta^{\rho}\Gamma(\rho))(x-\gamma)^{\rho-1} \\ &\quad \times \exp[-(x-\gamma)/\beta], \qquad \gamma < x < \infty, \rho > 0, \beta > 0, \\ &= 0, \qquad \text{elsewhere}, \end{aligned} \tag{5.5.1}$$

where γ is the threshold parameter, ρ is the shape parameter, and β is a scale parameter. As an alternate shape parameter, it is often preferable to employ $\alpha_3 = 2/\sqrt{\rho}$. The expected value (mean) and variance are

$$E(X) = \gamma + \rho\beta \quad \text{and} \quad V(X) = \rho\beta^2. \tag{5.5.2}$$

In standard units, where $Z = [X - E(X)]/\sqrt{V(X)}$, the pdf becomes

$$\begin{aligned} f(z; 0, 1, \alpha_3) &= \frac{1}{\Gamma(4/\alpha_3^2)}\left(\frac{2}{\alpha_3}\right)^{4/\alpha_3^2}\left(z + \frac{2}{\alpha_3}\right)^{(4/\alpha_3^2)-1} \\ &\quad \times \exp\left[-\frac{2}{\alpha_3}\left(z + \frac{2}{\alpha_3}\right)\right], \qquad (-2/\alpha_3) < z < \infty, \\ &= 0, \qquad \text{elsewhere}. \end{aligned} \tag{5.5.3}$$

When $\rho > 1$ and then $\alpha_3 < 2$, the gamma distribution is bell-shaped with a discernible mode at $x = \gamma + \beta(\rho - 1)$. Otherwise it is reverse J-shaped. When $\rho = 1$, it becomes the exponential distribution. Additional details concerning characteristics of the gamma distribution are given in Chapter 8.

The likelihood function of a random sample consisting of observations $\{x_i\}$, $i = 1, 2, \ldots, n$ from the distribution is

$$L(x_1, x_2, \ldots, x_n; \gamma, \rho, \beta) = 1/(\beta^\rho \Gamma(\rho))^n \left[\prod_{i=1}^{n} (x_i - \gamma)^{\rho-1} \right] \times \exp\left\{ -\sum_{i=1}^{n} \left(\frac{x_i - \gamma}{\beta} \right) \right\}. \tag{5.5.4}$$

Maximum likelihood estimation breaks down when $\rho \leq 1$, but for the bell-shaped case ($\rho > 1$), estimating equations may be obtained in the usual way by taking logarithms of (5.5.4) and equating partial derivatives of $\ln L$ to zero. Thus we obtain

$$\begin{aligned}
\frac{\partial \ln L}{\partial \gamma} &= \frac{n}{\beta} - (\rho - 1) \sum_{1}^{n} (x_i - \gamma)^{-1} = 0, \\
\frac{\partial \ln L}{\partial \rho} &= -n\psi(\rho) - n \ln \beta + \sum_{1}^{n} \ln(x_i - \gamma) = 0, \\
\frac{\partial \ln L}{\partial \beta} &= \frac{-n\rho}{\beta} + \frac{1}{\beta^2} \sum_{1}^{n} (x_i - \gamma) = 0,
\end{aligned} \tag{5.5.5}$$

where $\psi(\rho)$ is the digamma function, $\psi(\rho) = \partial \ln \Gamma(\rho)/\partial \rho$. Maximum likelihood estimates, $\hat{\gamma}$, $\hat{\rho}$, and $\hat{\beta}$ are obtained as the simultaneous solution of the three equations of (5.5.5). Although the computation is more or less straightforward, as pointed out by Johnson and Kotz (1970), difficulties might be encountered when ρ is near to 1 even though it actually exceeds 1. As a result they (Johnson and Kotz) recommend that the MLE be employed only if $\rho > 2.5$ ($\alpha_3 < 1.265$). Note that the third equation of (5.5.5) reduces to $\gamma + \rho\beta = \bar{x}$.

A degree of simplification can be achieved by eliminating β from the first and third equations of (5.5.5) to obtain

$$\begin{aligned}
\hat{\rho} &= \left[1 - n \Big/ \left\{ (\bar{x} - \hat{\gamma}) \sum_{1}^{n} (x_i - \hat{\gamma})^{-1} \right\} \right]^{-1}, \\
\hat{\beta} &= (\bar{x} - \hat{\gamma})/\hat{\rho}.
\end{aligned} \tag{5.5.6}$$

Any one of various iterative procedures might be employed to solve the three equations of (5.5.5) simultaneously for estimates $\hat{\gamma}$, $\hat{\rho}$, and $\hat{\beta}$. However, the "trial-and-error" technique that has been used in similar situations might be preferred. Accordingly, we select a first approximation $\gamma_1 < x_1$ and use the first equation of (5.5.6) to calculate a corresponding approximation ρ_1. Approximations γ_1 and ρ_1 are then substituted into the second equation of

(5.5.6) to calculate a corresponding approximation β_1. Approximations γ_1, ρ_1, and β_1 are substituted into the second equation of (5.5.5). If this result is zero, no further calculations are needed. Otherwise, we select a second approximation γ_2 and repeat the cycle of computations until we find two values γ_i and γ_j in a sufficiently narrow interval and such that $[\partial \ln L/\partial\rho]_i \gtrless 0 \gtrless [\partial \ln L/\partial\rho]_j$. Final estimates follow by linear interpolation. As a first approximation γ_1, we might choose the modified moment estimate that is included in Chapter 8.

A. *Two-Parameter Special Case*

With $\gamma = 0$, the second and third equations of (5.5.5) become

$$\ln\hat{\rho} - \psi(\hat{\rho}) = \ln\bar{x} - \frac{1}{n}\sum_{i=1}^{n}\ln x_i, \tag{5.5.7}$$

$$\hat{\beta} = \bar{x}/\hat{\rho}. \tag{5.5.8}$$

The first of these equations can be solved iteratively for $\hat{\rho}$, following which $\hat{\beta}$ is calculated from the second equation.

In the three-parameter gamma distribution, when $\rho \le 1$ and thus when $\alpha_3 \ge 2$, the likelihood functions become infinite as $\gamma \to x_{1:n}$. Although we might set $\gamma = x_1$ and $E(X) = \bar{x}$, it follows that $\partial \ln \mathrm{L}/\partial\rho \to -\infty$ as $\gamma \to x_1$, and then $\hat{\rho}$ and $\hat{\beta}$ fail to exist. In this case it is therefore necessary to use other estimators such as the MME of Chapter 8.

B. *Asymptotic Variances and Covariances*

Asymptotic variances and covariances can be obtained in the usual manner by inverting the Fisher information matrix. Unfortunately, these results are valid only if $\rho > 2$ and might prove difficult to calculate unless $\rho \ge 2.5$. Following the inversion, the variances and covariances of estimates may be expressed as given by Cohen and Whitten (1988) as

$$\begin{aligned}
\phi_{11} &= \frac{\rho\psi'(\rho)-1}{\rho M}, & \phi_{22} &= \frac{(\rho-1)^2\psi'(\rho)-(\rho-2)}{\rho(\rho-1)^2(\rho-2)M},\\
\phi_{33} &= \frac{2}{\rho(\rho-2)M}, & \phi_{12} &= \frac{1-(\rho-2)\psi'(\rho)}{\rho(\rho-1)M},\\
\phi_{13} &= \frac{-1}{\rho(\rho-1)M}, & \phi_{23} &= \frac{-1}{\rho(\rho-1)(\rho-2)M},
\end{aligned} \tag{5.5.9}$$

TABLE 5.5.1
Variance–Covariance Factors for Maximum Likelihood Estimates of Gamma Distribution Parameters[a]

α_3	ρ	ϕ_{11}	ϕ_{22}	ϕ_{33}	ϕ_{12}	ϕ_{13}	ϕ_{23}
0.32	39.0625	2026.5043	1.4746	8471.1637	53.7036	-4124.3022	-111.2797
0.34	34.6021	1564.8417	1.4714	6579.9761	47.0271	-3192.0778	-97.9103
0.36	30.8642	1223.9253	1.4679	5179.0026	41.4353	-2502.7921	-86.7092
0.38	27.7008	968.1494	1.4643	4124.3405	36.7061	-1984.9378	-77.2324
0.40	25.0000	773.5243	1.4605	3318.9110	32.6715	-1590.3115	-69.1440
0.42	22.6757	623.5494	1.4565	2695.8522	29.2026	-1285.7402	-62.1859
0.44	20.6612	506.6635	1.4523	2208.2496	26.1991	-1047.9671	-56.1577
0.46	18.9036	414.6290	1.4480	1822.6273	23.5819	-860.4125	-50.9012
0.48	17.3611	341.4876	1.4435	1514.7322	21.2883	-711.0755	-46.2906
0.50	16.0000	282.8694	1.4387	1266.7474	19.2675	-591.1488	-42.2249
0.52	14.7929	235.5289	1.4339	1065.4173	17.4785	-494.0867	-38.6219
0.54	13.7174	197.0275	1.4288	900.7639	15.8878	-414.9674	-35.4146
0.56	12.7551	165.5134	1.4236	765.1954	14.4677	-350.0503	-32.5474
0.58	11.8906	139.5666	1.4182	652.8780	13.1951	-296.4646	-29.9744
0.60	11.1111	118.0880	1.4127	559.2873	12.0509	-251.9866	-27.6571
0.62	10.4058	100.2198	1.4069	480.8844	11.0188	-214.8791	-25.5631
0.64	9.7656	85.2873	1.4011	414.8790	10.0851	-183.7744	-23.6651
0.66	9.1827	72.7559	1.3951	359.0545	9.2382	-157.5875	-21.9398
0.68	8.6505	62.1988	1.3889	311.6381	8.4681	-135.4519	-20.3671
0.70	8.1633	53.2735	1.3826	271.2024	7.7662	-116.6711	-18.9301
0.72	7.7160	45.7032	1.3761	236.5910	7.1252	-100.6816	-17.6138
0.74	7.3046	39.2632	1.3695	206.8618	6.5386	-87.0253	-16.4056
0.76	6.9252	33.7698	1.3628	181.2431	6.0009	-75.3273	-15.2942
0.78	6.5746	29.0721	1.3560	159.0995	5.5072	-65.2798	-14.2700
0.80	6.2500	25.0460	1.3490	139.9051	5.0532	-56.6283	-13.3243
0.82	5.9488	21.5883	1.3419	123.2229	4.6351	-49.1618	-12.4497
0.84	5.6689	18.6135	1.3347	108.6881	4.2497	-42.7046	-11.6395
0.86	5.4083	16.0499	1.3274	95.9948	3.8939	-37.1095	-10.8879
0.88	5.1653	13.8376	1.3200	84.8856	3.5653	-32.2532	-10.1896
0.90	4.9383	11.9260	1.3125	75.1431	3.2614	-28.0315	-9.5401
0.95	4.4321	8.1991	1.2933	55.6236	2.5968	-19.7084	-8.1034
1.00	4.0000	5.5950	1.2738	41.3553	2.0475	-13.7851	-6.8925
1.05	3.6281	3.7725	1.2539	30.8345	1.5927	-9.5510	-5.8663
1.10	3.3058	2.4987	1.2338	23.0221	1.2161	-6.5188	-4.9922
1.15	3.0246	1.6128	1.2137	17.1884	0.9047	-4.3493	-4.2449
1.20	2.7778	1.0021	1.1936	12.8143	0.6484	-2.8031	-3.6040
1.25	2.5600	0.5869	1.1738	9.5254	0.4386	-1.7097	-3.0530
1.30	2.3669	0.3104	1.1544	7.0487	0.2685	-.9459	-2.5784
1.35	2.1948	0.1318	1.1356	5.1835	0.1325	-.4225	-2.1692
1.40	2.0408	0.0219	1.1175	3.7806	0.0256	-.0741	-1.8162

[a] Valid only if $\rho > 2$.

Reproduced from Cohen and Whitten (1988), Table 6.2, page 100, by courtesy of Marcel Dekker, Inc.

where

$$M = \frac{2(\rho-1)^2\psi'(\rho)-(2\rho-3)}{(\rho-1)^2(\rho-2)}. \tag{5.5.10}$$

The foregoing results are in agreement with corresponding variances and covariances given by Johnson and Kotz (1970). They are valid only when $\rho > 2$, but as already noted, the MLEs are of doubtful utility unless $\rho > 2.5$. Consequently, when $\rho \le 2.5$ it becomes important to consider other estimators such as the MME. When they are applicable, the variances of (5.5.8) enable us to calculate approximate $100(1-\alpha)\%$ confidence intervals on the parameters as: estimate $\pm z_{\alpha/2}\sqrt{\mathrm{V(Est)}}$, where $z_{\alpha/2}$ is the upper $\alpha/2$ percentage point of a standard normal variate.

Note that the ϕ_{ij} are functions of ρ alone. In order to facilitate calculation of the variances and covariances, we give an abridged table (Table 5.5.1) of the ϕ_{ij} as functions of α_3 where $\alpha_3 = 2/\sqrt{\rho}$. We also note that for a two-parameter gamma distribution with $\gamma = 0$ known, that

$$\mathrm{V}(\hat{\beta}) = \frac{\beta^2}{n}\left(\frac{\psi'(\rho)}{\rho\psi'(\rho)-1}\right), \qquad \mathrm{V}(\hat{\rho}) = \frac{1}{n}\left(\frac{\rho}{\rho\psi'(\rho)-1}\right),$$
$$\mathrm{Cov}(\hat{\beta}, \hat{\rho}) = \frac{\sigma^2}{n\beta\rho}\left(\frac{-1}{\rho\psi'(\rho)-1}\right) = \frac{-\beta}{n}\left(\frac{1}{\rho\psi'(\rho)-1}\right). \tag{5.5.11}$$

In the case of the exponential distribution with $\rho = 1$ and $\gamma = 0$, then

$$\mathrm{V}(\hat{\beta}) = \beta^2/n. \tag{5.5.12}$$

5.6. The Rayleigh Distribution

The Rayleigh distribution is a special case of the Weibull distribution in which $\delta = 2$ and $\beta^2 = 2\sigma^2$ (cf. Eq. (5.2.1)). This distribution is of particular interest to engineers and physicists as a model for distributions that involve wave propagation, radiation and related phenomena. The pdf of the two-parameter Rayleigh distribution is

$$f(x;\gamma,\sigma) = \frac{(x-\gamma)}{\sigma^2}\exp\left\{-\frac{1}{2\sigma^2}(x-\gamma)^2\right\}, \qquad \gamma < x < \infty,$$
$$= 0, \qquad \text{elsewhere}. \tag{5.6.1}$$

The cdf is

$$F(x;\gamma,\sigma) = 1-\exp\left\{-\frac{1}{2\sigma^2}(x-\gamma)^2\right\}. \tag{5.6.2}$$

The kth moment of the distribution about the origin is

$$\mu'_{k:(x-\gamma)} = (\sigma\sqrt{2})^k \left(\frac{k}{2}\right)\Gamma\left(\frac{k}{2}\right), \tag{5.6.3}$$

and other characteristics of interest are

$$\begin{aligned} \mu = E(X) &= \gamma + 1.253314\sigma, \\ V(X) &= 0.429204\sigma^2, \\ \alpha_3(X) &= 0.631110, \\ \alpha_4(X) &= 3.245089, \\ M_0(X) &= \gamma + \sigma \quad \text{and} \quad ME(X) = \gamma + 1.17741\sigma. \end{aligned} \tag{5.6.4}$$

The distribution of the first order statistic in a random sample of size n from a Rayleigh distribution (cf. Cohen and Whitten (1988)) is another Rayleigh distribution with pdf

$$\begin{aligned} f_{1:n}(x_{1:n}) &= \frac{n}{\sigma^2}(x_{1:n} - \gamma)\exp\left\{-\frac{n}{2\sigma^2}(x_{1:n} - \gamma)^2\right\}, \qquad \gamma < x_{1:n} < \infty, \\ &= 0, \qquad \text{elsewhere.} \end{aligned} \tag{5.6.5}$$

The likelihood function of a random sample of size n from the distribution is

$$L(x_1, x_2, \ldots, x_n; \gamma, \sigma) = \frac{1}{\sigma^{2n}} \prod_{i=1}^{n} (x_i - \gamma)\left\{\exp\left[-\frac{1}{2\sigma^2}\sum_{i=1}^{n}(x_i - \gamma)^2\right]\right\}. \tag{5.6.6}$$

Maximum likelihood estimating equations follow as

$$\begin{aligned} \frac{\partial \ln L}{\partial \sigma} &= -\frac{2n}{\sigma} + \frac{1}{\sigma^3}\sum_{i=1}^{n}(x_i - \gamma)^2 = 0, \\ \frac{\partial \ln L}{\partial \gamma} &= -\sum_{i=1}^{n}\left(\frac{1}{x_i - \gamma}\right) + \frac{1}{\sigma^2}\sum_{i=1}^{n}(x_i - \gamma) = 0. \end{aligned} \tag{5.6.7}$$

We eliminate σ between these two equations and then solve the first equation for σ^2 to obtain

$$\begin{aligned} &\frac{n(\bar{x} - \hat{\gamma})}{\sum_{i=1}^{n}(x_i - \hat{\gamma})^{-1}} - \frac{1}{2n}\sum_{i=1}^{n}(x_i - \hat{\gamma})^2 = 0, \\ &\hat{\sigma}^2 = \frac{1}{2n}\sum_{i=1}^{n}(x_i - \hat{\gamma})^2. \end{aligned} \tag{5.6.8}$$

The first equation of (5.6.8) can be solved iteratively for $\hat{\gamma}$, and $\hat{\sigma}^2$ subsequently follows from the second of these equations.

Moment estimators, which employ the estimating equations $E(X)=\bar{x}$ and $V(X)=s^2$, yield the explicit estimators

$$\sigma^* = s\sqrt{\frac{2}{4-\pi}}, \text{ and } \gamma^* = \bar{x} - \sigma^*\sqrt{\frac{\pi}{2}}. \tag{5.6.9}$$

The estimate γ^* is subject to the restriction that $\gamma^* < x_1$.

Modified moment estimators, which employ the estimating equation $E(X_{1:n}) = x_{1:n}$ and $E(X)=\bar{x}$, yield the explicit estimators

$$\begin{aligned} \hat{\sigma} &= \frac{\bar{x} - x_{1:n}}{\sqrt{\pi/2} - \sqrt{\pi/2n}}, \\ \hat{\gamma} &= \bar{x} - \hat{\sigma}\sqrt{\frac{\pi}{2}}. \end{aligned} \tag{5.6.10}$$

In applications where it is considered important to calculate the maximum likelihood estimator, either the ME or the MME will provide good first approximations from which one might begin an iterative solution of the first equation of (5.6.8). In most applications, however, the MME of (5.6.10) will likely be the preferred estimator.

The *one-parameter Rayleigh distribution* with $\gamma = 0$ is, of course, much easier to deal with. In this case, the following estimators of σ are applicable:

$$\text{MLE:} \quad \hat{\sigma} = \sqrt{\frac{1}{2n}\sum_{i=1}^{n} x_i^2}; \tag{5.6.11}$$

$$\text{ME:} \quad \sigma^* = \bar{x}\sqrt{2/\pi}; \tag{5.6.12}$$

$$\text{First order statistic estimator:} \quad \tilde{\sigma} = x_{1:n}\sqrt{2n/\pi}. \tag{5.6.13}$$

A. *Truncation in the One-Parameter Rayleigh Distribution*

We consider singly right truncation at $X = T$ in the one-parameter Rayleigh distribution. The sample consists of n random observations, each of which lies in the interval $0 \le x \le T$. Cohen and Whitten (1988) have shown that when σ is replaced by z_0 as the unknown parameter, where

$$z_0 = T/\sigma, \tag{5.6.14}$$

the MLE of z_0 becomes

$$\left.\begin{aligned} J_2(\hat{z}_0) &= \frac{1}{nT^2}\sum_{i=1}^{n} x_i^2, \\ \text{where} \qquad J_2(z_0) &= \frac{2}{z_0^2} - \frac{\phi(z_0)}{\phi(0)-\phi(z_0)} \end{aligned}\right\} \tag{5.6.15}$$

and where $\phi(\cdot)$ is the pdf of the standard normal distribution (0, 1). To facilitate the calculation of $\hat{z}_0$ and thus of $\hat{\sigma}$ in practical applications, Cohen and Whitten (1988) provided tables and a graph of the function $J_2(z)$. A graph of $J_2(z)$ is included here as Fig. 5.6.1. With $J_2(\hat{z}_0)$ calculated from the sample data (cf. Eq. (5.6.15)), we read $\hat{z}_0$ from Fig. 5.6.1, and $\hat{\sigma}$ follows from (5.6.14) as $\hat{\sigma} = T/\hat{z}_0$.

B. Singly Right Censored Samples from the One-Parameter Distribution

We consider censoring such that N is the total sample size and that $N = n + c$, where n is the number of complete observations $\leq T$, while c is the number of censored observations $> T$. For Type I censoring, T is a fixed constant,

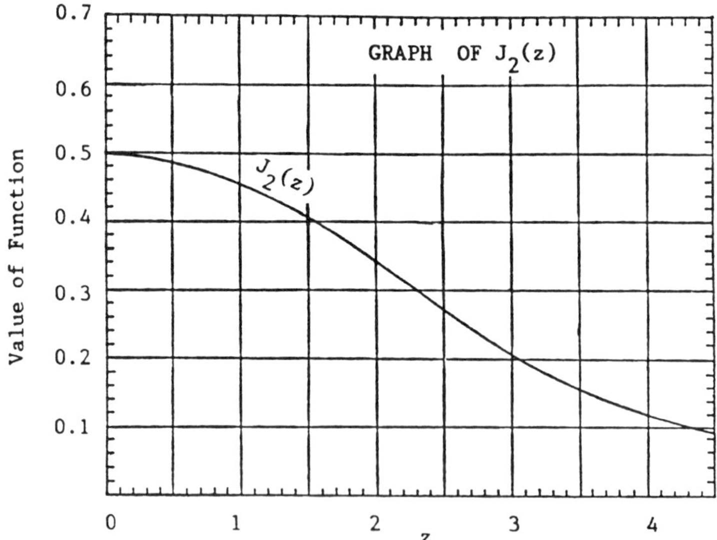

FIGURE 5.6.1. Truncated sample estimating function for Rayleigh distribution. Adapted from Cohen (1965), Figure 1, page 1127, with permission of the American Statistical Association.

whereas in Type II censoring, $T = x_{n:N}$. For samples of this type, Cohen and Whitten (1988) derived the explicit MLE

$$\hat{\sigma} = \sqrt{\frac{1}{2n}\left[\sum_{i=1}^{n} x_i^2 + cT^2\right]}. \tag{5.6.16}$$

C. *Reliability of Estimates*

For complete samples from the Rayleigh distribution, Cohen and Whitten (1988) pointed out that $2n\hat{\sigma}^2/\sigma^2$ has a chi-square distribution with $2n$ degrees of freedom. Thus, confidence intervals on σ can be determined as

$$P\left\{\sqrt{\frac{2n\hat{\sigma}^2}{\chi_2^2}} < \sigma < \sqrt{\frac{2n\hat{\sigma}^2}{\chi_1^2}}\right\} = 1 - \alpha, \tag{5.6.17}$$

where $\hat{\sigma}$ is given by (5.6.8) with $\gamma = 0$, and χ_1^2 and χ_2^2 can be obtained from standard chi-square tables with d.f. $= 2n$ such that $\Pr[\chi^2 > \chi_1^2] = 1 - \alpha/2$, and $\Pr[\chi^2 > \chi_2^2] = \alpha/2$.

Since $2n\hat{\sigma}^2/\sigma^2$ has a chi-square distribution, then $\hat{\sigma}\sqrt{2n}/\sigma$ has a chi distribution, also with d.f. $= 2n$. Kendall (1948, p. 294) gives moments of the chi distribution, and by using his results, we obtain the exact variance of $\hat{\sigma}$ as

$$V(\hat{\sigma}) = \frac{\sigma^2}{4n}\left[4n - \left(\frac{2\Gamma((2n+1)/2)}{\Gamma(n)}\right)^2\right]. \tag{5.6.18}$$

An expansion, also given by Kendall, permits (5.6.18) to be written as

$$V(\hat{\sigma}) = \frac{\sigma^2}{4n}\left[1 - \frac{1}{8n} + \cdots\right]. \tag{5.6.19}$$

For the mean of $\hat{\sigma}$, he obtains

$$E(\hat{\sigma}) = \sigma\left[1 - \frac{1}{8n} + \frac{1}{128n^2} + \cdots\right]. \tag{5.6.20}$$

Although similar exact results are not available for truncated and censored samples, the asymptotic variances in these cases can be calculated from

$$\text{Asy } V(\hat{\sigma}) = -\left[E\left(\frac{\partial^2 \ln L}{\partial \sigma^2}\right)\right]^{-1}. \tag{5.6.21}$$

Following are some specific results obtained from this equation:

For *truncated samples,*

$$\text{Asy } V(\hat{\sigma}) = \frac{\sigma^2}{4n}\left\{1 - \frac{z_0^4}{4}\left[\frac{\phi(0)\phi(z_0)}{[\phi(0) - \phi(z_0)]^2}\right]\right\}^{-1}. \tag{5.6.22}$$

For *Type I censored samples,*

$$\text{Asy V}(\hat{\sigma}) = \frac{\sigma^2}{4n[1-\sqrt{2\pi}\phi(z_0)]}. \tag{5.6.23}$$

For *Type II censored samples,*

$$\text{Asy V}(\hat{\sigma}) = \frac{\sigma^2}{4n}. \tag{5.6.24}$$

Although the Rayleigh distribution was introduced here as a special case of the Weibull distribution, it might also have been introduced as a two-dimensional special case of the more general p-dimensional Rayleigh distribution that was considered by Cohen and Whitten (1988). In this general form, the Rayleigh distribution might be considered as the distribution of the distance X from the origin to a point $(Y_1, Y_2, \ldots, Y_p)$ in a p-dimensional Euclidean space where the components Y_j, $j = 1, 2, \ldots, p$, are independent random variables, each of which is normally distributed $(0, \sigma^2)$. Variances given here for estimates obtained in the two-dimensional Rayleigh distribution can be readily expanded to apply in the general p-dimensional case, in agreement with results given by Cohen and Whitten.

5.7. The Exponential Distribution

The likelihood function of a random sample of size n from the two-parameter exponential distribution is

$$L(x_1, \ldots, x_n; \gamma, \beta) = \left(\frac{1}{\beta}\right) \exp -\sum_{i=1}^{n} [(x_i - \gamma)/\beta], \qquad \gamma \leq x < \infty. \tag{5.7.1}$$

$L(\ )$ is thus an increasing function of γ, subject to the restriction $\gamma \leq x_{1:n}$. Therefore the MLE for γ can only be

$$\hat{\gamma} = x_{1:n}. \tag{5.7.2}$$

With $\hat{\gamma}$ thus determined, the estimating equation for β is obtained in the usual manner as $\partial \ln L/\partial\beta = 0$. Thereby we have

$$\frac{\partial \ln L}{\partial \beta} = -\frac{n}{\beta} + \sum_{i=1}^{n} \left(\frac{x_i - \gamma}{\beta^2}\right) = 0, \tag{5.7.3}$$

and

$$\hat{\beta} = \frac{1}{n} \sum_{i=1}^{n} (x_i - x_{1:n}) = \bar{x} - x_{1:n}. \tag{5.7.4}$$

Estimators (5.7.2) and (5.7.4) are not unbiased. In fact,

$$E(\hat{\gamma}) = \frac{\beta}{n} + \gamma \quad \text{and} \quad E(\hat{\beta}) = \beta\left(\frac{n-1}{n}\right). \tag{5.7.5}$$

It is thus evident that both $\hat{\gamma}$ and $\hat{\beta}$ are severely biased in small samples. However, $\text{Lim}_{n\to\infty} E(\hat{\beta}) = \beta$, and the bias thus becomes negligible as n increases.

A further account of estimation in the exponential distribution is given in Chapter 8. There it is shown that modified moment estimation with estimating equations $E(X) = \bar{x}$ and $E(X_{1:n}) = x_{1:n}$ lead to the minimum variance unbiased estimators; see Chapter 4,

$$\tilde{\gamma} = \frac{nx_{1:n} - \bar{x}}{n-1} \quad \text{and} \quad \tilde{\beta} = \frac{n(\bar{x} - x_{1:n})}{n-1}. \tag{5.7.6}$$

These estimators are not only the MME; they are also BLUE (best linear unbiased estimators) and MVUE (minimum variance unbiased estimators).

In the one-parameter exponential distribution with $\gamma = 0$, the MLE $\hat{\beta}$ becomes

$$\hat{\beta} = \bar{x}. \tag{5.7.7}$$

This is also the moment estimator in this case.

A. *Singly Right Censored Samples*

We consider a sample that is censored on the right at $x = T$, for which n observations $\leq T$ are fully measured, whereas c observations are known only to exceed T. The total sample size is $N = n + c$, and the likelihood function for a sample of this type is

$$L(\ ) = k\left(\frac{1}{\beta}\right)^n \exp\left\{-\frac{1}{\beta}\left[\sum_{i=1}^{n}(x_i - \gamma) + c(T - \gamma)\right]\right\}, \tag{5.7.8}$$

where k is an ordering constant that depends on neither γ nor β. In Type I censoring, T is a fixed constant. In Type II censoring, T is the nth order statistic; i.e., $T = x_{n:N}$.

The same arguments that gave us the ML estimate (5.7.2) for the complete sample enable us to estimate γ for the censored sample as

$$\hat{\gamma} = x_{1:N}. \tag{5.7.9}$$

The estimating equation for β becomes

$$\frac{\partial \ln L}{\partial \beta} = -\frac{n}{\beta} + \frac{1}{\beta^2}\left[\sum_{i=1}^{n}(x_i - \gamma) + c(T - \gamma)\right] = 0. \tag{5.7.10}$$

With $\hat{\gamma}$ given by (5.7.9), it follows from (5.7.10) that

$$\hat{\beta}=\frac{1}{n}[ST-Nx_{1:N}], \tag{5.7.11}$$

where

$$ST=\sum_{i=1}^{n} x_i+cT. \tag{5.7.12}$$

Maximum likelihood estimators (MLE) in this case are thus given by (5.7.9) and (5.7.11). We note that here as for complete samples, $\hat{\gamma}$ is not unbiased. The expected value of $\hat{\gamma}$ is

$$E(\hat{\gamma})=E(X_{1:N})=\gamma+\frac{\beta}{N}. \tag{5.7.13}$$

Thus β/N is the bias, and for small samples, it might be substantial.

B. Modified Maximum Likelihood Estimators

In order to avoid the bias present in the MLE, we turn to modified maximum likelihood estimators (MMLE) with estimating equations: $E(X_{1:n})=x_{1:N}$ and $\partial \ln L/\partial\beta=0$. Thus it follows that

$$\tilde{\gamma}=\frac{nx_{1:N}-ST/N}{n-1},$$
$$\tilde{\beta}=\frac{ST-Nx_{1:N}}{n-1}. \tag{5.7.14}$$

For a complete sample, with $c=0$ and $n=N$, then $ST=n\bar{x}$, and the estimators of (5.7.14) are reduced to those given in (5.7.6).

5.8. Complete, Truncated, and Censored Samples from the Normal Distribution

The pdf of the normal distribution with mean μ and variance σ^2 is

$$f(x;\mu,\sigma^2)=\frac{1}{\sigma\sqrt{2\pi}}\exp-\frac{1}{2\sigma^2}(x-\mu)^2, \qquad -\infty<x<\infty, \sigma>0. \tag{5.8.1}$$

It is easy to show that the MLE for a *complete random sample* of size n from this distribution are

$$\hat{\mu} = \bar{x} = \sum_{i=1}^{n} x_i/n,$$
$$\hat{\sigma} = s = \sqrt{\sum_{i=1}^{n} (x_i - \bar{x})^2/(n-1)}, \tag{5.8.2}$$

and the asymptotic variances and covariance of these estimators are

$$V(\bar{x}) = \sigma^2/n, \qquad V(s) = \sigma^2/2n, \qquad \text{Cov}(\bar{x}, s) = 0. \tag{5.8.3}$$

We further note that the MLE and the moment estimators (ME) in this case are identical.

A. *The Truncated Normal Distribution*

We begin with one-sided (single) truncation on the left as considered by Cohen (1959, 1961), and suppose that a sample consists of n random observations, each of which $\geq T$, where T is a known point of truncation. The pdf of the truncated distribution is

$$f_T(x; \mu, \sigma | T) = \frac{1}{\sigma\sqrt{2\pi}[1-\Phi(\xi)]} \exp\left[-\frac{1}{2}\left(\frac{x-\mu}{\sigma}\right)^2\right], \qquad T \leq x < \infty,$$
$$= 0, \qquad \text{elsewhere}, \tag{5.8.4}$$

where ξ is the standardized point of truncation

$$\xi = (T - \mu)/\sigma, \tag{5.8.5}$$

and $\Phi(\ )$ is the cdf of the standard normal (0, 1). The likelihood function of a random sample of size n from the truncated distribution is

$$L = [1-\Phi(\xi)]^{-n}(2\pi\sigma^2)^{-n/2} \exp -\frac{1}{2}\sum_{i=1}^{n}\left(\frac{x_i-\mu}{\sigma}\right)^2. \tag{5.8.6}$$

Maximum likelihood estimating equations become

$$\frac{\partial \ln L}{\partial \mu} = \frac{-n\phi(\xi)}{\sigma[1-\Phi(\xi)]} + \frac{1}{\sigma^2}\sum_{i=1}^{n}(x_i - \mu) = 0,$$
$$\frac{\partial \ln L}{\partial \sigma} = \frac{-n\xi\phi(\xi)}{\sigma[1-\Phi(\xi)]} - \frac{n}{\sigma} + \frac{1}{\sigma^3}\sum_{i=1}^{n}(x_i - \mu)^2 = 0, \tag{5.8.7}$$

where $\phi(\ )$ is the pdf of the standard normal (0, 1). Let

$$\bar{x}_n = \sum_{i=1}^{n} x_i/n, \qquad s_n^2 = \sum_{i=1}^{n} (x_i - \bar{x}_n)^2/n \quad \text{and} \quad Q(\hat{\xi}) = \frac{\phi(\hat{\xi})}{1 - \Phi(\hat{\xi})}, \tag{5.8.8}$$

and the estimating equations of (5.8.7) become

$$\begin{aligned} \bar{x}_n - \hat{\mu} &= \hat{\sigma} Q(\hat{\xi}), \\ s_n^2 + (\bar{x}_n - \hat{\mu})^2 &= \hat{\sigma}^2[1 + \hat{\xi} Q(\hat{\xi})], \end{aligned} \tag{5.8.9}$$

where $\hat{\xi} = (T - \hat{\mu})/\hat{\sigma}$. We eliminate $(\bar{x}_n - \hat{\mu})$ between these two equations and write $\hat{Q}$ for $Q(\hat{\xi})$ to obtain

$$\hat{\sigma}^2 = s_n^2 + \hat{\sigma}^2 \hat{Q}(\hat{Q} - \hat{\xi}). \tag{5.8.10}$$

However, from (5.8.5) and the first equation of (5.8.9), it follows that

$$\hat{\sigma}\hat{\xi} = T - \hat{\mu} = T - \bar{x}_n + \hat{\sigma}\hat{Q}, \tag{5.8.11}$$

so that

$$\hat{\sigma} = \frac{\bar{x}_n - T}{\hat{Q} - \hat{\xi}}. \tag{5.8.12}$$

We substitute (5.8.12) into (5.8.10) to obtain

$$\hat{\sigma}^2 = s_n^2 + \left(\frac{\hat{Q}}{\hat{Q} - \hat{\xi}}\right)(\bar{x}_n - T)^2. \tag{5.8.13}$$

Let us define

$$\hat{\hat{\theta}} = \hat{\theta}(\hat{\xi}) = \frac{\hat{Q}}{\hat{Q} - \hat{\xi}}, \tag{5.8.14}$$

and from (5.8.13) and (5.8.14), we have

$$\begin{aligned} \hat{\sigma}^2 &= s_n^2 + \hat{\hat{\theta}}(\bar{x}_n - T)^2, \\ \hat{\mu} &= \bar{x}_n - \hat{\hat{\theta}}(\bar{x}_n - T). \end{aligned} \tag{5.8.15}$$

From (5.8.12) and (5.8.13) it follows that

$$\frac{s_n^2}{(\bar{x}_n - T)^2} = \frac{1 - \hat{Q}(\hat{Q} - \hat{\xi})}{(\hat{Q} - \hat{\xi})^2} = \alpha(\hat{\xi}) = \hat{\alpha}. \tag{5.8.16}$$

In order to calculate $\hat{\sigma}^2$ and $\hat{\mu}$ from (5.8.15), we need $\hat{\hat{\theta}}$, which is a function of $\hat{\xi}$ and in turn, a function of $\hat{\alpha}$. A graph and a table of $\hat{\theta}$ as a function of α were provided by Cohen (1959, 1961). These are reproduced here as Fig. 5.8.1 and Table 5.8.1, respectively. With $\hat{\alpha} = s_n^2/(\bar{x}_n - T)^2$ calculated

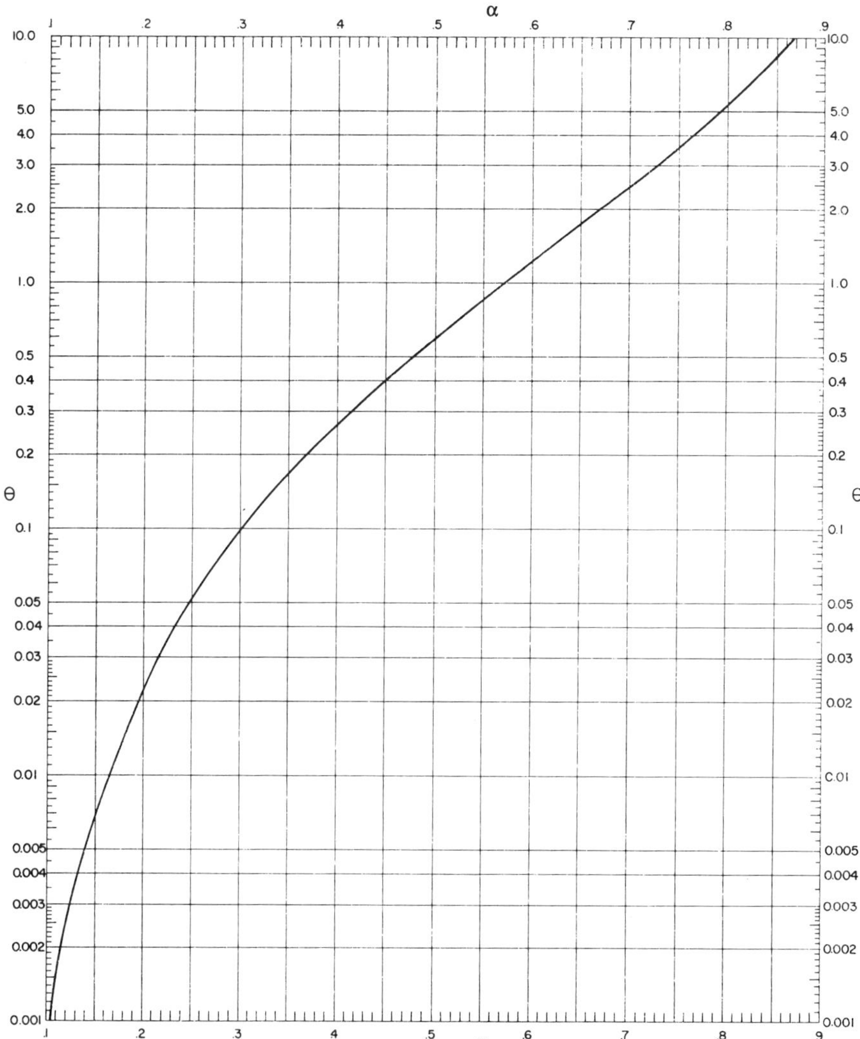

FIGURE 5.8.1. Estimation curve for singly truncated normal samples. Reproduced from Cohen (1959), Figure 1, p. 226, with permission of the Technometrics Management Committee.

TABLE 5.8.1
Auxiliary Estimation Function θ for Singly Truncated Normal Samples

α	.000	.001	.002	.003	.004	.005	.006	.007	.008	.009	α
0.05	.00000	.00000	.00000	.00001	.00001	.00001	.00001	.00001	.00002	.00002	0.05
0.06	.00002	.00003	.00003	.00003	.00004	.00004	.00005	.00006	.00007	.00007	0.06
0.07	.00008	.00009	.00010	.00011	.00013	.00014	.00016	.00017	.00019	.00020	0.07
0.08	.00022	.00024	.00026	.00028	.00031	.00033	.00036	.00039	.00042	.00045	0.08
0.09	.00048	.00051	.00055	.00059	.00062	.00067	.00071	.00075	.00080	.00085	0.09
0.10	.00090	.00095	.00101	.00106	.00112	.00118	.00125	.00131	.00138	.00145	0.10
0.11	.00153	.00160	.00168	.00176	.00184	.00193	.00202	.00211	.00220	.00230	0.11
0.12	.00240	.00250	.00261	.00271	.00283	.00294	.00305	.00317	.00330	.00342	0.12
0.13	.00355	.00369	.00382	.00396	.00410	.00425	.00440	.00455	.00470	.00486	0.13
0.14	.00503	.00519	.00536	.00553	.00571	.00589	.00608	.00626	.00646	.00665	0.14
0.15	.00685	.00705	.00726	.00747	.00769	.00790	.00813	.00835	.00859	.00882	0.15
0.16	.00906	.00930	.00955	.00980	.01006	.01032	.01058	.01085	.01112	.01140	0.16
0.17	.01168	.01197	.01226	.01256	.01286	.01316	.01347	.01378	.01410	.01443	0.17
0.18	.01475	.01509	.01542	.01577	.01611	.01647	.01682	.01718	.01755	.01792	0.18
0.19	.01830	.01868	.01907	.01946	.01986	.02026	.02067	.02108	.02150	.02193	0.19
0.20	.02236	.02279	.02323	.02367	.02413	.02458	.02504	.02551	.02599	.02647	0.20
0.21	.02695	.02744	.02793	.02844	.02894	.02946	.02998	.03050	.03103	.03157	0.21
0.22	.03211	.03266	.03322	.03378	.03435	.03492	.03550	.03609	.03668	.03728	0.22
0.23	.03788	.03849	.03911	.03973	.04036	.04100	.04164	.04230	.04295	.04362	0.23
0.24	.04429	.04496	.04565	.04634	.04703	.04774	.04845	.04917	.04989	.05062	0.24
0.25	.05136	.05211	.05286	.05362	.05439	.05516	.05594	.05673	.05753	.05834	0.25
0.26	.05915	.05997	.06079	.06163	.06247	.06332	.06417	.06504	.06591	.06679	0.26
0.27	.06768	.06857	.06948	.07039	.07131	.07223	.07317	.07411	.07507	.07603	0.27
0.28	.07700	.07797	.07895	.07995	.08095	.08196	.08298	.08400	.08504	.08609	0.28
0.29	.08714	.08820	.08927	.09035	.09143	.09253	.09364	.09475	.09588	.09701	0.29
0.30	.09815	.09930	.10046	.10163	.10281	.10400	.10520	.10640	.10762	.10885	0.30
0.31	.11009	.11133	.11258	.11385	.11512	.11641	.11770	.11901	.12032	.12165	0.31
0.32	.12298	.12433	.12568	.12704	.12842	.12981	.13120	.13261	.13403	.13546	0.32
0.33	.13690	.13834	.13980	.14127	.14276	.14425	.14575	.14727	.14880	.15033	0.33
0.34	.15188	.15344	.15501	.15659	.15819	.15979	.16141	.16304	.16468	.16633	0.34
0.35	.16800	.16967	.17136	.17306	.17477	.17650	.17823	.17998	.18174	.18352	0.35
0.36	.18531	.18710	.18891	.19074	.19257	.19442	.19629	.19816	.20005	.20195	0.36
0.37	.20387	.20580	.20774	.20969	.21166	.21364	.21564	.21765	.21967	.22171	0.37
0.38	.22376	.22582	.22790	.22999	.23210	.23422	.23636	.23851	.24067	.24285	0.38
0.39	.24505	.24725	.24948	.25171	.25397	.25624	.25852	.26082	.26314	.26547	0.39
0.40	.26782	.27018	.27255	.27494	.27735	.27978	.28222	.28468	.28715	.28964	0.40

0.41	.29215	.29467	.29721	.29976	.30233	.30492	.30753	.31015	.31280	.31546	0.41
0.42	.31813	.32082	.32353	.32626	.32901	.33177	.33455	.33736	.34018	.34301	0.42
0.43	.34587	.34874	.35163	.35454	.35747	.36042	.36339	.36638	.36939	.37242	0.43
0.44	.37546	.37853	.38161	.38471	.38784	.39098	.39415	.39734	.40054	.40377	0.44
0.45	.40702	.41029	.41358	.41688	.42022	.42357	.42645	.43034	.43376	.43721	0.45
0.46	.44067	.44415	.44765	.45118	.45473	.45831	.46190	.46553	.46917	.47284	0.46
0.47	.47653	.48024	.48398	.48774	.49152	.49533	.49916	.50302	.50691	.51082	0.47
0.48	.51475	.51870	.52268	.52669	.53072	.53478	.53887	.54298	.54712	.55129	0.48
0.49	.55548	.55969	.56393	.56820	.57250	.57682	.58118	.58556	.58997	.59441	0.49
0.50	.59888	.60336	.60788	.61243	.61701	.62162	.62626	.63093	.63563	.64036	0.50
0.51	.64512	.64990	.65472	.65956	.66444	.66936	.67430	.67928	.68429	.68933	0.51
0.52	.69440	.69950	.70463	.70980	.71500	.72023	.72550	.73081	.73615	.74152	0.52
0.53	.74693	.75237	.75784	.76334	.76889	.77447	.78009	.78575	.79144	.79717	0.53
0.54	.80294	.80873	.81456	.82044	.82635	.83230	.83829	.84433	.85040	.85651	0.54
0.55	.86266	.86884	.87507	.88133	.88764	.89399	.90038	.90682	.91330	.91982	0.55
0.56	.92638	.93298	.93962	.94630	.95303	.95981	.96663	.97350	.98042	.98738	0.56
0.57	.99439	1.0014	1.0085	1.0157	1.0228	1.0301	1.0374	1.0447	1.0521	1.0595	0.57
0.58	1.0670	1.0745	1.0821	1.0897	1.0974	1.1051	1.1129	1.1208	1.1287	1.1366	0.58
0.59	1.1446	1.1526	1.1607	1.1689	1.1771	1.1854	1.1937	1.2021	1.2105	1.2190	0.59
0.60	1.2276	1.2361	1.2448	1.2535	1.2623	1.2711	1.2801	1.2890	1.2981	1.3071	0.60
0.61	1.3163	1.3255	1.3348	1.3441	1.3535	1.3630	1.3725	1.3821	1.3918	1.4015	0.61
0.62	1.4113	1.4212	1.4311	1.4411	1.4512	1.4613	1.4715	1.4818	1.4922	1.5027	0.62
0.63	1.5132	1.5237	1.5344	1.5451	1.5559	1.5668	1.5777	1.5888	1.5999	1.6111	0.63
0.64	1.6224	1.6337	1.6452	1.6567	1.6683	1.6800	1.6917	1.7036	1.7156	1.7276	0.64
0.65	1.7397	1.7519	1.7642	1.7766	1.7890	1.8016	1.8143	1.8270	1.8399	1.8528	0.65
0.66	1.8659	1.8790	1.8922	1.9055	1.9189	1.9324	1.9461	1.9598	1.9736	1.9876	0.66
0.67	2.0016	2.0157	2.0300	2.0443	2.0588	2.0733	2.0880	2.1028	2.1178	2.1328	0.67
0.68	2.1480	2.1632	2.1785	2.1940	2.2096	2.2254	2.2412	2.2572	2.2733	2.2896	0.68
0.69	2.3059	2.3224	2.3390	2.3557	2.3726	2.3896	2.4067	2.4240	2.4414	2.4590	0.69
0.70	2.4767	2.4945	2.5125	2.5306	2.5489	2.5673	2.5859	2.6046	2.6235	2.6426	0.70
0.71	2.662	2.681	2.701	2.720	2.740	2.760	2.780	2.800	2.821	2.842	0.71
0.72	2.863	2.884	2.905	2.926	2.948	2.969	2.991	3.013	3.036	3.058	0.72
0.73	3.081	3.104	3.127	3.150	3.173	3.197	3.221	3.245	3.270	3.294	0.73
0.74	3.319	3.344	3.369	3.394	3.420	3.446	3.472	3.498	3.525	3.552	0.74
0.75	3.579	3.606	3.634	3.661	3.690	3.718	3.747	3.775	3.805	3.834	0.75
0.76	3.864	3.894	3.924	3.955	3.985	4.017	4.048	4.080	4.112	4.144	0.76
0.77	4.177	4.210	4.243	4.277	4.311	4.345	4.380	4.415	4.450	4.486	0.77
0.78	4.52	4.56	4.60	4.63	4.67	4.71	4.75	4.79	4.82	4.86	0.78
0.79	4.90	4.94	4.99	5.03	5.07	5.11	5.15	5.20	5.24	5.28	0.79
0.80	5.33	5.37	5.42	5.46	5.51	5.56	5.61	5.65	5.70	5.75	0.80
0.81	5.80	5.85	5.90	5.95	6.01	6.06	6.11	6.17	6.22	6.28	0.81
0.82	6.33	6.39	6.45	6.50	6.56	6.62	6.68	6.74	6.81	6.87	0.82
0.83	6.93	7.00	7.06	7.13	7.19	7.26	7.33	7.40	7.47	7.54	0.83
0.84	7.61	7.68	7.76	7.83	7.91	7.98	8.06	8.14	8.22	8.30	0.84
0.85	8.39	8.47	8.55	8.64	8.73	8.82	8.90	9.00	9.09	9.18	0.85

Reproduced from Cohen (1961), Table 1, page 537, with permission of the Technometrics Management Committee.

from sample data, $\hat{\hat{\theta}}$ can be read from the graph or obtained by interpolation from the table. Estimates $\hat{\mu}$ and $\hat{\sigma}$ can then be calculated from (5.8.15). Although the estimators of (5.8.15) were derived for a left truncated distribution, these estimators are also applicable when truncation is on the right because the normal distribution is symmetrical about its mean.

B. *Type I Censoring*

For censoring on the left, the sample consists of n complete observations, each of which equals or exceeds T, plus c observations, each of which is less than T, where T is a fixed (known) point of censoring. Let N designate the total of the randomly selected sample specimens where $N = n + c$. The likelihood function in this case is

$$L = k[\Phi(\xi)]^c (2\pi\sigma^2)^{-n/2} \exp\left\{ -\frac{1}{2} \sum_{i=1}^{n} \left(\frac{x_i - \mu}{\sigma}\right)^2 \right\}, \tag{5.8.17}$$

where k is an ordering constant.

Maximum likelihood estimating equations follow as

$$\begin{aligned} \frac{\partial \ln L}{\partial \mu} &= \frac{-c\phi(\xi)}{\sigma\Phi(\xi)} + \frac{1}{\sigma^2} \sum_{i=1}^{n} (x_i - \mu) = 0, \\ \frac{\partial \ln L}{\partial \sigma} &= \frac{-c\xi\phi(\xi)}{\sigma\Phi(\xi)} - \frac{n}{\sigma} + \frac{1}{\sigma^3} \sum_{i=1}^{n} (x_i - \mu)^2 = 0. \end{aligned} \tag{5.8.18}$$

With $\bar{x}_n$ and s_n^2 defined as in (5.8.8) and with

$$h = c/N \quad \text{and} \quad \Omega(h, \xi) = \frac{h}{1-h} Q(-\xi), \tag{5.8.19}$$

the estimating equations of (5.8.18) become

$$\begin{aligned} \bar{x} - \hat{\mu} &= \hat{\sigma}\Omega(h, \hat{\xi}), \\ s_n^2 + (\bar{x}_n - \hat{\mu})^2 &= \hat{\sigma}^2[1 + \hat{\xi}\Omega(h, \hat{\xi})]. \end{aligned} \tag{5.8.20}$$

The two equations of (5.8.20) differ from those of (5.8.9) only in that $Q(\hat{\xi})$ has been replaced by $\Omega(h, \hat{\xi})$. Consequently derivations from this point to the final estimators are completely analogous to those for truncated samples with $\hat{\hat{\theta}}$ replaced by $\hat{\lambda}$, where

$$\hat{\lambda} = \lambda(h, \hat{\xi}) = \frac{\hat{\Omega}}{\hat{\Omega} - \hat{\xi}} = \lambda(h, \hat{\alpha}), \tag{5.8.21}$$

and where

$$\hat{\alpha} = \alpha(\hat{\xi}) = \frac{1 - \hat{\Omega}(\hat{\Omega} - \hat{\xi})}{(\hat{\Omega} - \hat{\xi})^2} = \frac{s_n^2}{(\bar{x}_n - T)^2}.$$

Accordingly, the final estimators become

$$\begin{aligned} \hat{\sigma}^2 &= s_n^2 + \hat{\lambda}(\bar{x}_n - T)^2, \\ \hat{\mu} &= \bar{x}_n - \hat{\lambda}(\bar{x}_n - T). \end{aligned} \tag{5.8.22}$$

These equations differ from (5.8.15) for the truncated case only in that $\hat{\theta}$ has been replaced by $\hat{\lambda}$. In order to facilitate the practical application of these estimators, tables of λ as a function of h and α, prepared by Cohen (1961), are reproduced here as Table 5.8.2. With $h = c/N$, and $\hat{\alpha} = s_n^2/(\bar{x}_n - T)^2$, $\hat{\lambda}$ can be obtained by interpolation in this table. Estimates $\hat{\sigma}^2$ and $\hat{\mu}$ can then be calculated from the estimators of (5.8.22).

C. *Type II Censoring*

In Type II censored samples, $T = x_{(c+1):N}$ or $T = x_{n:N}$, depending on whether censoring occurs on the left or on the right. Otherwise, derivations and final estimators are identical with those derived for Type I censored samples. For both Type I and Type II censored samples, estimators in (5.8.22) apply to right censored as well as to left censored samples.

D. *Variances and Covariance*

Asymptotic variances and covariance obtained by inverting the Fisher information matrix may be expressed as

$$\begin{aligned} V(\hat{\mu}) &= \frac{\sigma^2}{E(n)}\left[\frac{\phi_{22}}{\phi_{11}\phi_{22} - \phi_{12}^2}\right], \\ V(\hat{\sigma}) &= \frac{\sigma^2}{E(n)}\left[\frac{\phi_{11}}{\phi_{11}\phi_{22} - \phi_{12}^2}\right], \\ \mathrm{Cov}(\hat{\mu}, \hat{\sigma}) &= \frac{\sigma^2}{E(n)}\left[\frac{-\phi_{12}}{\phi_{11}\phi_{22} - \phi_{12}^2}\right], \end{aligned} \tag{5.8.23}$$

where $E(n)$ is the expected value of the number of completely measured (uncensored) observations. In truncated and in Type II censored samples, n is fixed, and therefore $E(n) = n$. In samples that are Type I censored,

TABLE 5.8.2

Auxiliary Estimation Function $\lambda(h, \alpha)$ for Singly Censored Normal Samples

α \ h	.01	.02	.03	.04	.05	.06	.07	.08	.09	.10	.15	.20	h / α
.00	.010100	.020400	.030902	.041583	.052507	.063627	.074953	.086488	.09824	.11020	.17342	.24268	.00
.05	.010551	.021294	.032225	.043350	.054670	.066189	.077909	.089834	.10197	.11431	.17935	.25033	.05
.10	.010950	.022082	.033398	.044902	.056596	.068483	.080568	.092852	.10534	.11804	.18479	.25741	.10
.15	.011310	.022798	.034466	.046318	.058356	.070586	.083009	.095629	.10845	.12148	.18985	.26405	.15
.20	.011642	.023459	.035453	.047629	.059990	.072539	.085280	.098216	.11135	.12469	.19460	.27031	.20
.25	.011952	.024076	.036377	.048858	.061522	.074372	.087413	.10065	.11408	.12772	.19910	.27626	.25
.30	.012243	.024658	.037249	.050018	.062969	.076106	.089433	.10295	.11667	.13059	.20338	.28193	.30
.35	.012520	.025211	.038077	.051120	.064345	.077756	.091355	.10515	.11914	.13333	.20747	.28737	.35
.40	.012784	.025738	.038866	.052173	.065660	.079332	.093193	.10725	.12150	.13595	.21139	.29260	.40
.45	.013036	.026243	.039624	.053182	.066921	.080845	.094958	.10926	.12377	.13847	.21517	.29765	.45
.50	.013279	.026728	.040352	.054153	.068135	.082301	.096657	.11121	.12595	.14090	.21882	.30253	.50
.55	.013513	.027196	.041054	.055089	.069306	.083708	.098298	.11308	.12806	.14325	.22235	.30725	.55
.60	.013739	.027649	.041733	.055995	.070439	.085068	.099887	.11490	.13011	.14552	.22578	.31184	.60
.65	.013958	.028087	.042391	.056874	.071538	.086388	.10143	.11666	.13209	.14773	.22910	.31630	.65
.70	.014171	.028513	.043030	.057726	.072605	.087670	.10292	.11837	.13402	.14987	.23234	.32065	.70
.75	.014378	.028927	.043652	.058556	.073643	.088917	.10438	.12004	.13590	.15196	.23550	.32489	.75
.80	.014579	.029330	.044258	.059364	.074655	.090133	.10580	.12167	.13773	.15400	.23858	.32903	.80
.85	.014775	.029723	.044848	.060153	.075642	.091319	.10719	.12325	.13952	.15599	.24158	.33307	.85
.90	.014967	.030107	.045425	.060923	.076606	.092477	.10854	.12480	.14126	.15793	.24452	.33703	.90
.95	.015154	.030483	.045989	.061676	.077549	.093611	.10987	.12632	.14297	.15983	.24740	.34091	.95
1.00	.015338	.030850	.046540	.062413	.078471	.094720	.11116	.12780	.14465	.16170	.25022	.34471	1.00

α \ h	.25	.30	.35	.40	.45	.50	.55	.60	.65	.70	.80	.90	h / α
.00	.31862	.4021	.4941	.5961	.7096	.8368	.9808	1.145	1.336	1.561	2.176	3.283	.00
.05	.32793	.4130	.5066	.6101	.7252	.8540	.9994	1.166	1.358	1.585	2.203	3.314	.05
.10	.33662	.4233	.5184	.6234	.7400	.8703	1.017	1.185	1.379	1.608	2.229	3.345	.10
.15	.34480	.4330	.5296	.6361	.7542	.8860	1.035	1.204	1.400	1.630	2.255	3.376	.15
.20	.35255	.4422	.5403	.6483	.7678	.9012	1.051	1.222	1.419	1.651	2.280	3.405	.20
.25	.35993	.4510	.5506	.6600	.7810	.9158	1.067	1.240	1.439	1.672	2.305	3.435	.25
.30	.36700	.4595	.5604	.6713	.7937	.9300	1.083	1.257	1.457	1.693	2.329	3.464	.30
.35	.37379	.4676	.5699	.6821	.8060	.9437	1.098	1.274	1.476	1.713	2.353	3.492	.35
.40	.38033	.4755	.5791	.6927	.8179	.9570	1.113	1.290	1.494	1.732	2.376	3.520	.40
.45	.38665	.4831	.5880	.7029	.8295	.9700	1.127	1.306	1.511	1.751	2.399	3.547	.45
.50	.39276	.4904	.5967	.7129	.8408	.9826	1.141	1.321	1.528	1.770	2.421	3.575	.50
.55	.39870	.4976	.6051	.7225	.8517	.9950	1.155	1.337	1.545	1.788	2.443	3.601	.55
.60	.40447	.5045	.6133	.7320	.8625	1.007	1.169	1.351	1.561	1.806	2.465	3.628	.60
.65	.41008	.5114	.6213	.7412	.8729	1.019	1.182	1.366	1.577	1.824	2.486	3.654	.65
.70	.41555	.5180	.6291	.7502	.8832	1.030	1.195	1.380	1.593	1.841	2.507	3.679	.70
.75	.42090	.5245	.6367	.7590	.8932	1.042	1.207	1.394	1.608	1.858	2.528	3.705	.75
.80	.42612	.5308	.6441	.7676	.9031	1.053	1.220	1.408	1.624	1.875	2.548	3.730	.80
.85	.43122	.5370	.6515	.7761	.9127	1.064	1.232	1.422	1.639	1.892	2.568	3.754	.85
.90	.43622	.5430	.6586	.7844	.9222	1.074	1.244	1.435	1.653	1.908	2.588	3.779	.90
.95	.44112	.5490	.6656	.7925	.9314	1.085	1.255	1.448	1.668	1.924	2.607	3.803	.95
1.00	.44592	.5548	.6724	.8005	.9406	1.095	1.267	1.461	1.682	1.940	2.626	3.827	1.00

For all values $0 \leq \alpha \leq 1$, $\lambda(0, \alpha) = 0$.

Reproduced from Cohen (1961), Table 2, page 538, with permission of the Technometrics Management Committee.

TABLE 5.8.3
Variance Factors for Singly Truncated and Singly Censored Normal Samples[a]

η	For Truncated Samples				For Censored Samples				Percent Rest.	η
	μ_{11}	μ_{12}	μ_{22}	ρ	μ_{11}	μ_{12}	μ_{22}	ρ		
-4.0	1.00054	-.001143	.502287	-.001613	1.00000	-.000006	.500030	-.000001	0.00	-4.0
-3.5	1.00313	-.005922	.510366	-.008277	1.00001	-.000052	.500208	-.000074	0.02	-3.5
-3.0	1.01460	-.024153	.536283	-.032744	1.00010	-.000335	.501180	-.000473	0.13	-3.0
-2.5	1.05738	-.081051	.602029	-.101586	1.00056	-.001712	.505280	-.002407	0.62	-2.5
-2.4	1.07437	-.101368	.622786	-.123924	1.00078	-.002312	.506935	-.003247	0.82	-2.4
-2.3	1.09604	-.126136	.646862	-.149803	1.00107	-.003099	.509030	-.004341	1.07	-2.3
-2.2	1.12365	-.156229	.674663	-.179434	1.00147	-.004121	.511658	-.005757	1.39	-2.2
-2.1	1.15880	-.192688	.706637	-.212937	1.00200	-.005438	.514926	-.007571	1.79	-2.1
-2.0	1.20350	-.236743	.743283	-.250310	1.00270	-.007123	.518960	-.009875	2.28	-2.0
-1.9	1.26030	-.289860	.785158	-.291388	1.00363	-.009266	.523899	-.012778	2.87	-1.9
-1.8	1.33246	-.353771	.832880	-.335818	1.00485	-.011971	.529899	-.016405	3.59	-1.8
-1.7	1.42405	-.430531	.887141	-.383041	1.00645	-.015368	.537141	-.020901	4.46	-1.7
-1.6	1.54024	-.522564	.948713	-.432293	1.00852	-.019610	.545827	-.026431	5.48	-1.6
-1.5	1.68750	-.632733	1.01846	-.482644	1.01120	-.024884	.556186	-.033181	6.68	-1.5
-1.4	1.87398	-.764405	1.09734	-.533054	1.01467	-.031410	.568471	-.041358	8.08	-1.4
-1.3	2.10982	-.921533	1.18642	-.582464	1.01914	-.039460	.582981	-.051193	9.68	-1.3
-1.2	2.40764	-1.10874	1.28690	-.629889	1.02488	-.049355	.600046	-.062937	11.51	-1.2
-1.1	2.78311	-1.33145	1.40009	-.674498	1.03224	-.061491	.620049	-.076861	13.57	-1.1
-1.0	3.25557	-1.59594	1.52746	-.715676	1.04168	-.076345	.643438	-.093252	15.87	-1.0
-0.9	3.84879	-1.90952	1.67064	-.753044	1.05376	-.094501	.670724	-.112407	18.41	-0.9
-0.8	4.59189	-2.28066	1.83140	-.786452	1.06923	-.116674	.702513	-.134620	21.19	-0.8
-0.7	5.52036	-2.71911	2.01172	-.815942	1.08904	-.143744	.739515	-.160175	24.20	-0.7
-0.6	6.67730	-3.23612	2.21376	-.841703	1.11442	-.176798	.782574	-.189317	27.43	-0.6
-0.5	8.11482	-3.84458	2.43990	-.864019	1.14696	-.217183	.832691	-.222233	30.85	-0.5
-0.4	9.89562	-4.55921	2.69271	-.883229	1.18876	-.266577	.891077	-.259011	34.46	-0.4
-0.3	12.0949	-5.39683	2.97504	-.899688	1.24252	-.327080	.959181	-.299607	38.21	-0.3
-0.2	14.8023	-6.37653	3.28997	-.913744	1.31180	-.401326	1.03877	-.343800	42.07	-0.2
-0.1	18.1244	-7.51996	3.64083	-.925727	1.40127	-.492641	1.13198	-.391156	46.02	-0.1

0.0	22.1875	-8.85155	4.03126	-.935932	1.51709	-.605233	1.24145	-.441013	50.00	0.0
0.1	27.1403	-10.3988	4.46517	-.944623	1.66743	-.744459	1.37042	-.492483	53.98	0.1
0.2	33.1573	-12.1927	4.94678	-.952028	1.86310	-.917165	1.52288	-.544498	57.93	0.2
0.3	40.4428	-14.2679	5.48068	-.958345	2.11857	-1.13214	1.70381	-.595891	61.79	0.3
0.4	49.2342	-16.6628	6.07169	-.963742	2.45318	-1.40071	1.91942	-.645504	65.54	0.4
0.5	59.8081	-19.4208	6.72512	-.968361	2.89293	-1.73757	2.17751	-.692299	69.15	0.5
0.6	72.4834	-22.5896	7.44658	-.972322	3.47293	-2.16185	2.48793	-.735459	72.57	0.6
0.7	87.6276	-26.2220	8.24204	-.975727	4.24075	-2.69858	2.86318	-.774443	75.80	0.7
0.8	105.66	-30.376	9.1178	-.97866	5.2612	-3.3807	3.3192	-.80899	78.81	0.8
0.9	127.07	-35.117	10.081	-.98119	6.6229	-4.2517	3.8765	-.83912	81.59	0.9
1.0	152.40	-40.515	11.138	-.98338	8.4477	-5.3696	4.5614	-.86502	84.13	1.0
1.1	182.29	-46.650	12.298	-.98529	10.903	-6.8116	5.4082	-.88703	86.43	1.1
1.2	217.42	-53.601	13.567	-.98694	14.224	-8.6818	6.4616	-.90557	88.49	1.2
1.3	258.61	-61.465	14.954	-.98838	18.735	-11.121	7.7804	-.92109	90.32	1.3
1.4	306.78	-70.347	16.471	-.98964	24.892	-14.319	9.4423	-.93401	91.92	1.4
1.5	362.91	-80.350	18.124	-.99074	33.339	-18.539	11.550	-.94473	93.32	1.5
1.6	428.11	-91.586	19.922	-.99171	44.986	-24.139	14.243	-.95361	94.52	1.6
1.7	503.57	-104.17	21.874	-.99256	61.132	-31.616	17.706	-.96097	95.54	1.7
1.8	591.03	-118.31	24.003	-.99332	83.638	-41.664	22.193	-.96706	96.41	1.8
1.9	691.78	-134.10	26.311	-.99398	115.19	-55.252	28.046	-.97211	97.13	1.9
2.0	807.71	-151.73	28.813	-.99457	159.66	-73.750	35.740	-.97630	97.72	2.0
2.1	940.38	-171.30	31.511	-.99509	222.74	-99.100	45.930	-.97979	98.21	2.1
2.2	1091.4	-192.92	34.405	-.99555	312.73	-134.08	59.526	-.98270	98.61	2.2
2.3	1265.4	-217.17	37.575	-.99596	441.92	-182.68	77.810	-.98514	98.93	2.3
2.4	1458.6	-243.23	40.858	-.99632	628.58	-250.68	102.59	-.98718	99.18	2.4
2.5	1677.8	-271.99	44.392	-.99665	899.99	-346.53	136.44	-.98890	99.38	2.5

[a] When truncation or Type I censoring occurs on the left, entries in this table corresponding to $\eta = \xi$ are applicable. For right truncated or type I right censored samples, read entries corresponding to $\eta = -\xi$, but delete negative signs from μ_{12} and ρ. For both Type II left censored and Type II right censored samples, read entries corresponding to percent restriction $= 100h$, but for right censoring delete negative signs from μ_{12} and ρ.

Reproduced from Cohen (1961), Table 3, page 539, with permission of the Technometrics Management Committee.

$E(n)=N[1-\Phi(\xi)]$ for censoring on the left and $E(n)=N\Phi(\xi)$ for censoring on the right. The ϕ_{ij} are given below.

For truncated samples:

$$\phi_{11}(\xi)=1-Q(\xi)[Q(\xi)-\xi],$$
$$\phi_{12}(\xi)=Q(\xi)\{1-\xi[Q(\xi)-\xi]\},$$
$$\phi_{22}(\xi)=2+\xi\phi_{12}(\xi),$$

For Type I censored samples:

$$\phi_{11}(\xi)=1+Q(\xi)[Q(-\xi)+\xi],$$
$$\phi_{12}(\xi)=Q(\xi)\{1+\xi[Q(-\xi)+\xi]\},$$
$$\phi_{22}(\xi)=2+\xi\phi_{12}(\xi),$$

(5.8.24)

For Type II censored samples:

$$\phi_{11}(h,\xi)=1+\Omega(h,\xi)[Q(-\xi)+\xi],$$
$$\phi_{12}(h,\xi)=\Omega(h,\xi)\{1+\xi[Q(-\xi)+\xi]\},$$
$$\phi_{22}(h,\xi)=2+\xi\phi_{12}(h,\xi).$$

As $N\to\infty$, the ϕ_{ij} for Type II censored samples approach the ϕ_{ij} for Type I samples. Likewise, $\text{Lim}_{N\to\infty}(n/N)=[1-\Phi(\hat{\xi})]$. Therefore, in this sense, limiting values of estimate variances and covariance for both types of censored samples are equal. For samples that are restricted *on the left*, estimate variances and covariance may be calculated by substituting applicable values of $\phi_{ij}(\xi)$ or $\phi_{ij}(h,\xi)$ from (5.8.24) directly into (5.8.23). For samples that are restricted *on the right*, $\phi_{ij}(-\xi)$ or $\phi_{ij}(h,-\xi)$ from (5.8.24) is substituted into (5.8.23). In order to facilitate these calculations a table of variance factors μ_{11}, μ_{12}, and μ_{22} plus the correlation coefficient $\rho_{\hat{\mu},\hat{\sigma}}$, given by Cohen (1961), is reproduced here as Table 5.8.3, where

$$\mu_{11}=\phi_{22}/[\phi_{11}\phi_{22}-\phi_{12}^2], \qquad \mu_{12}=-\phi_{12}/[\phi_{11}\phi_{22}-\phi_{12}^2],$$
$$\mu_{22}=\phi_{11}/[\phi_{11}\phi_{22}-\phi_{12}^2]. \tag{5.8.25}$$

With the μ_{ij} available from this table, asymptotic variances, covariance, and correlation coefficient become

$$\mathrm{V}(\hat{\mu})=\frac{\sigma^2}{N}\mu_{11}, \qquad \mathrm{V}(\hat{\sigma})=\frac{\sigma^2}{N}\mu_{22},$$
$$\mathrm{Cov}(\hat{\mu},\hat{\sigma})=\frac{\sigma^2}{N}\mu_{12}, \qquad \rho_{\hat{\mu},\hat{\sigma}}=\frac{\mu_{12}}{\sqrt{\mu_{11}\mu_{22}}}. \tag{5.8.26}$$

One must remember that for truncated samples, $N=n$.

E. *Illustrative Examples*

Example 5.1. Results of a life test on 10 mice following inoculation with a uniform culture of human tuberculosis were presented by Gupta (1952).

The sample was censored (i.e., the test was terminated) with the death of the seventh specimen. Survival time, Y, in days from inoculation to death and logarithms of those times for the seven complete observations as reported are:

y	41	44	46	54	55	58	60
$x=\log_{10} y$	1.613	1.644	1.663	1.732	1.740	1.763	1.778

This is a Type II right censored sample for which Gupta assumed the logarithms to be distributed normally (μ, σ^2). Sample data are summarized as follows: $N=10$, $n=7$, $c=3$, $h=0.3$, $T=x_{7:10}=1.778$, $\bar{x}_7=1.7047$, $s_7^2=\sum_{i=1}^{7}(x_i-\bar{x}_7)^2/7=0.003514$, and $\hat{\alpha}=s_7^2/(\bar{x}_7-1.778)^2=0.654$. We enter Table 5.8.2 with $h=0.3$ and $\hat{\alpha}=0.654$ and interpolate to obtain $\hat{\lambda}=0.512$. From (5.8.22) we calculate

$$\hat{\mu}=1.742 \quad \text{and} \quad \hat{\sigma}=0.079.$$

From Table 5.8.3, we obtain (approximately)

$$\mu_{11}=1.14, \qquad \mu_{12}=0.20, \qquad \mu_{22}=0.82, \qquad \rho=0.21.$$

Following substitution in (5.8.26), we calculate standard errors as

$$\text{S.E. of } \hat{\mu}=0.079\sqrt{1.14/10}=0.027,$$
$$\text{S.E. of } \hat{\sigma}=0.079\sqrt{0.82/10}=0.023.$$

This example was considered previously in Section 4.4 of Chapter 4, where tables of Sarhan and Greenberg (1962) were used to calculate best linear unbiased estimators.

Example 5.2. A random selection of $N=300$ electric light bulbs were life tested until $n=119$ failed. Life spans x (in hours) were recorded for each failure, with the result that $\bar{x}_{119}=1304.832$, $s_{119}^2=12128.250$, and $T=x_{119:30}=1450.00$. Thus $c=181$, $h=181/300=0.6033$, and $\hat{\alpha}=s_{119}^2/(\bar{x}_{119}-x_{119:200})^2=0.5755$. This is another Type II right censored sample, assumed to be from a distribution that is normal (μ, σ^2), and it was also given by Gupta (1952). Two-way linear interpolation in Table 5.8.2 with h and $\hat{\alpha}$ as given above gives $\hat{\lambda}=1.36$, and from (5.8.22), we calculate

$$\hat{\mu}=1.502 \quad \text{and} \quad \hat{\sigma}=202.$$

We enter Table 5.8.3 with percent restriction $=100\text{ h}=60.33\%$, and interpolate linearly to obtain

$$\mu_{11}=2.022, \qquad \mu_{12}=1.051, \qquad \mu_{22}=1.635 \quad \text{and} \quad \rho_{\hat{\mu},\hat{\sigma}}=0.58.$$

When these values are substituted into equations (5.8.26) with $\hat{\sigma}=202$, we calculate standard errors as

$$\text{S.E. of } \hat{\mu}=16.58 \quad \text{and} \quad \text{S.E. of } \hat{\sigma}=14.91.$$

Chapter 6 Approximate Maximum Likelihood Estimation

6.1. Preliminary Remarks

The maximum likelihood estimation of the parameters (location and scale, in particular) based on complete and censored samples has been discussed in great detail in the last chapter. It has been noted there that in most cases, but for a few exceptions such as the exponential and the Rayleigh distributions, the maximum likelihood method does not provide explicit estimators. Hence, it is desirable (if possible) to develop approximations to this maximum likelihood method of estimation which would provide us with estimators for the location and scale parameters that are explicit functions of order statistics and also possess certain "optimal" properties.

In this chapter, we describe the approximate maximum likelihood estimation method that has been developed recently by Balakrishnan (1989a, b, 1990b, c, d), Balakrishnan and Varadan (1990), Balakrishnan and Wong (1990a, b), and Chan (1989). We explain this method of estimation by considering the Rayleigh, normal, logistic, extreme value, Type I generalized logistic, and half logistic distributions. We derive the estimators in all these cases by considering a general Type-II censored sample. Approximate expressions for the variances and covariance of these estimators are derived, and some properties are discussed. Comparisons are made with the corresponding BLUEs and the MLEs (whenever available), and some comments are made about the efficiency of these approximate maximum likelihood estimators (AMLEs) for each individual case. Several illustrative examples are also presented in order to facilitate the reader's understanding of the method of estimation described in this chapter.

6.2. Estimation for the Rayleigh Distribution

Let us consider the Rayleigh distribution with density function

$$f(x;\sigma)=\frac{x}{\sigma^2}e^{-x^2/2\sigma^2}, \qquad x>0,\ \sigma>0, \tag{6.2.1}$$

and cumulative distribution function

$$F(x;\sigma)=1-e^{-x^2/2\sigma^2}, \qquad x>0,\ \sigma>0. \tag{6.2.2}$$

The above-given Rayleigh distribution is very useful in communication engineering and is a special case of a two-parameter Weibull distribution with $\delta=2$ and $\beta^2=2\sigma^2$ (Mann, 1968); also see Chapter 8. The Rayleigh distribution in (6.2.2) has a linearly increasing failure rate, and this property, as rightly pointed out by Polovko (1968), makes it quite suitable as a model for a component that possibly has no manufacturing defects but that ages rapidly.

Let us consider an experiment in which n Rayleigh components are put to test simultaneously at time $x=0$, and the failure times of these components are recorded. Suppose some initial observations are censored (possibly because of some failures during the time when some checks and adjustments are being made on the devices) and some final observations are also censored (possibly because the experimenter terminates the experiment before all components have failed). Then, let

$$X_{r+1:n}\le X_{r+2:n}\le\cdots\le X_{n-s:n} \tag{6.2.3}$$

be the available Type-II censored sample from the Rayleigh distribution with pdf as in (6.2.1), where the first r and the last s observations are censored. Based on the censored sample in (6.2.3), we present in this section the AMLE of σ derived by Balakrishnan (1989a). It is important to mention here that when $r=0$ (case of right censoring only), Harter and Moore (1965a) have given an explicit form for the MLE of σ and have worked out its bias and variance. When $r\neq 0$, Harter and Moore (1965b) have provided an iterative procedure for determining the MLE of σ. As mentioned in Chapter 4, Dyer and Whisenand (1973a) have worked out the BLUE of σ based on complete and censored samples. The problem of estimation of

σ based on selected order statistics has been discussed by several authors including Quayle (1963), Dubey (1967), and Dyer and Whisenand (1973b). The details on this problem are provided in the next chapter.

The likelihood function based on the censored sample in (6.2.3) is given by

$$L = \frac{n!}{r!\,s!}\{F(X_{r+1:n};\,\sigma)\}^r\{1 - F(X_{n-s:n};\,\sigma)\}^s \prod_{i=r+1}^{n-s} f(X_{i:n};\,\sigma),$$

which, upon denoting $Z_{i:n} = X_{i:n}/\sigma$, can be written down as

$$L = \frac{n!}{r!\,s!}\sigma^{-A}\{F(Z_{r+1:n})\}^r\{1 - F(Z_{n-s:n})\}^s \prod_{i=r+1}^{n-s} f(Z_{i:n}), \tag{6.2.4}$$

where $A = n - r - s$ is the size of the censored sample in (6.2.3), and $f(z) = z\,e^{-z^2/2}$ and $F(z) = 1 - e^{-z^2/2}$ are the pdf and cdf of the standard Rayleigh distribution, respectively. Now by realizing that $f(z) = z\{1 - F(z)\}$, we obtain from (6.2.4) the likelihood equation for σ to be

$$\frac{d\ln L}{d\sigma} = -\frac{1}{\sigma}\left[2A + rZ_{r+1:n}\frac{f(Z_{r+1:n})}{F(Z_{r+1:n})} - sZ_{n-s:n}^2 - \sum_{i=r+1}^{n-s} Z_{i:n}^2\right]$$

$$= 0. \tag{6.2.5}$$

Equation (6.2.5) does not admit an explicit solution for σ unless $r = 0$; also see Lee *et al.* (1980). But, we may expand the function $f(Z_{r+1:n})/F(Z_{r+1:n})$ appearing in (6.2.5) in a Taylor series around the point $F^{-1}(p_{r+1}) = \{-2\ln q_{r+1}\}^{1/2}$ and then approximate it by

$$\frac{f(Z_{r+1:n})}{F(Z_{r+1:n})} \simeq \alpha - \beta Z_{r+1:n}, \tag{6.2.6}$$

where

$$p_{r+1} = (r+1)/(n+1), \qquad q_{r+1} = 1 - p_{r+1},$$

$$\alpha = q_{r+1}\{-2\ln q_{r+1}\}^{3/2}/p_{r+1}^2,$$

and

$$\beta = -q_{r+1}\left\{1 + \frac{2}{p_{r+1}}\ln q_{r+1}\right\}\bigg/ p_{r+1}.$$

One may refer to David (1981) and Arnold and Balakrishnan (1989) for justification of such an expansion. By using (6.2.6), we may approximate the likelihood equation in (6.2.5) by

$$\frac{d \ln L}{d\sigma} \simeq \frac{d \ln L^*}{d\sigma}$$
$$= -\frac{1}{\sigma}\left[2A + rZ_{r+1:n}(\alpha - \beta Z_{r+1:n}) - sZ^2_{n-s:n} - \sum_{i=r+1}^{n-s} Z^2_{i:n}\right]$$
$$= 0. \tag{6.2.7}$$

Upon solving Eq. (6.2.7) for σ, we derive the AMLE of σ as

$$\hat{\sigma} = \{-B + (B^2 + 8AC)^{1/2}\}/4A, \tag{6.2.8}$$

where

$$B = r\alpha X_{r+1:n}$$

and

$$C = \sum_{i=r+1}^{n-s} X^2_{i:n} + sX^2_{n-s:n} + r\beta X^2_{r+1:n}.$$

It must be mentioned here that upon solving (6.2.7) for σ, we get a quadratic equation in σ that has two roots, one of which drops out because $\beta > 0$ and, hence, $C > 0$.

Since the estimator $\hat{\sigma}$ in (6.2.8) is the solution of the approximate likelihood equation in (6.2.7), it follows that $\hat{\sigma}$ is asymptotically normally distributed with mean σ and variance $1/E(-d^2 \ln L^*/d\sigma^2)$. For a proof of this result, one may refer to Kendall and Stuart (1973) or Mood *et al.* (1974). From (6.2.7) we derive

$$E\left(-\frac{d^2 \ln L^*}{d\sigma^2}\right) = D/\sigma^2, \tag{6.2.9}$$

where

$$D = 3\left\{\sum_{i=r+1}^{n-s} E(Z^2_{i:n}) + sE(Z^2_{n-s:n}) + r\beta E(Z^2_{r+1:n})\right\} - 2r\alpha E(Z_{r+1:n}) - 2A.$$

Therefore, the variance of $\hat{\sigma}$ may be approximated from the above expressions by using the formulas

$$E(Z_{i:n}) = \sqrt{\frac{\pi}{2}} \frac{n!}{(i-1)!\,(n-i)!} \sum_{j=0}^{i-1} (-1)^{i-1-j} \binom{i-1}{j} \Big/ (n-j)^{3/2}$$

and

$$E(Z^2_{i:n}) = 2\frac{n!}{(i-1)!\,(n-i)!} \sum_{j=0}^{i-1} (-1)^{i-1-j} \binom{i-1}{j} \Big/ (n-j)^2,$$

derived by Lieblein (1955).

By noting that the estimator $\hat{\sigma}$ in (6.2.8) simply becomes $d\hat{\sigma}$ when $X_{i:n}$ in (6.2.3) gets replaced by $dX_{i:n}$, Balakrishnan (1989a) has simulated the values of the bias and the variance of $\hat{\sigma}$ (based on 10,000 Monte Carlo runs) for $n = 10, 20, 30$ and $r, s = 0(1)4$ when $\sigma = 1$, and then has used them to compute the values of (1) $E(\hat{\sigma} - \sigma)/\sigma$ and (2) $\text{Var}(\hat{\sigma})/\sigma^2$. These values are presented in Table 6.2.1. Also presented in this table are the values of (3) approximate $\text{Var}(\hat{\sigma})/\sigma^2$ computed from (6.2.9), and (4) (variance of the BLUE of σ)/σ^2 for $n = 10$ taken from the tables of Dyer and Whisenand (1973a). The value of the (variance of the BLUE of σ)/σ^2 for $n = 10$ and $r = s = 0$ has been taken from Table II of Dyer and Whisenand (1973b). It should be pointed out here that $\hat{\sigma}$ in (6.2.8) is the same as the MLE of σ for the case when $r = 0$, and hence no comparison is needed. When $r \neq 0$, however, a comparison cannot be made because the MLE of σ can be found only by an iterative method.

From Table 6.2.1, we observe that the estimator $\hat{\sigma}$ is as efficient as the BLUE of σ and, at the same time, does not require the construction of any special tables by using the means, variances, and covariances of order statistics as does the BLUE. Further, we also observe that the formula in (6.2.9) yields a very good approximation to $\text{Var}(\hat{\sigma})/\sigma^2$.

Example 6.2.1. Let us consider the following Type-II censored sample from the Rayleigh distribution (see Example 4.2.1), simulated with $\sigma = 12.38$, given by Dyer and Whisenand (1973a) (with $n = 12$, $r = 3$, $s = 2$) as

$$7.2113, 13.0397, 13.5380, 15.0124, 16.1683, 18.1613, 23.2829.$$

In Example 4.2.1, we have worked out the BLUE of σ to be 12.27 and the variance of this estimate to be $0.025\sigma^2$.

We have, in this case,

$$\alpha = 4.61206, \qquad \beta = 3.12797,$$

$$A = 7, \qquad B = 99.77684, \qquad C = 3336.2033,$$

and

$$\hat{\sigma} = \{-B + (B^2 + 8AC)^{1/2}\}/4A = 12.2795.$$

Also, the variance of this estimate is computed from Eq. (6.2.9) to be $0.0244\sigma^2$. It is of interest to note here that these values are almost the same as those based on the BLUE.

TABLE 6.2.1

Bias, Variance, and Approximate Variance of $\hat{\sigma}$, and Variance of the BLUE of σ for the Rayleigh Distribution

		$n = 10$				$n = 20$			$n = 30$		
r	s	(1)	(2)	(3)	(4)	(1)	(2)	(3)	(1)	(2)	(3)
0	0	−.00997	.02509	.02500	.02536	−.00361	.01234	.01250	−.00205	.00845	.00833
0	1	−.01053	.02759	.02778	.02818	−.00362	.01306	.01316	−.00202	.00874	.00862
0	2	−.01219	.03091	.03125	.03174	−.00383	.01367	.01389	−.00182	.00903	.00893
0	3	−.01550	.03579	.03571	.03633	−.00460	.01449	.01471	−.00215	.00930	.00926
0	4	−.01981	.04169	.04167	.04248	−.00489	.01544	.01563	−.00268	.00964	.00962
1	0	−.00491	.02542	.02463	.02539	−.00089	.01241	.01242	−.00018	.00848	.00830
1	1	−.00506	.02798	.02733	.02821	−.00077	.01313	.01307	−.00009	.00878	.00858
1	2	−.00624	.03139	.03068	.03178	−.00083	.01375	.01379	.00017	.00907	.00889
1	3	−.00902	.03643	.03497	.03638	−.00146	.01459	.01459	−.00009	.00935	.00922
1	4	−.01275	.04252	.04066	.04255	−.00159	.01554	.01550	−.00055	.00969	.00957
2	0	−.00388	.02551	.02447	.02541	−.00001	.01244	.01238	.00040	.00849	.00828
2	1	−.00404	.02806	.02713	.02824	.00014	.01316	.01303	.00051	.00878	.00857
2	2	−.00529	.03148	.03043	.03182	.00010	.01379	.01374	.00079	.00908	.00887
2	3	−.00825	.03655	.03465	.03644	−.00049	.01463	.01454	.00054	.00936	.00920
2	4	−.01237	.04261	.04022	.04262	−.00058	.01559	.01544	.00010	.00970	.00955
3	0	−.00358	.02568	.02439	.02551	.00023	.01245	.01236	.00065	.00850	.00828
3	1	−.00386	.02826	.02703	.02837	.00037	.01316	.01301	.00076	.00879	.00856
3	2	−.00532	.03170	.03030	.03198	.00033	.01380	.01372	.00104	.00910	.00886
3	3	−.00866	.03676	.03448	.03665	−.00028	.01464	.01452	.00079	.00937	.00919
3	4	−.01344	.04287	.04000	.04291	−.00040	.01561	.01541	.00035	.00972	.00954
4	0	−.00349	.02596	.02438	.02572	.00035	.01248	.01235	.00075	.00851	.00827
4	1	−.00390	.02856	.02701	.02866	.00049	.01319	.01299	.00085	.00881	.00855
4	2	−.00560	.03208	.03029	.03231	.00042	.01384	.01371	.00114	.00911	.00886
4	3	−.00934	.03718	.03446	.03708	−.00019	.01467	.01450	.00089	.00939	.00918
4	4	−.01493	.04334	.03997	.04350	−.00034	.01565	.01539	.00045	.00974	.00953

From N. Balakrishnan "Approximate MLE of the scale parameter of the Rayleigh distribution with censoring," *IEEE Trans. on Reliab.* **28**, 255–257 © 1989 IEEE.

6.3. Estimation for the Normal Distribution

In this section we shall consider the normal distribution with cdf $F(x; \mu, \sigma)$ and pdf

$$f(x; \mu, \sigma) = \frac{1}{\sqrt{2\pi}\sigma} e^{-(x-\mu)^2/2\sigma^2}, \qquad -\infty < x < \infty. \tag{6.3.1}$$

Let

$$X_{r+1:n} \leq X_{r+2:n} \leq \cdots \leq X_{n-s:n} \tag{6.3.2}$$

be a Type-II censored sample available from the above normal population.

As mentioned in Chapter 4, Sarhan and Greenberg (1956, 1958a, 1962) have worked out the BLUEs of μ and σ from singly and doubly censored samples for sample sizes up to 20, and these tables have been recently extended by Balakrishnan (1990e) for sample sizes up to 40. Dixon (1957, 1960) has derived some simplified estimators for complete as well as censored samples that are almost as efficient as the BLUEs. Saw (1959) has derived simplified unbiased estimators from a singly censored sample for sample sizes up to 20.

The maximum likelihood estimation of μ and σ, based on the censored sample in (6.3.2), has already been discussed in great detail in the last chapter. By considering the likelihood equations for μ and σ based on a Type-II right-censored sample, Gupta (1952) has determined asymptotic variances and covariance of the estimators. The asymptotic properties of the MLEs based on singly censored samples have been studied by Halperin (1952) and Breakwell (1953). Plackett (1958) has shown that the MLEs are asymptotically linear, and that the BLUEs are asymptotically efficient and normally distributed. He has also proposed a linearized MLE of σ and has compared its performance with the BLUE from censored samples. By conducting extensive Monte Carlo studies, Isida and Tagami (1959) and Harter and Moore (1966b) have simulated the values of the bias and mean square error of the MLEs of μ and σ based on singly and doubly Type-II censored samples, and these results are also reproduced by Harter (1970b). By modifying the likelihood equations based on doubly Type-II censored samples, Tiku (1967) has derived the modified MLEs of μ and σ. Tiku (1970) has also carried out a Monte Carlo comparative study of various simplified estimators of μ and σ based on censored samples. A detailed account of the properties of the modified MLEs and their applications in developing robust inference procedures may be found in the recent monograph by Tiku *et al.* (1986).

In this section we present the derivation of the AMLEs of μ and σ based on the censored sample in (6.3.2) due to Balakrishnan (1989b). The likelihood function based on the censored sample in (6.3.2) is given by

$$L=\frac{n!}{r!\,s!}\{F(X_{r+1:n};\,\mu,\sigma)\}^r\{1-F(X_{n-s:n};\,\mu,\sigma)\}^s\prod_{i=r+1}^{n-s}f(X_{i:n};\,\mu,\sigma),$$

which, upon denoting $Z_{i:n}=(X_{i:n}-\mu)/\sigma$, can be written down as

$$L=\frac{n!}{r!\,s!}\sigma^{-A}\{F(Z_{r+1:n})\}^r\{1-F(Z_{n-s:n})\}^s\prod_{i=r+1}^{n-s}f(Z_{i:n}), \qquad (6.3.3)$$

where $A=n-r-s$, and $f(z)$ and $F(z)$ are, respectively, the pdf and cdf of the standard normal distribution. Now by realizing that $f'(z)=-zf(z)$, we obtain the likelihood equations for μ and σ from (6.3.3) to be

$$\frac{\partial \ln L}{\partial\mu}=-\frac{1}{\sigma}\left[r\frac{f(Z_{r+1:n})}{F(Z_{r+1:n})}-s\frac{f(Z_{n-s:n})}{1-F(Z_{n-s:n})}-\sum_{i=r+1}^{n-s}Z_{i:n}\right]=0 \qquad (6.3.4)$$

and

$$\frac{\partial \ln L}{\partial\sigma}=-\frac{1}{\sigma}\left[A+rZ_{r+1:n}\frac{f(Z_{r+1:n})}{F(Z_{r+1:n})}-sZ_{n-s:n}\frac{f(Z_{n-s:n})}{1-F(Z_{n-s:n})}-\sum_{i=r+1}^{n-s}Z_{i:n}^2\right]$$
$$=0. \qquad (6.3.5)$$

The likelihood equations (6.3.4) and (6.3.5) do not admit explicit solutions except in the complete sample case. But, we may expand the functions $f(Z_{r+1:n})/F(Z_{r+1:n})$ and $f(Z_{n-s:n})/\{1-F(Z_{n-s:n})\}$ appearing in (6.3.4) and (6.3.5) in Taylor series around the points $\xi_{r+1}=F^{-1}(p_{r+1})$ and $\xi_{n-s}=F^{-1}(p_{n-s})$, respectively, and approximate them by

$$\frac{f(Z_{r+1:n})}{F(Z_{r+1:n})}\simeq\alpha-\beta Z_{r+1:n} \qquad (6.3.6)$$

and

$$\frac{f(Z_{n-s:n})}{1-F(Z_{n-s:n})}\simeq\gamma+\delta Z_{n-s:n}, \qquad (6.3.7)$$

where

$$p_i=\frac{i}{n+1}, \qquad q_i=1-p_i,$$

$$\alpha=f(\xi_{r+1})\{1+\xi_{r+1}^2+\xi_{r+1}f(\xi_{r+1})/p_{r+1}\}/p_{r+1}, \qquad (6.3.8)$$

$$\beta=f(\xi_{r+1})\{f(\xi_{r+1})+p_{r+1}\xi_{r+1}\}/p_{r+1}^2, \qquad (6.3.9)$$

$$\gamma=f(\xi_{n-s})\{1+\xi_{n-s}^2-\xi_{n-s}f(\xi_{n-s})/q_{n-s}\}/q_{n-s}, \qquad (6.3.10)$$

and

$$\delta = f(\xi_{n-s})\{f(\xi_{n-s}) - q_{n-s}\xi_{n-s}\}/q_{n-s}^2. \tag{6.3.11}$$

It is very easy to see from Eq. (6.3.9) that $\beta > 0$ whenever $p_{r+1} \geq \frac{1}{2}$. Also, when $p_{r+1} < \frac{1}{2}$, we have $\xi_{r+1} < 0$ and

$$|\xi_{r+1}p_{r+1}| = |\xi_{r+1}F(\xi_{r+1})| = -\xi_{r+1}\int_{-\infty}^{\xi_{r+1}} f(z)\,dz$$

$$\leq \int_{-\infty}^{\xi_{r+1}} -zf(z)\,dz = f(\xi_{r+1}),$$

and, consequently, $\beta > 0$. By using a similar argument, we can show from Eq. (6.3.11) that $\delta > 0$. By using (6.3.6) and (6.3.7), we may approximate the likelihood equations (6.3.4) and (6.3.5) by

$$\frac{\partial \ln L}{\partial \mu} \simeq \frac{\partial \ln L^*}{\partial \mu} = -\frac{1}{\sigma}\left[r\alpha - s\gamma - r\beta Z_{r+1:n} - s\delta Z_{n-s:n} - \sum_{i=r+1}^{n-s} Z_{i:n}\right]$$
$$= 0, \tag{6.3.12}$$

and

$$\frac{\partial \ln L}{\partial \sigma} \simeq \frac{\partial \ln L^*}{\partial \sigma} = -\frac{1}{\sigma}\left[A + r\alpha Z_{r+1:n} - s\gamma Z_{n-s:n} - r\beta Z_{r+1:n}^2 - s\delta Z_{n-s:n}^2\right.$$
$$\left. - \sum_{i=r+1}^{n-s} Z_{i:n}^2\right] = 0. \tag{6.3.13}$$

Upon solving Eqs. (6.3.12) and (6.3.13), we derive the AMLEs of μ and σ as

$$\hat{\mu} = B - \hat{\sigma}C \tag{6.3.14}$$

and

$$\hat{\sigma} = \{-D + (D^2 + 4AE)^{1/2}\}/2A, \tag{6.3.15}$$

where

$$m = A + r\beta + s\delta,$$

$$B = \left\{\sum_{i=r+1}^{n-s} X_{i:n} + r\beta X_{r+1:n} + s\delta X_{n-s:n}\right\}\Big/ m,$$

$$C = (r\alpha - s\gamma)/m,$$

$$D = r\alpha(X_{r+1:n} - B) - s\gamma(X_{n-s:n} - B)$$
$$= r\alpha X_{r+1:n} - s\gamma X_{n-s:n} - mBC,$$

and

$$E = \sum_{i=r+1}^{n-s} (X_{i:n} - B)^2 + r\beta(X_{r+1:n} - B)^2 + s\delta(X_{n-s:n} - B)^2$$

$$= \sum_{i=r+1}^{n-s} X_{i:n}^2 + r\beta X_{r+1:n}^2 + s\delta X_{n-s:n}^2 - mB^2.$$

It should be mentioned here that upon solving (6.3.13), we get a quadratic equation in σ that has two roots; however, one of them drops out since $\beta > 0$ and $\delta > 0$ and, hence, $E > 0$.

Furthermore, for the case of symmetric censoring (i.e., $r = s$), we observe easily from Eqs. (6.3.8)–(6.3.11) that $\alpha = \gamma$ and $\beta = \delta$ so that $C = 0$; in this case, therefore, $\hat{\mu}$ is an unbiased estimator of μ.

The conditional bias of $\hat{\mu}$ can be computed exactly from (6.3.14). But it is not possible to determine the conditional bias of $\hat{\sigma}$ exactly. However, it may be evaluated approximately by $E(\partial \ln L^*/\partial\sigma)/E(-\partial^2 \ln L^*/\partial\sigma^2)$; for a proof, one may refer to Kendall and Stuart (1973). Moreover, we derive from Eqs. (6.3.12) and (6.3.13)

$$E\left(-\frac{\partial^2 \ln L^*}{\partial\mu^2}\right) = m/\sigma^2, \tag{6.3.16}$$

$$E\left(-\frac{\partial^2 \ln L^*}{\partial\mu\,\partial\sigma}\right) = mV_1/\sigma^2, \tag{6.3.17}$$

and

$$E\left(-\frac{\partial^2 \ln L^*}{\partial\sigma^2}\right) = mV_2/\sigma^2, \tag{6.3.18}$$

where

$$V_1 = \frac{2}{m}\left\{\sum_{i=r+1}^{n-s} E(Z_{i:n}) + r\beta E(Z_{r+1:n}) + s\delta E(Z_{n-s:n})\right\} - C \tag{6.3.19}$$

and

$$V_2 = \frac{1}{m}\left[3\left\{\sum_{i=r+1}^{n-s} E(Z_{i:n}^2) + r\beta E(Z_{r+1:n}^2) + s\delta E(Z_{n-s:n}^2)\right\}\right.$$

$$\left. - 2\{r\alpha E(Z_{r+1:n}) - s\gamma E(Z_{n-s:n})\} - A\right]. \tag{6.3.20}$$

From the above expressions we obtain

$$\mathrm{Var}(\hat{\mu}) = \frac{\sigma^2}{m}\left\{\frac{V_2}{V_2 - V_1^2}\right\}, \tag{6.3.21}$$

$$\mathrm{Var}(\hat{\sigma}) = \frac{\sigma^2}{m}\left\{\frac{1}{V_2 - V_1^2}\right\}, \tag{6.3.22}$$

and

$$\mathrm{Cov}(\hat{\mu}, \hat{\sigma}) = -\frac{\sigma^2}{m}\left\{\frac{V_1}{V_2 - V_1^2}\right\}. \tag{6.3.23}$$

In the case of symmetric censoring (i.e., $r = s$), we have already noted that $C = 0$. Further, since $\beta = \delta$ and $E(Z_{i:n}) = -E(Z_{n-i+1:n})$, we have $V_1 = 0$ and hence, in this case, we find the estimators $\hat{\mu}$ and $\hat{\sigma}$ to be uncorrelated.

By using all these formulas, Balakrishnan (1989b) has computed the values of:

(1) {Exact conditional bias of $\hat{\mu}$}/σ,
(2) {Approximate conditional bias of $\hat{\sigma}$}/σ,
(3) {Approximate variance of $\hat{\mu}$}/σ^2,
(4) {Approximate variance of $\hat{\sigma}$}/σ^2,
(5) {Approximate covariance of $\hat{\mu}$ and $\hat{\sigma}$}/σ^2,
(6) {Approximate conditional variance of $\hat{\mu}$}/σ^2, and
(7) {Approximate conditional variance of $\hat{\sigma}$}/σ^2,

for various sample sizes and over various choices of censoring. In Table 6.3.1, we have presented these values for sample size $n = 10$ and 20. Upon comparing these values with the corresponding values of the MLEs (based on Monte Carlo simulations) given by Harter and Moore (1966b) and those of the BLUEs given by Sarhan and Greenberg (1962), we observe that the AMLEs $\hat{\mu}$ and $\hat{\sigma}$ are both remarkably efficient. Moreover, by comparing the values in Table 6.3.1 with the corresponding values of the MMLEs given by Tiku (1967), we find the AMLEs $\hat{\mu}$ and $\hat{\sigma}$ to be as efficient as his estimators. As pointed out by Balakrishnan (1989b), this is not surprising, since the two estimators are quite similar with the only difference occurring in the linear approximation method adopted. The linear approximation used here, however, has an advantage in the sense that it could be easily extended to Type-I censoring, multiple censoring, truncation, grouped data, and even to the case of estimation based on selected order statistics, as will be explained in the next chapter.

TABLE 6.3.1
Bias, Variances, and Covariance of Estimators $\hat{\mu}$ and $\hat{\sigma}$ for the Normal Distribution

n	r	s	(1)	(2)	(3)	(4)	(5)	(6)	(7)
10	0	1	−0.00133	−0.00086	0.10250	0.05827	0.00453	0.10215	0.05807
	0	2	−0.00276	−0.00122	0.10721	0.06814	0.01137	0.10531	0.06694
	0	3	−0.00460	−0.00116	0.11546	0.08055	0.02158	0.10968	0.07652
	0	4	−0.00713	−0.00035	0.12989	0.09681	0.03707	0.11570	0.08623
	1	1	0.00000	−0.00206	0.10439	0.06926	0.00000	0.10439	0.06926
	1	2	−0.00142	−0.00273	0.10834	0.08274	0.00734	0.10769	0.08224
	1	3	−0.00324	−0.00292	0.11585	0.10030	0.01894	0.11227	0.09720
	1	4	−0.00576	−0.00215	0.12993	0.12428	0.03756	0.11858	0.11342
	2	2	0.00000	−0.00370	0.11121	0.10122	0.00000	0.11121	0.10122
	2	3	−0.00182	−0.00417	0.11735	0.12622	0.01257	0.11610	0.12487
	2	4	−0.00435	−0.00342	0.13014	0.16204	0.03433	0.12287	0.15298
	3	3	0.00000	−0.00493	0.12144	0.16295	0.00000	0.12144	0.16295
	3	4	−0.00254	−0.00411	0.13159	0.21889	0.02445	0.12886	0.21435
	4	4	0.00000	−0.00254	0.13725	0.31308	0.00000	0.13725	0.31308
20	0	1	−0.00038	−0.00032	0.05045	0.02705	0.00096	0.05042	0.02703
	0	2	−0.00069	−0.00049	0.05114	0.02922	0.00217	0.05098	0.02913
	0	3	−0.00102	−0.00061	0.05209	0.03161	0.00368	0.05166	0.03135
	0	4	−0.00139	−0.00070	0.05337	0.03426	0.00553	0.05248	0.03369
	0	5	−0.00180	−0.00074	0.05508	0.03725	0.00780	0.05345	0.03614
	1	1	0.00000	−0.00071	0.05085	0.02942	0.00000	0.05085	0.02942
	1	2	−0.00031	−0.00092	0.05146	0.03195	0.00125	0.05141	0.03192
	1	3	−0.00064	−0.00109	0.05234	0.03475	0.00281	0.05211	0.03460
	1	4	−0.00100	−0.00122	0.05354	0.03790	0.00476	0.05294	0.03748
	1	5	−0.00142	−0.00132	0.05517	0.04148	0.00719	0.05393	0.04054
	2	2	0.00000	−0.00118	0.05199	0.03489	0.00000	0.05199	0.03489
	2	3	−0.00033	−0.00139	0.05277	0.03817	0.00160	0.05270	0.03812
	2	4	−0.00069	−0.00156	0.05387	0.04189	0.00363	0.05355	0.04164
	2	5	−0.00110	−0.00170	0.05539	0.04615	0.00619	0.05456	0.04546
	3	3	0.00000	−0.00164	0.05343	0.04201	0.00000	0.05343	0.04201
	3	4	−0.00036	−0.00186	0.05440	0.04641	0.00207	0.05431	0.04633
	3	5	−0.00077	−0.00204	0.05578	0.05151	0.00474	0.05535	0.05110
	4	4	0.00000	−0.00213	0.05522	0.05163	0.00000	0.05522	0.05163
	4	5	−0.00041	−0.00237	0.05642	0.05776	0.00273	0.05629	0.05763
	5	5	0.00000	−0.00267	0.05740	0.06522	0.00000	0.05740	0.06522

Balakrishnan (1989b) has also simulated (based on 10,000 Monte Carlo runs) the values of

(8) {Bias of $\hat{\mu}$}/σ,
(9) {Bias of $\hat{\sigma}$}/σ,
(10) {Variance of $\hat{\mu}$}/σ^2,
(11) {Variance of $\hat{\sigma}$}/σ^2, and
(12) {Covariance of $\hat{\mu}$ and $\hat{\sigma}$}/σ^2,

TABLE 6.3.2
Simulated Values of Bias, Variances, and Covariance of $\hat{\mu}$ and $\hat{\sigma}$ for the Normal Distribution

n	r	s	(8)	(9)	(10)	(11)	(12)
10	0	1	−0.00806	−0.08964	0.10293	0.05603	0.00230
	0	2	−0.01783	−0.10151	0.10797	0.06570	0.00919
	0	3	−0.03584	−0.12160	0.11704	0.07693	0.01916
	1	1	0.00003	−0.10545	0.10522	0.06520	−0.00227
	1	2	−0.01050	−0.12169	0.10944	0.07821	0.00501
	1	3	−0.03016	−0.14900	0.11772	0.09344	0.01610
	2	2	0.00072	−0.14565	0.11239	0.09516	−0.00189
	2	3	−0.01987	−0.18327	0.11952	0.11571	0.01011
	3	3	0.00022	−0.23787	0.12386	0.14379	−0.00064
20	0	1	−0.00166	−0.03964	0.04864	0.02729	0.00131
	0	2	−0.00380	−0.04270	0.04919	0.02940	0.00239
	0	3	−0.00681	−0.04677	0.05010	0.03147	0.00373
	0	4	−0.00997	−0.05062	0.05095	0.03352	0.00506
	0	5	−0.01319	−0.05421	0.05320	0.03629	0.00757
	1	1	0.00046	−0.04385	0.04906	0.02945	0.00033
	1	2	−0.00174	−0.04742	0.04955	0.03188	0.00142
	1	3	−0.00485	−0.05216	0.05038	0.03433	0.00283
	1	4	−0.00819	−0.05673	0.05116	0.03685	0.00422
	1	5	−0.01169	−0.06113	0.05330	0.04025	0.00693
	2	2	0.00068	−0.05216	0.05008	0.03468	0.00022
	2	3	−0.00247	−0.05763	0.05082	0.03760	0.00166
	2	4	−0.00594	−0.06302	0.05149	0.04070	0.00311
	2	5	−0.00965	−0.06829	0.05352	0.04477	0.00600
	3	3	0.00048	−0.06360	0.05142	0.04178	0.00009
	3	4	−0.00306	−0.06997	0.05199	0.04563	0.00156
	3	5	−0.00696	−0.07631	0.05385	0.05058	−0.00463
	4	4	0.00025	−0.07687	0.05282	0.05079	−0.00046
	4	5	−0.00374	−0.08446	0.05448	0.05682	0.00273
	5	5	0.00052	−0.09425	0.05528	0.06419	0.00036

Produced with kind permission from Gordon and Breach Science Publishers, Inc.

for various sample sizes and over various choices of censoring. In Table 6.3.2, we have presented these simulated values for sample size $n = 10$ and 20. By comparing these values with the corresponding entries in Table 6.3.1, we observe that the formulas (6.3.21) through (6.3.23) yield very good approximations to the variances and covariance of the AMLEs $\hat{\mu}$ and $\hat{\sigma}$.

Furthermore, for the case of symmetric censoring (i.e., $r = s$), we have given in Table 6.3.3 the values of {Approximate variance of $\hat{\mu}$}/σ^2 computed from (6.3.21), {Exact variance of $\hat{\mu}$}/σ^2 computed from (6.3.14) by using the table of variances and covariances of standard normal order statistics prepared by Tietjen *et al.* (1977), {Variance of the BLUE of μ}/σ^2 taken from the table of Sarhan and Greenberg (1962), and {Mean square error of the MLE of μ}/σ^2 taken from the table of Harter (1970b), for sample size $n = 10$ and 20 and proportion of censoring $q = r/n = 0.1$, 0.2, 0.3, and 0.4. It is very clear from Table 6.3.3 that the AMLE $\hat{\mu}$ of μ is almost as efficient as the MLE of μ and just as efficient as the BLUE of μ, even for samples of size as small as 10.

Example 6.3.1. Let us consider the following data (see Example 4.4.1) due to Gupta (1952), showing the number X' of days to death of the first seven in a sample of 10 mice after inoculation with a uniform culture of human tuberculosis:

X':	41	44	46	54	55	58	60
$X = \log X'$:	1.613	1.644	1.663	1.732	1.740	1.763	1.778

Gupta (1952) has taken $X = \log X'$ to be distributed as $N(\mu, \sigma^2)$.

In the last chapter, we obtained the MLEs of μ and σ from the above data to be

$$\tilde{\mu} = 1.742 \qquad \text{and} \qquad \tilde{\sigma} = 0.079$$

TABLE 6.3.3
Variances of the Estimators of μ for the Normal Distribution When $r = s$

	$n = 10$				$n = 20$			
r:	1	2	3	4	2	4	6	8
Appr. Var($\hat{\mu}$)/σ^2	0.104	0.111	0.121	0.137	0.052	0.055	0.060	0.068
Exact Var($\hat{\mu}$)/σ^2	0.104	0.111	0.122	0.138	0.052	0.055	0.060	0.068
Var (BLUE of μ)/σ^2	0.104	0.111	0.122	0.138	0.052	0.055	0.060	0.068
MSE (MLE of μ)/σ^2	0.101	0.108	0.119	0.134	0.049	0.052	0.057	0.064

Produced with kind permission of Gordon and Breach Science Publishers, Inc.

and the standard errors of these estimates to be

$$\text{S.E.}(\tilde{\mu}) = 0.027 \qquad \text{and} \qquad \text{S.E.}(\tilde{\sigma}) = 0.023.$$

In Chapter 4, by using Sarhan and Greenberg's (1962) tables, we obtained the BLUEs of μ and σ to be

$$\mu^* = 1.746 \qquad \text{and} \qquad \sigma^* = 0.091$$

and the standard errors of these estimates to be

$$\text{S.E.}(\mu^*) = 0.031 \qquad \text{and} \qquad \text{S.E.}(\sigma^*) = 0.029.$$

Also in Chapter 4, by applying Gupta's (1952) simplified linear estimation method, we obtained estimates of μ and σ to be

$$\mu^{**} = 1.748 \qquad \text{and} \qquad \sigma^{**} = 0.094$$

and the standard errors of these estimates to be

$$\text{S.E.}(\mu^{**}) = 0.033 \qquad \text{and} \qquad \text{S.E.}(\sigma^{**}) = 0.031.$$

For the determination of the AMLEs of μ and σ in this case, we have

$$n = 10, \qquad r = 0, \qquad s = 3, \qquad A = 7,$$

$$\alpha = 0.68239, \qquad \beta = 0.83724,$$

$$\gamma = 0.78611, \qquad \delta = 0.70532,$$

$$m = 9.11596, \qquad C = -0.25870,$$

$$B = 1.721725, \qquad D = -0.132769, \qquad \text{and} \qquad E = 0.033324,$$

so that

$$\hat{\sigma} = \{-D + (D^2 + 4AE)^{1/2}\}/2A = 0.079129$$

and

$$\hat{\mu} = B - \hat{\sigma}C = 1.742196.$$

From Table 6.3.2 we also obtain the standard errors of these estimates to be

$$\text{S.E.}(\hat{\mu}) = 0.027 \qquad \text{and} \qquad \text{S.E.}(\hat{\sigma}) = 0.022.$$

We note that these results are quite close to those obtained by ML method.

Example 6.3.2. Next, let us consider the well-known Darwin data (see Example 4.4.2) that represent the differences in heights between cross- and self-fertilized plants of the same pair grown together in one pot:

$$49, -67, 8, 16, 6, 23, 28, 41, 14, 29, 56, 24, 75, 60, -48.$$

Assuming a normal $N(\mu, \sigma^2)$ distribution for the above data, we may estimate μ and σ by

$$\bar{X} = 20.9333 \quad \text{and} \quad s = 37.744,$$

respectively, with standard errors

$$\text{S.E.}(\bar{X}) = 9.745 \quad \text{and} \quad \text{S.E.}(s) = 7.133.$$

In order to achieve robustness, let us censor the two smallest and the two largest observations as suggested by Tiku *et al.* (1986). The remaining 11 observations, arranged in increasing order of magnitude,

$$6, 8, 14, 16, 23, 24, 28, 29, 41, 49, 56,$$

constitute a symmetrically Type-II censored sample. In Chapter 4, by using Sarhan and Greenberg's (1962) tables, we obtained the BLUEs of μ and σ from the above censored sample to be

$$\mu^* = 27.695 \quad \text{and} \quad \sigma^* = 26.381$$

and

$$\text{S.E.}(\mu^*) = 7.009 \quad \text{and} \quad \text{S.E.}(\sigma^*) = 6.430.$$

For the determination of the AMLEs of μ and σ based on the above symmetrically censored sample, we have

$$n = 15, \quad r = 2, \quad s = 2, \quad A = 11,$$

$$\alpha = \gamma = 0.737162, \quad \beta = \delta = 0.787189,$$

$$m = 14.148754, \quad C = 0,$$

$$B = 27.678153, \quad D = -73.716204, \quad \text{and} \quad E = 4654.844569,$$

so that

$$\hat{\mu} = B = 27.678153$$

and

$$\hat{\sigma} = \{-D + (D^2 + 4AE)^{1/2}\}/2A = 24.192884,$$

and the standard errors of these estimates (based on Monte Carlo simulations) are reported by Balakrishnan (1989b) to be

$$\text{S.E.}(\hat{\mu}) = 6.432 \quad \text{and} \quad \text{S.E.}(\hat{\sigma}) = 5.536.$$

It is with interest that we note that the AMLEs of μ and σ both have smaller standard error than the corresponding BLUEs.

6.4. Estimation for the Logistic Distribution

Let

$$X_{r+1:n} \leq X_{r+2:n} \leq \cdots \leq X_{n-s:n} \tag{6.4.1}$$

be a Type-II censored sample from the logistic population with pdf

$$f(x;\, \mu, \sigma) = \frac{\pi}{\sigma\sqrt{3}} \frac{e^{-\pi(x-\mu)/\sigma\sqrt{3}}}{\{1+e^{-\pi(x-\mu)/\sigma\sqrt{3}}\}^2}, \qquad -\infty < x < \infty, \tag{6.4.2}$$

and cdf

$$F(x;\, \mu, \sigma) = 1/\{1+e^{-\pi(x-\mu)/\sigma\sqrt{3}}\}, \qquad -\infty < x < \infty. \tag{6.4.3}$$

As mentioned in Chapter 4, Gupta *et al.* (1967) have worked out the BLUEs of μ and σ based on the doubly censored sample in (6.4.1) for selected sample sizes up to 25 and for some selected choices of censoring. These tables have been recently extended by Balakrishnan (1990f) for sample sizes up to 40. Raghunandanan and Srinivasan (1970) have proposed some simple estimators of μ and σ based on a search made on some specific linear functions of order statistics. They have also set up the necessary tables for their estimators for sample sizes up to 20; however, their estimators do not extend to larger sample sizes.

The maximum likelihood estimation of μ and σ, based on the censored sample in (6.4.1), has been investigated by Harter and Moore (1967). They have simulated the bias and mean square error of these estimators (based on 1,000 Monte Carlo runs) for sample size $n = 10$ and 20 and over various choices of censoring; these values are reproduced by Harter (1970b). Tiku (1968) has derived the modified MLEs of μ and σ by modifying the likelihood equations based on the censored sample in (6.4.1). The estimation of μ and σ based on selected order statistics has been considered in great detail by several authors, including Beyer *et al.* (1976), Chan (1969), Chan and Cheng (1972, 1974), Chan *et al.* (1971), Gupta and Gnanadesikan (1966), and Hassanein (1969a). A detailed discussion of this topic is presented in the next chapter.

In this section we present the derivation of the AMLEs of μ and σ based on the doubly censored sample in (6.4.1) due to Balakrishnan (1990b). The likelihood function based on the censored sample in (6.4.1) is given by

$$L = \frac{n!}{r!\, s!} \{F(X_{r+1:n};\, \mu, \sigma)\}^r \{1 - F(X_{n-s:n};\, \mu, \sigma)\}^s \prod_{i=r+1}^{n-s} f(X_{i:n};\, \mu, \sigma),$$

which, by denoting $Z_{i:n} = \pi(X_{i:n} - \mu)/\sigma\sqrt{3}$, can be written as

$$L = \frac{n!}{r!\,s!}\left(\frac{\pi}{\sqrt{3}}\right)^{A} \sigma^{-A}\{F(Z_{r+1:n})\}^{r}\{1 - F(Z_{n-s:n})\}^{s} \prod_{i=r+1}^{n-s} f(Z_{i:n}), \tag{6.4.4}$$

where $A = n - r - s$ is the size of the censored sample in (6.4.1), $f(z) = e^{-z}/(1+e^{-z})^2$, and $F(z) = 1/(1+e^{-z})$. Now, by realizing that $f(z) = F(z)\{1 - F(z)\}$ we obtain the likelihood equations for μ and σ as

$$\frac{\partial \ln L}{\partial \mu} = -\frac{\pi}{\sigma\sqrt{3}}\left[r\{1 - F(Z_{r+1:n})\} - sF(Z_{n-s:n}) + \sum_{i=r+1}^{n-s} \{1 - 2F(Z_{i:n})\}\right] = 0, \tag{6.4.5}$$

and

$$\frac{\partial \ln L}{\partial \sigma} = -\frac{1}{\sigma}\left[A + rZ_{r+1:n}\{1 - F(Z_{r+1:n})\} - sZ_{n-s:n}F(Z_{n-s:n}) + \sum_{i=r+1}^{n-s} Z_{i:n}\{1 - 2F(Z_{i:n})\}\right] = 0. \tag{6.4.6}$$

Equations (6.4.5) and (6.4.6) do not admit explicit solutions. But, we may expand the function $F(Z_{i:n})$ appearing in (6.4.5) and (6.4.6) in a Taylor series around the point $F^{-1}(p_i) = \ln(p_i/q_i)$, and then approximate it by

$$F(Z_{i:n}) \simeq \alpha_i + \beta_i Z_{i:n}, \tag{6.4.7}$$

where

$$p_i = i/(n+1), \qquad q_i = 1 - p_i,$$

$$\alpha_i = p_i - p_i q_i \ln(p_i/q_i), \tag{6.4.8}$$

and

$$\beta_i = p_i q_i. \tag{6.4.9}$$

By using (6.4.7), we may approximate the likelihood equations (6.4.5) and (6.4.6) by

$$\frac{\partial \ln L}{\partial \mu} \simeq \frac{\partial \ln L^*}{\partial \mu} = -\frac{\pi}{\sigma\sqrt{3}}\left[r(1 - \alpha_{r+1}) - s\alpha_{n-s} + A - 2\sum_{i=r+1}^{n-s} \alpha_i - r\beta_{r+1}Z_{r+1:n} - s\beta_{n-s}Z_{n-s:n} - 2\sum_{i=r+1}^{n-s} \beta_i Z_{i:n}\right] = 0, \tag{6.4.10}$$

and

$$\frac{\partial \ln L}{\partial \sigma} \simeq \frac{\partial \ln L^*}{\partial \sigma}$$

$$= -\frac{1}{\sigma}\left[A + r(1-\alpha_{r+1})Z_{r+1:n} - s\alpha_{n-s}Z_{n-s:n} - \sum_{i=r+1}^{n-s}(2\alpha_i - 1)Z_{i:n}\right.$$

$$\left. - r\beta_{r+1}Z_{r+1:n}^2 - s\beta_{n-s}Z_{n-s:n}^2 - 2\sum_{i=r+1}^{n-s}\beta_i Z_{i:n}^2\right] = 0. \qquad (6.4.11)$$

Upon solving Eqs. (6.4.10) and (6.4.11), we derive the AMLEs of μ and σ as

$$\hat{\mu} = B - \frac{\sqrt{3}}{\pi}\hat{\sigma}C \qquad (6.4.12)$$

and

$$\hat{\sigma} = \frac{\pi}{\sqrt{3}}\{-D + (D^2 + 4AE)^{1/2}\}/2A, \qquad (6.4.13)$$

where

$$m = 2\sum_{i=r+1}^{n-s}\beta_i + r\beta_{r+1} + s\beta_{n-s},$$

$$B = \left\{2\sum_{i=r+1}^{n-s}\beta_i X_{i:n} + r\beta_{r+1}X_{r+1:n} + s\beta_{n-s}X_{n-s:n}\right\}\Big/ m,$$

$$C = \left\{n - s - r\alpha_{r+1} - s\alpha_{n-s} - 2\sum_{i=r+1}^{n-s}\alpha_i\right\}\Big/ m,$$

$$D = \sum_{i=r+1}^{n-s}(1-2\alpha_i)(X_{i:n} - B) + r(1-\alpha_{r+1})(X_{r+1:n} - B) - s\alpha_{n-s}(X_{n-s:n} - B)$$

$$= \sum_{i=r+1}^{n-s}(1-2\alpha_i)X_{i:n} + r(1-\alpha_{r+1})X_{r+1:n} - s\alpha_{n-s}X_{n-s:n} - mBC,$$

and

$$E = 2\sum_{i=r+1}^{n-s}\beta_i(X_{i:n} - B)^2 + r\beta_{r+1}(X_{r+1:n} - B)^2 + s\beta_{n-s}(X_{n-s:n} - B)^2$$

$$= 2\sum_{i=r+1}^{n-s}\beta_i X_{i:n}^2 + r\beta_{r+1}X_{r+1:n}^2 + s\beta_{n-s}X_{n-s:n}^2 - mB^2.$$

It is important to mention here that upon solving Eq. (6.4.11) we get a quadratic equation in σ that has two roots; however, one of them drops out, since we observe easily from (6.4.9) that $\beta_i > 0$ and, hence, $E > 0$.

Furthermore, for the case of symmetric censoring (i.e., $r=s$), we note from (6.4.8) that $\alpha_{n-i+1}=1-\alpha_i$ and $\sum_{i=r+1}^{n-r}\alpha_i=(n-2r)/2$ so that $C=0$; hence, in this case the estimator $\hat{\mu}$ in (6.4.12) is unbiased for μ.

The conditional bias of $\hat{\mu}$ can be computed exactly from (6.4.12). But it is difficult to find the conditional bias of $\hat{\sigma}$ exactly. However, as pointed out in the last section, we may approximate the conditional bias of $\hat{\sigma}$ by

$$E\left(\frac{\partial \ln L^*}{\partial \sigma}\right)\Big/E\left(-\frac{\partial^2 \ln L^*}{\partial \sigma^2}\right).$$

In addition, we derive from Eqs. (6.4.10) and (6.4.11) that

$$E\left(-\frac{\partial^2 \ln L^*}{\partial \mu^2}\right)=m\pi^2/(3\sigma^2), \tag{6.4.14}$$

$$E\left(-\frac{\partial^2 \ln L^*}{\partial \mu\,\partial \sigma}\right)=m\pi V_1/(\sqrt{3}\sigma^2), \tag{6.4.15}$$

and

$$E\left(-\frac{\partial^2 \ln L^*}{\partial \sigma^2}\right)=mV_2/\sigma^2, \tag{6.4.16}$$

where

$$V_1=\frac{2}{m}\left\{2\sum_{i=r+1}^{n-s}\beta_i E(Z_{i:n})+r\beta_{r+1}E(Z_{r+1:n})+s\beta_{n-s}E(Z_{n-s:n})\right\}-C \tag{6.4.17}$$

and

$$\begin{aligned}V_2=\frac{1}{m}\Bigg[&3\left\{2\sum_{i=r+1}^{n-s}\beta_i E(Z_{i:n}^2)+r\beta_{r+1}E(Z_{r+1:n}^2)+s\beta_{n-s}E(Z_{n-s:n}^2)\right\}\\&+2\left\{\sum_{i=r+1}^{n-s}(2\alpha_i-1)E(Z_{i:n})-r(1-\alpha_{r+1})E(Z_{r+1:n})\right.\\&\left.+s\alpha_{n-s}E(Z_{n-s:n})\right\}-A\Bigg].\end{aligned} \tag{6.4.18}$$

From these expressions we obtain

$$\operatorname{Var}(\hat{\mu})=\frac{3\sigma^2}{m\pi^2}\left\{\frac{V_2}{V_2-V_1^2}\right\}, \tag{6.4.19}$$

$$\operatorname{Var}(\hat{\sigma})=\frac{\sigma^2}{m}\left\{\frac{1}{V_2-V_1^2}\right\}, \tag{6.4.20}$$

and

$$\text{Cov}(\hat{\mu}, \hat{\sigma}) = -\frac{\sqrt{3}\sigma^2}{m\pi}\left\{\frac{V_1}{V_2 - V_1^2}\right\}. \tag{6.4.21}$$

In the case of symmetric censoring (i.e., $r = s$), we have noted already that $C = 0$. Further, since $\beta_i = \beta_{n-i+1}$ and $E(Z_{n-i+1:n}) = -E(Z_{i:n})$, we have from (6.4.17) that $V_1 = 0$ and hence, in this case, we find the estimators $\hat{\mu}$ and $\hat{\sigma}$ to be uncorrelated.

By making use of all the above formulas, Balakrishnan (1990b) has computed the values of

(1) {Exact conditional bias of $\hat{\mu}$}/σ,
(2) {Approximate conditional bias of $\hat{\sigma}$}/σ,
(3) {Approximate variance of $\hat{\mu}$}/σ^2,
(4) {Approximate variance of $\hat{\sigma}$}/σ^2,
(5) {Approximate covariance of $\hat{\mu}$ and $\hat{\sigma}$}/σ^2,
(6) {Approximate conditional variance of $\hat{\mu}$}/σ^2, and
(7) {Approximate conditional variance of $\hat{\sigma}$}/σ^2,

for various sample sizes and over various choices of censoring. In Table 6.4.1, we present these values for sample size $n = 10$ and 20 and $0 \le r \le s \le 4$. By comparing these values with the corresponding values of the MLEs (based on Monte Carlo simulations) given by Harter and Moore (1967) and those of the BLUEs given by Gupta *et al.* (1967), we find the AMLEs $\hat{\mu}$ and $\hat{\sigma}$ to be both remarkably efficient. In addition, by comparing the values in Table 6.4.1 with the corresponding values of the MMLEs given by Tiku (1968), we observe that the AMLEs $\hat{\mu}$ and $\hat{\sigma}$ are jointly more efficient than his estimators.

We have also simulated (based on 10,000 Monte Carlo runs) the values of

(8) {Bias of $\hat{\mu}$}/σ,
(9) {Bias of $\hat{\sigma}$}/σ,
(10) {Variance of $\hat{\mu}$}/σ^2,
(11) {Variance of $\hat{\sigma}$}/σ^2, and
(12) {Covariance of $\hat{\mu}$ and $\hat{\sigma}$}/σ^2,

for various sample sizes and over various choices of censoring. In Table 6.4.2, we present these simulated values for sample size $n = 10$ and 20. A comparison of these simulated values in Table 6.4.2 with the corresponding entries in Table 6.4.1 reveals that the formulas (6.4.19)–(6.4.21) yield very good approximation to the variances and covariance of the AMLEs $\hat{\mu}$ and $\hat{\sigma}$.

TABLE 6.4.1
Bias, Variances, and Covariance of Estimators $\hat{\mu}$ and $\hat{\sigma}$ for the Logistic Distribution

n	r	s	(1)	(2)	(3)	(4)	(5)	(6)	(7)
10	0	0	0.0000	0.0623	0.08359	0.04864	0.00000	0.08359	0.04864
	0	1	−0.0084	0.0560	0.08406	0.05578	0.00227	0.08397	0.05572
	0	2	−0.0164	0.0542	0.08579	0.06434	0.00647	0.08514	0.06385
	0	3	−0.0247	0.0550	0.09003	0.07522	0.01362	0.08756	0.07316
	0	4	−0.0340	0.0580	0.09917	0.08946	0.02541	0.09195	0.08295
	1	1	0.0000	0.0475	0.08436	0.06522	0.00000	0.08436	0.06522
	1	2	−0.0079	0.0439	0.08581	0.07688	0.00458	0.08554	0.07664
	1	3	−0.0160	0.0430	0.08978	0.09229	0.01285	0.08799	0.09045
	1	4	−0.0249	0.0448	0.09898	0.11343	0.02730	0.09241	0.10590
	2	2	0.0000	0.0388	0.08674	0.09290	0.00000	0.08674	0.09290
	2	3	−0.0080	0.0366	0.08999	0.11492	0.00909	0.08927	0.11400
	2	4	−0.0166	0.0374	0.09853	0.14670	0.02626	0.09383	0.13970
	3	3	0.0000	0.0330	0.09195	0.14753	0.00000	0.09195	0.14753
	3	4	−0.0085	0.0330	0.09872	0.19747	0.01954	0.09679	0.19360
	4	4	0.0000	0.0331	0.10217	0.28152	0.00000	0.10217	0.28152
20	0	0	0.0000	0.0353	0.04352	0.02848	0.00000	0.04352	0.02848
	0	1	−0.0023	0.0323	0.04355	0.03024	0.00032	0.04355	0.03024
	0	2	−0.0044	0.0309	0.04366	0.03217	0.00085	0.04364	0.03215
	0	3	−0.0065	0.0301	0.04388	0.03438	0.00162	0.04380	0.03432
	0	4	−0.0086	0.0298	0.04429	0.03692	0.00270	0.04409	0.03676
	1	1	0.0000	0.0289	0.04358	0.03222	0.00000	0.04358	0.03222
	1	2	−0.0021	0.0271	0.04367	0.03442	0.00054	0.04366	0.03441
	1	3	−0.0042	0.0260	0.04388	0.03695	0.00134	0.04383	0.03691
	1	4	−0.0063	0.0254	0.04428	0.03988	0.00249	0.04412	0.03974
	2	2	0.0000	0.0251	0.04375	0.03693	0.00000	0.04375	0.03693
	2	3	−0.0021	0.0237	0.04394	0.03983	0.00083	0.04392	0.03981
	2	4	−0.0041	0.0228	0.04431	0.04322	0.00202	0.04422	0.04313
	3	3	0.0000	0.0222	0.04409	0.04319	0.00000	0.04409	0.04319
	3	4	−0.0020	0.0211	0.04442	0.04715	0.00122	0.04439	0.04712
	4	4	0.0000	0.0197	0.04468	0.05183	0.00000	0.04468	0.05183

In addition, for the symmetrically censored sample case (i.e., $r = s$), we present in Table 6.4.3 the values of {Approximate variance of $\hat{\mu}$}/σ^2 computed from (6.4.19), {Exact variance of $\hat{\mu}$}/σ^2 computed from (6.4.12) by using the table of variances and covariances of logistic order statistics prepared by Gupta *et al.* (1967) and Balakrishnan and Malik (1990), {Variance of the BLUE of μ}/σ^2 taken from the table of Gupta *et al.* (1967), and {Mean square error of the MLE of μ}/σ^2 taken from the table of Harter (1970b), for sample size $n = 10$ and 20 and $r = s = 1(1)4$. It is quite apparent

TABLE 6.4.2
Simulated Values of Bias, Variances, and Covariance of $\hat{\mu}$ and $\hat{\sigma}$ for the Logistic Distribution

n	r	s	(8)	(9)	(10)	(11)	(12)
10	0	0	0.00235	−0.02152	0.09403	0.07517	0.00058
	0	1	−0.00733	−0.04210	0.09406	0.07984	0.00083
	0	2	−0.01781	−0.05837	0.09515	0.09012	0.00319
	0	3	−0.03193	−0.07685	0.09823	0.10363	0.01003
	1	1	0.00199	−0.06649	0.09401	0.08493	0.00095
	1	2	−0.00933	−0.08854	0.09515	0.09652	0.00421
	1	3	−0.02556	−0.11523	0.09830	0.11298	0.01100
	2	2	0.00279	−0.12062	0.09642	0.10869	0.00075
	2	3	−0.01487	−0.15894	0.09914	0.13022	0.00789
	3	3	0.00380	−0.22025	0.10189	0.15543	−0.00005
20	0	0	−0.00056	−0.00804	0.04652	0.03551	−0.00042
	0	1	−0.00290	−0.01377	0.04658	0.03699	−0.00018
	0	2	−0.00511	−0.01768	0.04665	0.03886	0.00015
	0	3	−0.00762	−0.02184	0.04686	0.04125	0.00079
	0	4	−0.01035	−0.02585	0.04706	0.04399	0.00153
	1	1	−0.00063	−0.01964	0.04655	0.03819	−0.00008
	1	2	−0.00289	−0.02409	0.04662	0.04015	0.00026
	1	3	−0.00549	−0.02888	0.04684	0.04260	0.00093
	1	4	−0.00837	−0.03362	0.04705	0.04550	0.00169
	2	2	−0.00049	−0.02972	0.04667	0.04200	0.00004
	2	3	−0.00313	−0.03120	0.04688	0.04468	0.00072
	2	4	−0.00612	−0.04072	0.04708	0.04798	0.00152
	3	3	−0.00039	−0.04192	0.04706	0.04765	0.00005
	3	4	−0.00345	−0.04836	0.04725	0.05134	0.00085
	4	4	−0.00023	−0.05665	0.04738	0.05570	0.00012

TABLE 6.4.3
Variances of the Estimators of μ for the Logistic Distribution When $r = s$

	$n = 10$				$n = 20$			
r:	1	2	3	4	1	2	3	4
Approximate $\mathrm{Var}(\hat{\mu})/\sigma^2$	0.084	0.087	0.092	0.102	0.044	0.044	0.044	0.045
Exact $\mathrm{Var}(\hat{\mu})/\sigma^2$	0.094	0.096	0.101	0.112	0.046	0.046	0.047	0.047
Var(BLUE of $\mu)/\sigma^2$	0.094	0.096	0.101	0.112	0.046	0.046	0.047	0.047
MSE(MLE of $\mu)/\sigma^2$	0.093	0.096	0.099	0.108	—	0.047	—	0.048

from Table 6.4.3 that the AMLE $\hat{\mu}$ of μ is almost as efficient as the MLE of μ and just as efficient as the BLUE of μ even for a sample of size as small as 10.

Example 6.4.1. Sarhan and Greenberg (1962) have presented data resulting from an experiment in which students were learning to measure strontium-90 concentrations in samples of milk. The test substance was supposed to contain 9.22 picocuries per liter. The measurements, each involving readings and calculations, were made, but, since the measurement error was known to be relatively larger at the extremes, especially the upper one, a decision was made to censor the two smallest and the three largest observations, leaving the following Type-II censored sample (see Example 4.4.5):

$$8.2, 8.4, 9.1, 9.8, 9.9.$$

In Example 4.4.5, by assuming a logistic distribution for the above censored sample, we obtained the BLUEs of μ and σ as

$$\mu^* = 9.3032 \qquad \text{and} \qquad \sigma^* = 1.8758$$

and their standard errors as

$$\text{S.E.}(\mu^*) = 0.5908 \qquad \text{and} \qquad \text{S.E.}(\sigma^*) = 0.8033.$$

The same Type-II censored sample has been used by Harter and Moore (1967) to calculate the MLEs of μ and σ as

$$\tilde{\mu} = 9.2718 \qquad \text{and} \qquad \tilde{\sigma} = 1.5678$$

and the standard errors of these estimates as

$$\text{S.E.}(\tilde{\mu}) = 1.5678(0.098)^{1/2} = 0.4908$$

and

$$\text{S.E.}(\tilde{\sigma}) = 1.5678(0.130)^{1/2} = 0.5653.$$

For this example, we have $n = 10$, $r = 2$, $s = 3$,

i	p_i	q_i	α_i	β_i
3	3/11	8/11	0.467272	0.198347
4	4/11	7/11	0.493134	0.231405
5	5/11	6/11	0.499749	0.247934
6	6/11	5/11	0.500251	0.247934
7	7/11	4/11	0.506866	0.231405

$A = 5$, $m = 3.404959$, $C = -0.114447$, $B = 9.168932$, $D = -2.228374$, and $E = 1.836713$. Assuming now that the censored sample has come from the logistic distribution in (6.4.2) and using the above computed values in Eqs. (6.4.12) and (6.4.13), we obtain the AMLEs of μ and σ to be

$$\hat{\mu} = 9.2683 \qquad \text{and} \qquad \hat{\sigma} = 1.5755.$$

Further, the standard errors of these estimates are computed from Table 6.4.2 to be

$$\text{S.E.}(\hat{\mu}) = 1.5755(0.09914)^{1/2} = 0.4961$$

and

$$\text{S.E.}(\hat{\sigma}) = 1.5755(0.13022)^{1/2} = 0.5685.$$

We find these values to be very close to those provided by the ML method.

Example 6.4.2. Davis (1952) has given lifetimes in hours of 417 40-watt incandescent lamps taken from 42 weekly forced-life test samples. As in Example 4.8.1, let us consider the first 20 observations from this data and for the sake of illustration assume that 50% censoring had occurred in this data and that only the following first 10 observations were available:

$$785, 855, 905, 918, 919, 920, 929, 936, 948, 950.$$

By taking $n = 20$, $r = 0$, and $s = 10$, and by using the tables of Balakrishnan (1990f), we obtain the BLUEs of μ and σ based on the above censored data as

$$\begin{aligned}\mu^* &= -0.04336(785) - 0.02968(855) - 0.01300(905) + 0.00454(918)\\ &\quad + 0.02200(919) + 0.03882(920) + 0.05460(929) + 0.06901(936)\\ &\quad + 0.08176(948) + 0.81530(950)\\ &= 956.28\end{aligned}$$

and

$$\begin{aligned}\sigma^* &= -0.16913(785) - 0.17466(855) - 0.16574(905) - 0.14860(918)\\ &\quad - 0.12598(919) - 0.09961(920) - 0.07070(929) - 0.04027(936)\\ &\quad - 0.00916(948) + 1.00385(950)\\ &= 65.67,\end{aligned}$$

and the standard errors of these estimates as

$$\text{S.E.}(\mu^*) = 65.67(0.06168)^{1/2} = 16.309$$

and

$$\text{S.E.}(\sigma^*) = 65.67(0.08679)^{1/2} = 19.346.$$

Based on the above censored data, Bain *et al.* (1990) have obtained the MLEs of μ and σ as

$$\tilde{\mu} = 948 \qquad \text{and} \qquad \tilde{\sigma} = 59$$

and their standard errors as

$$\text{S.E.}(\tilde{\mu}) = 14.452 \qquad \text{and} \qquad \text{S.E.}(\tilde{\sigma}) = 16.372.$$

For this example, we have:

i	p_i	q_i	α_i	β_i
1	1/21	20/21	0.183479	0.045351
2	2/21	19/21	0.289227	0.086168
3	3/21	18/21	0.362256	0.122449
4	4/21	17/21	0.413584	0.154195
5	5/21	16/21	0.449098	0.181406
6	6/21	15/21	0.472713	0.204082
7	7/21	14/21	0.487366	0.222222
8	8/21	13/21	0.495449	0.235828
9	9/21	12/21	0.499024	0.244898
10	10/21	11/21	0.499964	0.249433

$A = 10$, $m = 5.986394$, $C = -0.551912$, $B = 934.525771$, $D = 219.005212$, and $E = 4409.36436$. Assuming now that the censored sample has come from the logistic distribution in (6.4.2) and using the above computed values in Eqs. (6.4.12) and (6.4.13), we obtain the AMLEs of μ and σ as

$$\hat{\mu} = 953.6398 \qquad \text{and} \qquad \hat{\sigma} = 62.8163.$$

The standard errors of the above estimates are computed to be

$$\text{S.E.}(\hat{\mu}) = 62.8163(0.059640)^{1/2} = 15.341$$

and

$$\text{S.E.}(\hat{\sigma}) = 62.8163(0.077270)^{1/2} = 17.461.$$

6.5. Estimation for the Extreme Value Distribution

Let

$$X_{r+1:n} \leq X_{r+2:n} \leq \cdots \leq X_{n-s:n} \tag{6.5.1}$$

be a Type-II censored sample from the extreme value population with pdf

$$f(x; \mu, \sigma) = \frac{1}{\sigma} e^{(x-\mu)/\sigma} \exp\{-e^{(x-\mu)/\sigma}\}, \qquad -\infty < x < \infty, \tag{6.5.2}$$

and cdf

$$F(x; \mu, \sigma) = 1 - \exp\{-e^{(x-\mu)/\sigma}\}, \qquad -\infty < x < \infty. \tag{6.5.3}$$

Order statistics from the above population and their moments have been studied by Lieblein and Zelen (1956), Lieblein and Salzer (1957), and White (1969). The last two references give the expected values and variances of order statistics, respectively, for sample sizes up to 20, and the first reference gives the covariances for sample sizes up to six. The BLUEs of μ and σ have been worked out by Lieblein and Zelen (1956). Mann (1967a) has derived the best linear invariant estimators (BLIEs) of μ and σ, and the tables necessary for the calculation of these BLIEs for sample sizes up to 25 have been prepared by Mann (1967a, b) and Mann *et al.* (1974, p. 194–207). D'agostino (1971) has considered some approximations to the BLUEs and the BLIEs of μ and σ. Reference may be made to Mann and Fertig (1977) and Engelhardt and Bain (1977) for a survey of several such simplified linear estimators that have been proposed in literature that require fewer tables than the BLUEs and the BLIEs. Harter and Moore (1968) have presented an iterative procedure for obtaining the MLEs of μ and σ based on complete and Type-II censored samples. By using Monte Carlo simulations, Harter and Moore have also investigated the bias, variances and covariance of the MLEs. Recently, Balakrishnan *et al.* (1990a) have tabulated the means, variances, and covariances of order statistics for sample sizes up to 30. By using these values, Balakrishnan *et al.* (1990b) have also worked out the BLUEs of μ and σ for all doubly censored samples for sample sizes up to 20 and for all right-censored samples for $21 \le n \le 30$. The estimation of μ and σ based on optimally selected order statistics (to be discussed in detail in the next chapter) has been considered by Chan and Kabir (1969) and Chan and Mead (1971a).

In this section we present the derivation of the AMLEs of μ and σ based on the doubly censored sample in (6.5.1) due to Balakrishnan and Varadan (1990). The likelihood function based on the censored sample in (6.5.1) is given by

$$L = \frac{n!}{r!\,s!}\{F(X_{r+1:n}; \mu, \sigma)\}^r\{1 - F(X_{n-s:n}; \mu, \sigma)\}^s \prod_{i=r+1}^{n-s} f(X_{i:n}; \mu, \sigma),$$

which, by denoting $Z_{i:n} = (X_{i:n} - \mu)/\sigma$, can be written as

$$L = \frac{n!}{r!\,s!}\sigma^{-A}\{F(Z_{r+1:n})\}^r\{1 - F(Z_{n-s:n})\}^s \prod_{i=r+1}^{n-s} f(Z_{i:n}), \tag{6.5.4}$$

where $A = n - r - s$, $f(z) = e^z \exp(-e^z)$, and $F(z) = 1 - \exp(-e^z)$. From

(6.5.4) we obtain the likelihood equations for μ and σ as

$$\frac{\partial \ln L}{\partial \mu} = -\frac{1}{\sigma}\left[r\frac{f(Z_{r+1:n})}{F(Z_{r+1:n})} - s\frac{f(Z_{n-s:n})}{1-F(Z_{n-s:n})} + \sum_{i=r+1}^{n-s} \frac{f'(Z_{i:n})}{f(Z_{i:n})} \right]$$
$$= 0, \tag{6.5.5}$$

and

$$\frac{\partial \ln L}{\partial \sigma} = -\frac{1}{\sigma}\left[A + rZ_{r+1:n}\frac{f(Z_{r+1:n})}{F(Z_{r+1:n})} - sZ_{n-s:n}\frac{f(Z_{n-s:n})}{1-F(Z_{n-s:n})} \right.$$
$$\left. + \sum_{i=r+1}^{n-s} Z_{i:n}\frac{f'(Z_{i:n})}{f(Z_{i:n})} \right] = 0. \tag{6.5.6}$$

The likelihood equations (6.5.5) and (6.5.6) do not admit explicit solutions. But, we may expand the functions $f(Z_{r+1:n})/F(Z_{r+1:n}), f'(Z_{i:n})/f(Z_{i:n})$, and $f(Z_{n-s:n})/\{1-F(Z_{n-s:n})\}$ in Taylor series around the points $F^{-1}(p_{r+1}) = \ln(-\ln q_{r+1})$, $F^{-1}(p_i) = \ln(-\ln q_i)$, and $F^{-1}(p_{n-s}) = \ln(-\ln q_{n-s})$, respectively, and then approximate them by

$$\frac{f(Z_{r+1:n})}{F(Z_{r+1:n})} \simeq \gamma - \delta Z_{r+1:n}, \tag{6.5.7}$$

$$\frac{f'(Z_{i:n})}{f(Z_{i:n})} \simeq \alpha_i - \beta_i Z_{i:n}, \tag{6.5.8}$$

and

$$\frac{f(Z_{n-s:n})}{1-F(Z_{n-s:n})} \simeq 1 - \alpha_{n-s} + \beta_{n-s}Z_{n-s:n}, \tag{6.5.9}$$

where

$$p_i = i/(n+1), \qquad q_i = 1 - p_i,$$

$$\gamma = -\frac{q_{r+1}}{p_{r+1}} \ln q_{r+1}\{1 - \ln(-\ln q_{r+1})\}$$
$$+ \frac{q_{r+1}}{p_{r+1}^2} (\ln q_{r+1})^2 \ln(-\ln q_{r+1}), \tag{6.5.10}$$

$$\delta = \frac{q_{r+1}}{p_{r+1}} \ln q_{r+1}\left\{1 + \frac{1}{p_{r+1}} \ln q_{r+1}\right\}, \tag{6.5.11}$$

$$\alpha_i = 1 + \ln q_i\{1 - \ln(-\ln q_i)\}, \tag{6.5.12}$$

and

$$\beta_i = -\ln q_i. \tag{6.5.13}$$

By making use of the linear approximations in (6.5.7) through (6.5.9), we may approximate the likelihood equations (6.5.5) and (6.5.6) by

$$\frac{\partial \ln L}{\partial \mu} \simeq \frac{\partial \ln L^*}{\partial \mu}$$

$$= -\frac{1}{\sigma}\left[r\gamma - s(1-\alpha_{n-s}) + \sum_{i=r+1}^{n-s} \alpha_i - r\delta Z_{r+1:n} - s\beta_{n-s}Z_{n-s:n} - \sum_{i=r+1}^{n-s} \beta_i Z_{i:n}\right] = 0 \tag{6.5.14}$$

and

$$\frac{\partial \ln L}{\partial \sigma} \simeq \frac{\partial \ln L^*}{\partial \sigma}$$

$$= -\frac{1}{\sigma}\left[A + r\gamma Z_{r+1:n} - s(1-\alpha_{n-s})Z_{n-s:n} + \sum_{i=r+1}^{n-s} \alpha_i Z_{i:n} - r\delta Z_{r+1:n}^2 - s\beta_{n-s}Z_{n-s:n}^2 - \sum_{i=r+1}^{n-s} \beta_i Z_{i:n}^2\right] = 0. \tag{6.5.15}$$

Upon solving Eqs. (6.5.14) and (6.5.15), we derive the AMLEs of μ and σ as

$$\hat{\mu} = B - \hat{\sigma}C \tag{6.5.16}$$

and

$$\hat{\sigma} = \{-D + (D^2 + 4AE)^{1/2}\}/2A, \tag{6.5.17}$$

where

$$m = r\delta + s\beta_{n-s} + \sum_{i=r+1}^{n-s} \beta_i,$$

$$B = \left\{r\delta X_{r+1:n} + s\beta_{n-s}X_{n-s:n} + \sum_{i=r+1}^{n-s} \beta_i X_{i:n}\right\} \Big/ m,$$

$$C = \left\{r\gamma - s(1-\alpha_{n-s}) + \sum_{i=r+1}^{n-s} \alpha_i\right\} \Big/ m,$$

$$D = r\gamma(X_{r+1:n} - B) - s(1-\alpha_{n-s})(X_{n-s:n} - B) + \sum_{i=r+1}^{n-s} \alpha_i(X_{i:n} - B)$$

$$= r\gamma X_{r+1:n} - s(1-\alpha_{n-s})X_{n-s:n} + \sum_{i=r+1}^{n-s} \alpha_i X_{i:n} - mBC,$$

and

$$E = r\delta(X_{r+1:n} - B)^2 + s\beta_{n-s}(X_{n-s:n} - B)^2 + \sum_{i=r+1}^{n-s} \beta_i(X_{i:n} - B)^2$$

$$= r\delta X_{r+1:n}^2 + s\beta_{n-s}X_{n-s:n}^2 + \sum_{i=r+1}^{n-s} \beta_i X_{i:n}^2 - mB^2.$$

It is important to mention here that upon solving Eq. (6.5.15), we obtain a quadratic equation in σ that has two roots; however, one of them drops out, since we observe from (6.5.11) and (6.5.13) that $\delta > 0$ and $\beta_i > 0$ and, hence, $E > 0$.

The conditional bias of $\hat{\mu}$ can be computed exactly from (6.5.16) by using the expected values of order statistics tabulated by Lieblein and Salzer (1957) and White (1969). But it is not possible to determine the exact conditional bias of $\hat{\sigma}$. However, as mentioned in the last section, it may be approximated by $E(\partial \ln L^*/\partial\sigma)/E(-\partial^2 \ln L^*/\partial\sigma^2)$. Furthermore, we derive from Eqs. (6.5.14) and (6.5.15) that

$$E\left(-\frac{\partial^2 \ln L^*}{\partial\mu^2}\right) = m/\sigma^2, \tag{6.5.18}$$

$$E\left(-\frac{\partial^2 \ln L^*}{\partial\mu\,\partial\sigma}\right) = mV_1/\sigma^2, \tag{6.5.19}$$

and

$$E\left(-\frac{\partial^2 \ln L^*}{\partial\sigma^2}\right) = mV_2/\sigma^2, \tag{6.5.20}$$

where

$$V_1 = \frac{2}{m}\left\{r\delta E(Z_{r+1:n}) + s\beta_{n-s}E(Z_{n-s:n}) + \sum_{i=r+1}^{n-s} \beta_i E(Z_{i:n})\right\} - C \tag{6.5.21}$$

and

$$V_2 = \frac{1}{m}\Bigg[3\left\{r\delta E(Z_{r+1:n}^2) + s\beta_{n-s}E(Z_{n-s:n}^2) + \sum_{i=r+1}^{n-s} \beta_i E(Z_{i:n}^2)\right\}$$
$$-2\Bigg\{r\gamma E(Z_{r+1:n}) - s(1-\alpha_{n-s})E(Z_{n-s:n})$$
$$+ \sum_{i=r+1}^{n-s} \alpha_i E(Z_{i:n})\Bigg\} - A\Bigg]. \tag{6.5.22}$$

From the above expressions we obtain

$$\mathrm{Var}(\hat{\mu}) = \frac{\sigma^2}{m}\left\{\frac{V_2}{V_2 - V_1^2}\right\}, \tag{6.5.23}$$

$$\mathrm{Var}(\hat{\sigma}) = \frac{\sigma^2}{m}\left\{\frac{1}{V_2 - V_1^2}\right\}, \tag{6.5.24}$$

and

$$\mathrm{Cov}(\hat{\mu}, \hat{\sigma}) = -\frac{\sigma^2}{m}\left\{\frac{V_1}{V_2 - V_1^2}\right\}. \tag{6.5.25}$$

By using (6.5.16) and (6.5.17), Balakrishnan and Varadan (1990) have simulated (based on 10,000 Monte Carlo runs) the values of

(1) $\{\text{Bias of } \hat{\mu}\}/\sigma$,
(2) $\{\text{Bias of } \hat{\sigma}\}/\sigma$,
(3) $\{\text{Variance of } \hat{\mu}\}/\sigma^2$,
(4) $\{\text{Variance of } \hat{\sigma}\}/\sigma^2$, and
(5) $\{\text{Covariance of } \hat{\mu} \text{ and } \hat{\sigma}\}/\sigma^2$.

Furthermore, by making use of the formulas in (6.5.23) through (6.5.25), Balakrishnan and Varadan (1990) have also computed the values of

(6) $\{\text{Approximate variance of } \hat{\mu}\}/\sigma^2$,
(7) $\{\text{Approximate variance of } \hat{\sigma}\}/\sigma^2$,
(8) $\{\text{Approximate covariance of } \hat{\mu} \text{ and } \hat{\sigma}\}/\sigma^2$,
(9) $\{\text{Approximate conditional variance of } \hat{\mu}\}/\sigma^2$, and
(10) $\{\text{Approximate conditional variance of } \hat{\sigma}\}/\sigma^2$.

In Tables 6.5.1 and 6.5.2, we have presented these values for the cases in which $n = 10$, r, $s = 0(1)3$ and $n = 20$, r, $s = 0(1)4$, respectively. From these two tables we observe that the approximate expressions for the variances of $\hat{\mu}$ and $\hat{\sigma}$ given in (6.5.23) and (6.5.24) are both very close approximations, while that of the covariance of $\hat{\mu}$ and $\hat{\sigma}$ given in (6.5.25) is only a reasonable approximation for sample size up to 20. However, computations carried out for larger sample sizes reveal that the approximate expression of the covariance of $\hat{\mu}$ and $\hat{\sigma}$ in (6.5.25) also improves for large sample sizes. From Tables 6.5.1 and 6.5.2, we also observe that the bias of the estimators $\hat{\mu}$ and $\hat{\sigma}$ both converge to zero, for fixed r and s, as n increases. This is to be expected, as the estimators $\hat{\mu}$ and $\hat{\sigma}$ in (6.5.16) and (6.5.17) are approximate solutions to the likelihood equations (6.5.5) and (6.5.6), respectively,

TABLE 6.5.1
Values of Bias, Variances, and Covariance of $\hat{\mu}$ and $\hat{\sigma}$ for the Extreme Value Distribution When $n = 10$

r	s	(1)	(2)	(3)	(4)	(5)	(6)	(7)	(8)	(9)	(10)
0	0	−0.0845	−0.0661	0.1159	0.0628	−0.0257	0.1132	0.0572	−0.0054	0.1127	0.0569
0	1	−0.0892	−0.0725	0.1209	0.0767	−0.0174	0.1224	0.0662	0.0034	0.1212	0.0662
0	2	−0.1029	−0.0848	0.1325	0.0904	−0.0049	0.1393	0.0772	0.0168	0.1357	0.0751
0	3	−0.1245	−0.0999	0.1556	0.1062	0.0140	0.1683	0.0905	0.0364	0.1537	0.0826
1	0	−0.0741	−0.0854	0.1169	0.0660	−0.0275	0.1135	0.0642	−0.0071	0.1127	0.0638
1	1	−0.0807	−0.0959	0.1215	0.0818	−0.0191	0.1224	0.0757	0.0026	0.1223	0.0756
1	2	−0.0976	−0.1139	0.1327	0.0980	−0.0059	0.1393	0.0899	0.0179	0.1357	0.0876
1	3	−0.1247	−0.1367	0.1557	0.1168	0.0148	0.1695	0.1079	0.0412	0.1537	0.0979
2	0	−0.0637	−0.1021	0.1187	0.0719	−0.0307	0.1141	0.0730	−0.0098	0.1128	0.0721
2	1	−0.0715	−0.1170	0.1227	0.0904	−0.0223	0.1224	0.0877	0.0008	0.1224	0.0877
2	2	−0.0907	−0.1417	0.1332	0.1103	−0.0081	0.1319	0.1066	0.0182	0.1359	0.1042
2	3	−0.1228	−0.1746	0.1558	0.1341	0.0149	0.1701	0.1315	0.0461	0.1539	0.1191
3	0	−0.0498	−0.1224	0.1216	0.0784	−0.0350	0.1156	0.0845	−0.0142	0.1132	0.0828
3	1	−0.0584	−0.1439	0.1248	0.1001	−0.0268	0.1229	0.1040	−0.0029	0.1228	0.1039
3	2	−0.0799	−0.1790	0.1342	0.1259	−0.0118	0.1386	0.1300	0.0170	0.1364	0.1279
3	3	−0.1180	−0.2291	0.1560	0.1567	−0.0141	0.1700	0.1662	0.0507	0.1546	0.1511

TABLE 6.5.2

Values of Bias, Variances, and Covariance of $\hat{\mu}$ and $\hat{\sigma}$ for the Extreme Value Distribution When $n = 20$

r	s	(1)	(2)	(3)	(4)	(5)	(6)	(7)	(8)	(9)	(10)
0	0	−0.0424	−0.0334	0.0564	0.0311	−0.0128	0.0555	0.0298	−0.0070	0.0539	0.0289
0	1	−0.0417	−0.0332	0.0574	0.0353	−0.0107	0.0568	0.0325	−0.0053	0.0560	0.0320
0	2	−0.0430	−0.0349	0.0587	0.0389	−0.0085	0.0589	0.0354	−0.0029	0.0587	0.0352
0	3	−0.0462	−0.0383	0.0609	0.0427	−0.0056	0.0618	0.0385	0.0000	0.0618	0.0385
0	4	−0.0494	−0.0409	0.0636	0.0465	0.0025	0.0657	0.0418	0.0036	0.0654	0.0416
1	0	−0.0398	−0.0383	0.0566	0.0319	−0.0133	0.0557	0.0313	−0.0075	0.0539	0.0303
1	1	−0.0393	−0.0385	0.0575	0.0363	−0.0111	0.0569	0.0342	−0.0057	0.0560	0.0336
1	2	−0.0409	−0.0407	0.0588	0.0402	−0.0089	0.0589	0.0374	−0.0032	0.0587	0.0372
1	3	−0.0445	−0.0448	0.0610	0.0443	0.0059	0.0618	0.0409	−0.0001	0.0618	0.0409
1	4	−0.0482	−0.0482	0.0636	0.0485	−0.0026	0.0657	0.0447	0.0037	0.0654	0.0445
2	0	−0.0372	−0.0429	0.0568	0.0330	−0.0138	0.0559	0.0329	−0.0080	0.0539	0.0317
2	1	−0.0369	−0.0436	0.0577	0.0379	−0.0117	0.0571	0.0361	−0.0062	0.0560	0.0355
2	2	−0.0387	−0.0464	0.0590	0.0421	−0.0094	0.0590	0.0397	−0.0037	0.0587	0.0395
2	3	−0.0426	−0.0512	0.0611	0.0465	−0.0063	0.0618	0.0436	−0.0004	0.0618	0.0436
2	4	−0.0467	−0.0553	0.0636	0.0514	−0.0029	0.0657	0.0480	0.0036	0.0654	0.0478
3	0	−0.0346	−0.0470	0.0572	0.0345	−0.0145	0.0561	0.0347	0.0087	0.0539	0.0334
3	1	−0.0345	−0.0482	0.0580	0.0398	−0.0124	0.0572	0.0384	0.0069	0.0560	0.0375
3	2	−0.0365	−0.0515	0.0591	0.0445	−0.0100	0.0591	0.0424	−0.0042	0.0587	0.0421
3	3	−0.0407	−0.0571	0.0612	0.0494	−0.0069	0.0618	0.0468	−0.0008	0.0618	0.0468
3	4	−0.0452	−0.0621	0.0636	0.0549	−0.0032	0.0657	0.0518	0.0035	0.0654	0.0516
4	0	−0.0320	−0.0510	0.0576	0.0361	−0.0153	0.0565	0.0369	−0.0096	0.0540	0.0352
4	1	−0.0320	−0.0526	0.0583	0.0420	−0.0132	0.0575	0.0409	−0.0077	0.0561	0.0399
4	2	−0.0342	−0.0566	0.0593	0.0473	−0.0108	0.0593	0.0455	−0.0050	0.0587	0.0451
4	3	−0.0385	−0.0630	0.0613	0.0526	−0.0076	0.0619	0.0505	−0.0014	0.0619	0.0505
4	4	−0.0433	−0.0688	0.0637	0.0589	−0.0038	0.0656	0.0563	0.0032	0.0655	0.0562

and hence are asymptotically unbiased; see, for example, Kendall and Stuart (1973).

For the purpose of comparing these AMLEs with some other estimators, we have presented in Table 6.5.3 the values of

(11) {Bias of $\hat{\mu}$}/σ,
(12) {Mean square error of $\hat{\mu}$}/σ^2,
(13) {Bias of MLE $\tilde{\mu}$}/σ,
(14) {Mean square error of $\tilde{\mu}$}/σ^2,
(15) {Variance of BLUE μ^*}/σ^2,
(16) {Mean square error of BLIE μ^{**}}/σ^2,
(17) {Bias of $\hat{\sigma}$}/σ,
(18) {Mean square error of $\hat{\sigma}$}/σ^2,
(19) {Bias of MLE $\tilde{\sigma}$}/σ,
(20) {Mean square error of $\tilde{\sigma}$}/σ^2,
(21) {Variance of BLUE σ^*}/σ^2, and
(22) {Mean square error of BLIE σ^{**}}/σ^2.

These values are presented in Table 6.5.3 for $r=0$ and $s=0(1)3$ when $n=10$, and $r=0$ and $s=0(1)4$ when $n=20$. The values of (11), (12), (17) and (18) are based on 10,000 Monte Carlo runs and computed from the respective

TABLE 6.5.3
Comparison of the Bias and Mean Square Error of Various Estimators of μ and σ for the Extreme Value Distribution

	$n=10$				$n=20$				
$s=$	0	1	2	3	0	1	2	3	4
(11)	−0.085	−0.089	−0.103	−0.125	−0.042	−0.042	−0.043	−0.046	−0.049
(12)	0.123	0.129	0.143	0.171	0.058	0.059	0.061	0.063	0.066
(13)	−0.04	−0.05	−0.08	−0.11	−0.02	—	−0.02	—	−0.04
(14)	0.114	0.122	0.137	0.166	0.056	—	0.060	—	0.066
(15)	0.113	0.120	0.134	0.162	0.056	—	0.059	—	0.065
(16)	0.113	0.120	0.134	0.161	0.056	—	0.059	—	0.065
(17)	−0.066	−0.073	−0.085	−0.100	−0.033	−0.033	−0.035	−0.038	−0.041
(18)	0.067	0.082	0.098	0.116	0.032	0.036	0.040	0.044	0.048
(19)	−0.07	−0.08	−0.10	−0.12	−0.04	—	−0.04	—	−0.05
(20)	0.063	0.077	0.094	0.113	0.033	—	0.042	—	0.050
(21)	0.072	0.088	0.107	0.132	0.033	—	0.041	—	0.050
(22)	0.067	0.081	0.097	0.117	0.032	—	0.039	—	0.047

values in Tables 6.5.1 and 6.5.2, while the values reported in (13)-(16) and (19)-(22) have been taken from the tables of Harter and Moore (1968), Mann *et al.* (1974), Harter (1970b), and Balakrishnan *et al.* (1990b). As rightly pointed out by Balakrishnan and Varadan (1990), it is quite clear from Table 6.5.3 that the AMLEs are almost as efficient as the MLEs and just as efficient as the BLUEs and the BLIEs. Furthermore, it is important to emphasize here that the AMLEs of μ and σ developed in this section are explicit estimators (unlike the MLEs) and also do not need any special tables to be constructed using the means, variances, and covariances of order statistics (unlike the BLUEs and the BLIEs).

Example 6.5.1. Let us consider here the example of Mann and Fertig (1973), who give failure times of airplane components for a life test in which 13 components were placed on test, with the test terminating at the time of the 10th failure.

Failure times in hours were

$$0.22, 0.50, 0.88, 1.00, 1.32, 1.33, 1.54, 1.76, 2.50, 3.00.$$

By assuming that the above data arose from a Weibull distribution, in order to obtain estimates we transform the above data to extreme value form by taking the logarithms of the observations, which are as follows:

$$-1.541, -0.693, -0.128, 0.000, 0.278, 0.285, 0.432, 0.565, 0.916, 1.099.$$

Now by using the appropriate weights taken from Table 5.3 of Mann *et al.* (1974), we obtain the BLIEs of μ and σ as

$$\begin{aligned}\mu^{**} &= (-0.002927)(-1.541) + (0.005067)(-0.693) + (0.014356)(-0.128)\\ &\quad + (0.024891)(0.000) + (0.036816)(0.278) + (0.050389)(0.285)\\ &\quad + (0.065995)(0.432) + (0.084201)(0.565) + (0.105863)(0.916)\\ &\quad + (0.615348)(1.099)\\ &= 0.87308\end{aligned}$$

and

$$\begin{aligned}\sigma^{**} &= (-0.083170)(-1.541) + (-0.087085)(-0.693) + (-0.085792)(-0.128)\\ &\quad + (-0.080789)(0.000) + (-0.072325)(0.278) + (-0.060181)(0.285)\\ &\quad + (-0.043768)(0.432) + (-0.022048)(0.565) + (0.006715)(0.916)\\ &\quad + (0.528441)(1.099)\\ &= 0.71778.\end{aligned}$$

Also, by starting with the graphical estimate of 0.69 as an initial guess for σ, Lawless (1982) has applied Newton's method to solve the likelihood equations (6.5.5) and (6.5.6) iteratively in order to obtain the MLEs of μ and σ to be $\tilde{\mu} = 0.821$ and $\tilde{\sigma} = 0.706$.

In this case, we have

$$n = 13, \qquad r = 0, \qquad s = 3, \qquad A = 10,$$

$$\gamma = 3.28461, \qquad \delta = 0.03614,$$

i	p_i	q_i	α_i	β_i
1	1/14	13/14	0.73305	0.07411
2	2/14	12/14	0.55761	0.15415
3	3/14	11/14	0.41584	0.24116
4	4/14	10/14	0.29703	0.33647
5	5/14	9/14	0.19727	0.44183
6	6/14	8/14	0.11552	0.55962
7	7/14	7/14	0.05281	0.69315
8	8/14	6/14	0.01230	0.84730
9	9/14	5/14	0.00043	1.02962
10	10/14	4/14	0.02955	1.25276

$$m = 9.38846, \qquad C = -0.05325,$$

$$B = 0.77317, \qquad D = -4.23191,$$

and

$$E = 2.03734.$$

From (6.5.16) and (6.5.17), we then obtain the AMLEs of μ and σ as

$$\hat{\mu} = 0.81098 \qquad \text{and} \qquad \hat{\sigma} = 0.71010.$$

We observe that the estimates given by the AML method are very close to those provided by the ML method.

Example 6.5.2. Let us consider the following Type-II censored sample of size 10 (obtained from a sample of 20 by censoring the largest 10 observations) as given by Lawless (1982, p. 156):

$$-3.57, -2.55, -2.02, -1.66, -1.36, -1.15, -0.95, -0.77, -0.61, -0.45.$$

By assuming that the above censored data have come from an extreme value population with density function as in (6.5.2), Lawless (1982) has computed the maximum likelihood estimates of μ and σ to be -0.112 and 0.907, respectively; also, Lawless has used the tables of Mann *et al.* (1971) and computed the best linear invariant estimates of μ and σ to be -0.048 and 0.915, respectively.

In this case, we have

$$n = 20, \qquad r = 0, \qquad s = 10, \qquad A = 10,$$

$$m = 9.680292,$$

$$B = -0.6347905, \qquad C = -0.5722707,$$

$$D = -7.104816,$$

and

$$E = 1.829956.$$

From (6.5.16) and (6.5.17), we then obtain the AMLEs of μ and σ as

$$\hat{\mu} = -0.1133 \qquad \text{and} \qquad \hat{\sigma} = 0.9113.$$

Once again, we observe that the estimates given by the AML method are very close to those obtained by the ML method.

6.6. Estimation for the Type I Generalized Logistic Distribution

The approximate maximum likelihood estimation of the location and scale parameters of the logistic distribution has already been discussed in Section 4. In this section, we consider the Type I generalized logistic population with pdf

$$f(x;\mu,\sigma) = \frac{b}{\sigma}\frac{e^{-(x-\mu)/\sigma}}{\{1+e^{-(x-\mu)/\sigma}\}^{b+1}}, \qquad -\infty < x < \infty,\ b > 0, \tag{6.6.1}$$

and cdf

$$F(x;\mu,\sigma) = \{1+e^{-(x-\mu)/\sigma}\}^{-b}, \qquad -\infty < x < \infty,\ b > 0. \tag{6.6.2}$$

The above distribution has been derived by Balakrishnan and Leung (1988a) as one of three generalized forms of the standard logistic distribution. The density function in (6.6.1), for example, has been obtained by Balakrishnan and Leung (1988a) by compounding an extreme value distribution with a gamma distribution. They have shown that this is a family of positively skewed distributions with its coefficient of kurtosis greater than that of the logistic distribution, and hence this family provides us with a useful class of distributions since the distributions also possess mathematical tractability. Balakrishnan and Leung (1988a) have made a detailed study of order statistics from this distribution and have derived exact and explicit expressions for the first two single moments and product moments of order statistics in terms of the gamma function and its derivatives. In a subsequent paper, Balakrishnan and Leung (1988b) have tabulated the means, variances, and covariances of order statistics for some selected choices of b and for sample sizes up to 15, and have also computed the coefficients required for the BLUEs of μ and σ.

In this section, we present the derivation of the AMLEs of μ and σ due to Balakrishnan (1990c). Let

$$X_{r+1:n} \le X_{r+2:n} \le \cdots \le X_{n-s:n} \tag{6.6.3}$$

be a doubly Type-II censored sample available from the Type-I generalized logistic distribution with its pdf as in (6.6.1). Then the likelihood equations, based on the censored sample in (6.6.3), for μ and σ are given by

$$\frac{\partial \ln L}{\partial \mu} = -\frac{1}{\sigma}\left[r\frac{f(Z_{r+1:n})}{F(Z_{r+1:n})} - s\frac{f(Z_{n-s:n})}{1-F(Z_{n-s:n})} + \sum_{i=r+1}^{n-s} \frac{f'(Z_{i:n})}{f(Z_{i:n})} \right] = 0 \tag{6.6.4}$$

and

$$\frac{\partial \ln L}{\partial \sigma} = -\frac{1}{\sigma}\left[A + rZ_{r+1:n}\frac{f(Z_{r+1:n})}{F(Z_{r+1:n})} - sZ_{n-s:n}\frac{f(Z_{n-s:n})}{1-F(Z_{n-s:n})} + \sum_{i=r+1}^{n-s} Z_{i:n}\frac{f'(Z_{i:n})}{f(Z_{i:n})} \right] = 0, \tag{6.6.5}$$

where $A = n-r-s$, $Z_{i:n} = (X_{i:n}-\mu)/\sigma$, $F(z) = (1+e^{-z})^{-b}$, and $f(z) = b\,e^{-z}/(1+e^{-z})^{b+1}$. The likelihood equations (6.6.4) and (6.6.5) do not admit explicit solutions. However, by expanding the functions $f(Z_{r+1:n})/F(Z_{r+1:n})$, $f(Z_{n-s:n})/\{1-F(Z_{n-s:n})\}$, and $f'(Z_{i:n})/f(Z_{i:n})$ in Taylor series around the points ξ_{r+1}, ξ_{n-s}, and ξ_i, respectively, where $\xi_i = F^{-1}(p_i) = \ln\{p_i^{1/b}/(1-p_i^{1/b})\}$ with $p_i = 1-q_i = i/(n+1)$, we may then approximate these functions by

$$\frac{f(Z_{r+1:n})}{F(Z_{r+1:n})} \simeq \gamma_1 - \delta_1 Z_{r+1:n}, \tag{6.6.6}$$

$$\frac{f(Z_{n-s:n})}{1-F(Z_{n-s:n})} \simeq \gamma_2 + \delta_2 Z_{n-s:n}, \tag{6.6.7}$$

and

$$\frac{f'(Z_{i:n})}{f(Z_{i:n})} \simeq \alpha_i - \beta_i Z_{i:n}, \tag{6.6.8}$$

where

$$\delta_1 = b p_{r+1}^{1/b}(1 - p_{r+1}^{1/b}), \tag{6.6.9}$$

$$\gamma_1 = b(1 - p_{r+1}^{1/b}) + \delta_1 \xi_{r+1}, \tag{6.6.10}$$

$$\delta_2 = \frac{b}{q_{n-s}^2} p_{n-s}(1 - p_{n-s}^{1/b})\{b + p_{n-s}^{1+1/b} - (b+1)p_{n-s}^{1/b}\}, \tag{6.6.11}$$

$$\gamma_2 = \frac{b}{q_{n-s}} p_{n-s}(1 - p_{n-s}^{1/b}) - \delta_2 \xi_{n-s}, \tag{6.6.12}$$

$$\beta_i = (b+1)p_i^{1/b}(1 - p_i^{1/b}), \tag{6.6.13}$$

and

$$\alpha_i = b - (b+1)p_i^{1/b} + \beta_i \xi_i. \tag{6.6.14}$$

It is quite easy to see from Eqs. (6.6.9) and (6.6.13) that $\delta_1 > 0$ and $\beta_i > 0$, respectively. In order to show from (6.6.11) that $\delta_2 > 0$, let us consider the function

$$H(z) = 1 + b\,e^{-z}(1 + e^{-z})^b - (1 + e^{-z})^b \qquad \text{for } z \in \mathbb{R}.$$

It is then easy to verify that

$$H'(z) = -b(b+1)(e^{-z})^2(1 + e^{-z})^{b+1} < 0,$$

and, hence, $H(z)$ is a monotonically decreasing function of z. Now by using the facts that

$$\lim_{z \to -\infty} H(z) = \infty \qquad \text{and} \qquad \lim_{z \to \infty} H(z) = 0,$$

we realize that $H(z) > 0$ for $z \in \mathbb{R}$; consequently, we have from (6.6.11) that $\delta_2 > 0$.

By making use of the linear approximations in (6.6.6) through (6.6.8), we obtain from (6.6.4) and (6.6.5) the approximate likelihood equations for μ and σ as

$$\frac{\partial \ln L}{\partial \mu} \simeq \frac{\partial \ln L^*}{\partial \mu}$$

$$= -\frac{1}{\sigma}\left[r\gamma_1 - s\gamma_2 + \sum_{i=r+1}^{n-s} \alpha_i - r\delta_1 Z_{r+1:n} \right.$$

$$\left. - s\delta_2 Z_{n-s:n} - \sum_{i=r+1}^{n-s} \beta_i Z_{i:n} \right] = 0 \qquad (6.6.15)$$

and

$$\frac{\partial \ln L}{\partial \sigma} \simeq \frac{\partial \ln L^*}{\partial \sigma}$$

$$= -\frac{1}{\sigma}\left[A + r\gamma_1 Z_{r+1:n} - s\gamma_2 Z_{n-s:n} + \sum_{i=r+1}^{n-s} \alpha_i Z_{i:n} \right.$$

$$\left. - r\delta_1 Z_{r+1:n}^2 - s\delta_2 Z_{n-s:n}^2 - \sum_{i=r+1}^{n-s} \beta_i Z_{i:n}^2 \right] = 0. \qquad (6.6.16)$$

By solving Eqs. (6.6.15) and (6.6.16), we derive the AMLEs of μ and σ as

$$\hat{\mu} = B - \hat{\sigma} C \qquad (6.6.17)$$

and

$$\hat{\sigma} = \{-D + (D^2 + 4AE)^{1/2}\}/2A, \qquad (6.6.18)$$

where

$$m = r\delta_1 + s\delta_2 + \sum_{i=r+1}^{n-s} \beta_i,$$

$$B = \left\{ r\delta_1 X_{r+1:n} + s\delta_2 X_{n-s:n} + \sum_{i=r+1}^{n-s} \beta_i X_{i:n} \right\} \Big/ m,$$

$$C = \left\{ r\gamma_1 - s\gamma_2 + \sum_{i=r+1}^{n-s} \alpha_i \right\} \Big/ m,$$

$$D = r\gamma_1 (X_{r+1:n} - B) - s\gamma_2 (X_{n-s:n} - B) + \sum_{i=r+1}^{n-s} \alpha_i (X_{i:n} - B)$$

$$= r\gamma_1 X_{r+1:n} - s\gamma_2 X_{n-s:n} + \sum_{i=r+1}^{n-s} \alpha_i X_{i:n} - mBC,$$

and

$$E = r\delta_1(X_{r+1:n} - B)^2 + s\delta_2(X_{n-s:n} - B)^2 + \sum_{i=r+1}^{n-s} \beta_i(X_{i:n} - B)^2$$
$$= r\delta_1 X_{r+1:n}^2 + s\delta_2 X_{n-s:n}^2 + \sum_{i=r+1}^{n-s} \beta_i X_{i:n}^2 - mB^2.$$

As before, upon solving Eq. (6.6.16) for σ, we obtain a quadratic equation in σ that has two roots; however, one of them drops out, since $\delta_1 > 0$, $\delta_2 > 0$, and $\beta_i > 0$, and consequently, $E > 0$.

As mentioned in the last section, we can compute the conditional bias of $\hat{\mu}$ exactly from (6.6.17) and approximate the conditional bias of $\hat{\sigma}$ by

$$E\left(\frac{\partial \ln L^*}{\partial \sigma}\right) \Big/ E\left(-\frac{\partial^2 \ln L^*}{\partial \sigma^2}\right).$$

The variances and covariance of the estimators may be approximated by

$$\operatorname{Var}(\hat{\mu}) = \frac{\sigma^2}{m}\left\{\frac{V_2}{V_2 - V_1^2}\right\}, \tag{6.6.19}$$

$$\operatorname{Var}(\hat{\sigma}) = \frac{\sigma^2}{m}\left\{\frac{1}{V_2 - V_1^2}\right\}, \tag{6.6.20}$$

and

$$\operatorname{Cov}(\hat{\mu}, \hat{\sigma}) = -\frac{\sigma^2}{m}\left\{\frac{V_1}{V_2 - V_1^2}\right\}, \tag{6.6.21}$$

where

$$V_1 = \frac{2}{m}\left\{r\delta_1 E(Z_{r+1:n}) + s\delta_2 E(Z_{n-s:n}) + \sum_{i=r+1}^{n-s} \beta_i E(Z_{i:n})\right\} - C \tag{6.6.22}$$

and

$$V_2 = \frac{1}{m}\left[3\left\{r\delta_1 E(Z_{r+1:n}^2) + s\delta_2 E(Z_{n-s:n}^2) + \sum_{i=r+1}^{n-s} \beta_i E(Z_{i:n}^2)\right\}\right.$$
$$\left. -2\left\{r\gamma_1 E(Z_{r+1:n}) - s\gamma_2 E(Z_{n-s:n}) + \sum_{i=r+1}^{n-s} \alpha_i E(Z_{i:n})\right\} - A\right]. \tag{6.6.23}$$

Balakrishnan (1990c) has simulated (based on 2,000 Monte Carlo runs) the values of

(1) {Bias of $\hat{\mu}$}/σ,
(2) {Bias of $\hat{\sigma}$}/σ,
(3) {Variance of $\hat{\mu}$}/σ^2,
(4) {Variance of $\hat{\sigma}$}/σ^2, and
(5) {Covariance of $\hat{\mu}$ and $\hat{\sigma}$}/σ^2,

for various sample sizes, various choices of censoring, and different values of b. In Tables 6.6.1–6.6.3, we have presented these values for $n = 15$, r, $s = 0(1)4$, and $b = 1.0(1.0)3.0$. From the formulas in (6.6.19) through (6.6.21), the values of

(6) {Approximate variance of $\hat{\mu}$}/σ^2,
(7) {Approximate variance of $\hat{\sigma}$}/σ^2, and
(8) {Approximate covariance of $\hat{\mu}$ and $\hat{\sigma}$}/σ^2,

have been computed and are also presented in Tables 6.6.1–6.6.3. By comparing these approximate values with the corresponding simulated values, we note from Tables 6.6.1–6.6.3 that the formulas in (6.6.19) through (6.6.21) yield very good approximation to the variances and covariance of $\hat{\mu}$ and $\hat{\sigma}$.

For the complete sample case (i.e., $r = s = 0$), Balakrishnan and Leung (1988b) have tabulated the coefficients required for the BLUEs of μ and σ for sample sizes up to 15 and for selected choices of b, and also the values of variances and covariance of these estimators. In this case, for $n = 15$ we have computed the values of

(9) {Mean Square Error of $\hat{\mu}$}/σ^2 and
(10) {Mean square error of $\hat{\sigma}$}/σ^2

from Tables 6.6.1–6.6.3 and have presented these values in Table 6.6.4 for $b = 1.0(1.0)3.0$. For comparison purposes we have also given in Table 6.6.4 the corresponding values of

(11) {Variance of BLUE μ^*}/σ^2 and
(12) {Variance of BLUE σ^*}/σ^2,

which have been taken from the tables of Balakrishnan and Leung (1988b). The efficiencies of the estimators $\hat{\mu}$ and $\hat{\sigma}$, relative to μ^* and σ^*, respectively, have been computed, and the values of

(13) Efficiency of $\hat{\mu}$ relative to μ^*,
(14) Efficiency of $\hat{\sigma}$ relative to σ^*, and
(15) Joint efficiency of $(\hat{\mu}, \hat{\sigma})$ relative to (μ^*, σ^*)

are also presented in Table 6.6.4 for $b = 1.0(1.0)3.0$. Here, the joint efficiency of the estimators $\hat{\mu}$ and $\hat{\sigma}$ has been taken to be the trace efficiency. From Table 6.6.4, we note that the AMLEs $\hat{\mu}$ and $\hat{\sigma}$ are just as efficient as the BLUEs of μ and σ even for a sample of size as small as 15.

TABLE 6.6.1

Bias, Variances, and Covariance of $\hat{\mu}$ and $\hat{\sigma}$ for the Generalized Logistic Distribution When $n = 15$ and $b = 1.0$

r	s	(1)	(2)	(3)	(4)	(5)	(6)	(7)	(8)
0	0	0.00411	−0.01703	0.21498	0.04778	0.00262	0.18824	0.03585	0.00000
	1	−0.00374	−0.02770	0.21446	0.05002	0.00197	0.18856	0.03903	0.00133
	2	−0.01206	−0.03638	0.21470	0.05404	0.00309	0.18965	0.04261	0.00357
	3	−0.02285	−0.04675	0.21615	0.05846	0.00542	0.19210	0.04685	0.00704
	4	−0.03436	−0.05579	0.22162	0.06520	0.01132	0.19678	0.05192	0.01213
1	0	0.01160	−0.02652	0.21585	0.04905	0.00152	0.18856	0.03903	−0.00133
	1	0.00377	−0.03868	0.21536	0.05115	0.00070	0.18879	0.04280	0.00000
	2	−0.00481	−0.04899	0.21554	0.05531	0.00167	0.18975	0.04711	0.00234
	3	−0.01621	−0.06155	0.21676	0.06039	0.00389	0.19203	0.05228	0.00606
	4	−0.02886	−0.07303	0.22195	0.06800	0.00997	0.19655	0.05857	0.01165
2	0	0.01907	−0.03382	0.21625	0.05231	0.00034	0.18965	0.04261	−0.00357
	1	0.01157	−0.04759	0.21586	0.05460	−0.00056	0.18975	0.04711	−0.00234
	2	0.00304	−0.05965	0.21601	0.05940	0.00031	0.19058	0.05232	0.00000
	3	−0.00856	−0.07452	0.21720	0.06508	0.00259	0.19245	0.05866	0.00387
	4	−0.02193	−0.08859	0.22227	0.07343	0.00896	0.19659	0.06650	0.00989
3	0	0.02790	−0.04129	0.21739	0.05700	−0.00175	0.19210	0.04685	−0.00704
	1	0.02109	−0.05689	0.21715	0.05934	−0.00273	0.19203	0.05228	−0.00606
	2	0.01292	−0.07095	0.21743	0.06426	−0.00194	0.19245	0.05865	−0.00387
	3	0.00150	−0.08871	0.21851	0.07080	0.00032	0.19394	0.06653	0.00000
	4	−0.01230	−0.10629	0.22315	0.08089	0.00711	0.19744	0.07648	0.00631
4	0	0.03790	−0.04844	0.22254	0.06327	−0.00719	0.19678	0.05192	−0.01214
	1	0.03227	−0.06617	0.22255	0.06537	−0.00825	0.19655	0.05857	−0.01165
	2	0.02495	−0.08274	0.22280	0.07081	−0.00777	0.19659	0.06650	−0.00989
	3	0.01432	−0.10423	0.22341	0.07924	−0.00597	0.19744	0.07648	−0.00631
	4	0.00064	−0.12658	0.22704	0.09227	0.00079	0.20000	0.08936	0.00000

TABLE 6.6.2
Bias, Variances, and Covariance of $\hat{\mu}$ and $\hat{\sigma}$ for the Generalized Logistic Distribution When $n = 15$ and $b = 2.0$

r	s	(1)	(2)	(3)	(4)	(5)	(6)	(7)	(8)
0	0	0.02657	−0.01104	0.14786	0.04323	−0.02261	0.13799	0.03546	−0.01942
	1	0.02625	−0.01786	0.14839	0.04655	−0.02391	0.13816	0.03819	−0.02025
	2	0.02568	−0.02646	0.14856	0.04879	−0.02437	0.13814	0.04131	−0.02084
	3	0.02408	−0.03310	0.14895	0.05349	−0.02609	0.13798	0.04502	−0.02112
	4	0.02170	−0.04001	0.14921	0.05807	−0.02585	0.13776	0.04949	−0.02095
1	0	0.04257	−0.02165	0.14905	0.04575	−0.02449	0.14441	0.03971	−0.02467
	1	0.04323	−0.02983	0.14971	0.04949	−0.02604	0.14487	0.04314	−0.02604
	2	0.04377	−0.04013	0.14978	0.05162	−0.02633	0.14507	0.04709	−0.02720
	3	0.04303	−0.04858	0.15044	0.05682	−0.02836	0.14506	0.05186	−0.02811
	4	0.04147	−0.05774	0.15057	0.06216	−0.02827	0.14487	0.05771	−0.02862
2	0	0.05562	−0.02920	0.15769	0.05155	−0.03143	0.15378	0.04414	−0.03113
	1	0.05756	−0.03873	0.15876	0.05608	−0.03360	0.15474	0.04836	−0.03324
	2	0.05970	−0.05089	0.15908	0.05901	−0.03445	0.15539	0.05328	−0.03522
	3	0.06030	−0.06131	0.16032	0.06545	−0.03739	0.15576	0.05931	−0.03707
	4	0.06015	−0.07292	0.16052	0.07224	−0.03805	0.15584	0.06683	−0.03866
3	0	0.06839	−0.03591	0.17158	0.05664	−0.03978	0.16728	0.04914	−0.03937
	1	0.07195	−0.04684	0.17307	0.06160	−0.04246	0.16912	0.05434	−0.04255
	2	0.07622	−0.06106	0.17369	0.06496	−0.04381	0.17061	0.06051	−0.04574
	3	0.07872	−0.07366	0.17583	0.07234	−0.04784	0.17180	0.06816	−0.04902
	4	0.08080	−0.08830	0.17670	0.08154	−0.05035	0.17265	0.07789	−0.05234
4	0	0.09146	−0.04814	0.19164	0.06440	−0.05222	0.18646	0.05492	−0.04994
	1	0.09750	−0.06121	0.19313	0.06948	−0.05497	0.18980	0.06140	−0.05466
	2	0.10523	−0.07871	0.19468	0.07399	−0.05753	0.19283	0.06918	−0.05966
	3	0.11110	−0.09511	0.19866	0.08349	−0.06374	0.19566	0.07901	−0.06515
	4	0.11705	−0.11469	0.20014	0.09416	−0.06753	0.19825	0.09178	−0.07124

TABLE 6.6.3
Bias, Variances, and Covariance of $\hat{\mu}$ and $\hat{\sigma}$ for the Generalized Logistic Distribution When $n = 15$ and $b = 3.0$

r	s	(1)	(2)	(3)	(4)	(5)	(6)	(7)	(8)
0	0	0.04696	−0.01721	0.14891	0.04480	−0.04113	0.13869	0.03604	−0.03320
	1	0.05137	−0.02633	0.15029	0.04677	−0.04276	0.14008	0.03872	−0.03517
	2	0.05389	−0.03350	0.15155	0.04906	−0.04445	0.14125	0.04178	−0.03714
	3	0.05521	−0.03927	0.15341	0.05235	−0.04407	0.14224	0.04543	−0.03916
	4	0.05772	−0.04906	0.15405	0.05667	−0.04872	0.14302	0.04985	−0.04120
1	0	0.06541	−0.02703	0.15733	0.04805	−0.04622	0.15175	0.04054	−0.04085
	1	0.07174	−0.03769	0.15872	0.04994	−0.04781	0.15405	0.04393	−0.04368
	2	0.07601	−0.04647	0.16037	0.05251	−0.04984	0.15617	0.04784	−0.04662
	3	0.07889	−0.05392	0.16298	0.05636	−0.05313	0.15816	0.05257	−0.04979
	4	0.08379	−0.06630	0.16416	0.06150	−0.05558	0.15999	0.05838	−0.05321
2	0	0.08272	−0.03558	0.16768	0.05205	−0.05263	0.16906	0.04521	−0.04983
	1	0.09154	−0.04799	0.16969	0.05435	−0.05475	0.17271	0.04941	−0.05378
	2	0.09808	−0.05858	0.17208	0.05749	−0.05747	0.17626	0.05432	−0.05802
	3	0.10295	−0.67825	0.17489	0.06137	−0.06080	0.17984	0.06033	−0.06275
	4	0.11110	−0.08316	0.17644	0.06714	−0.06385	0.18343	0.06783	−0.06808
3	0	0.09930	−0.04267	0.18658	0.05812	−0.06329	0.19226	0.05044	−0.06084
	1	0.11118	−0.05698	0.18919	0.06083	−0.06592	0.19797	0.05564	−0.06632
	2	0.12071	−0.06966	0.19334	0.06520	−0.07016	0.20377	0.06180	−0.07236
	3	0.12839	−0.08111	0.19807	0.07037	−0.07514	0.20993	0.06946	−0.07932
	4	0.14109	−0.10008	0.20054	0.07736	−0.07941	0.21654	0.07920	−0.08747
4	0	0.12414	−0.05332	0.21690	0.06373	−0.07632	0.22338	0.05644	−0.07451
	1	0.14042	−0.07027	0.22137	0.06712	−0.08019	0.23226	0.06292	−0.08213
	2	0.15440	−0.08594	0.22844	0.07275	−0.08651	0.24165	0.07072	−0.09075
	3	0.16663	−0.10078	0.23581	0.07904	−0.09334	0.25207	0.08058	−0.10097
	4	0.18690	−0.12555	0.24310	0.08928	−0.10197	0.26385	0.09339	−0.11339

TABLE 6.6.4
Comparison of $\hat{\mu}$ and $\hat{\sigma}$ with BLUEs μ^* and σ^* for the Generalized Logistic Distribution

b	(9)	(10)	(11)	(12)	(13)	(14)	(15)
1.0	0.21500	0.04807	0.20337	0.04958	94.5907	103.1413	96.1531
2.0	0.14857	0.04335	0.14782	0.04692	99.4952	108.2353	101.4694
3.0	0.15112	0.04510	0.14609	0.04608	97.3332	102.1729	98.4456

Example 6.6.1. Let us consider the sample of size 15 simulated by Balakrishnan (1990c) from the Type I generalized logistic distribution in (6.6.1) with $\mu = 25$, $\sigma = 0.5$, and $b = 2.0$. The data, arranged in increasing order of magnitude, are as follows:

$$24.110, 24.850, 25.022, 25.030, 25.176, 25.336, 25.431, 25.728,$$
$$25.848, 25.874, 25.899, 26.021, 26.286, 26.595, 26.717.$$

By using the tables prepared by Balakrishnan and Leung (1988b), we obtain the BLUEs of μ and σ from the above data to be

$$\begin{aligned}\mu^* &= (0.10773)(24.110) + (0.13292)(24.850) + (0.13220)(25.022)\\ &\quad + (0.12487)(25.030) + (0.11434)(25.176) + (0.10194)(25.336)\\ &\quad + (0.08843)(25.431) + (0.07424)(25.728) + (0.05961)(25.848)\\ &\quad + (0.04518)(25.874) + (0.03037)(25.899) + (0.01621)(26.021)\\ &\quad + (0.00257)(26.286) - (0.01002)(26.595) - (0.02057)(26.717)\\ &= 25.12377\end{aligned}$$

and

$$\begin{aligned}\sigma^* &= (-0.10700)(24.110) + (-0.08864)(24.850) + (-0.06630)(25.022)\\ &\quad + (-0.04583)(25.030) + (-0.02736)(25.176) + (-0.01077)(25.336)\\ &\quad + (0.00402)(25.431) + (0.01707)(25.728) + (0.02864)(25.848)\\ &\quad + (0.03750)(25.874) + (0.04579)(25.899) + (0.05138)(26.021)\\ &\quad + (0.05477)(26.286) + (0.05532)(26.595) + (0.05142)(26.717)\\ &= 0.50931;\end{aligned}$$

the standard errors of these estimates are obtained from Table 5 of Balakrishnan and Leung (1988b) to be

$$\text{S.E.}(\mu^*) = \sigma^*(0.14782)^{1/2} = 0.19582$$

and

$$\text{S.E.}(\sigma^*) = \sigma^*(0.04692)^{1/2} = 0.11032.$$

Next, by censoring the r smallest and the s largest observations in the sample, we computed the AMLEs of μ and σ based on the resulting Type-II censored sample. These estimates have been computed for various choices of r and s. In Table 6.6.5, we have presented the values of $\hat{\mu}$ and $\hat{\sigma}$ for r, $s = 0(1)4$. Furthermore, by using the simulated values of {Variance of $\hat{\mu}$}/σ^2 and {Variance of $\hat{\sigma}$}/σ^2 given in Table 6.6.2, we have also determined the standard errors of these estimates and have presented them in Table 6.6.5. From this table, we observe that the AMLEs $\hat{\mu}$ and $\hat{\sigma}$ are just as efficient as the BLUEs of μ and σ even for a sample size as small as 15. Moreover,

TABLE 6.6.5
Values of $\hat{\mu}$ and $\hat{\sigma}$ and Their Standard Errors for Censored Samples from the Generalized Logistic Distribution

r	s	$\hat{\mu}$	$\hat{\sigma}$	S.E.($\hat{\mu}$)	S.E.($\hat{\sigma}$)
0	0	25.13504	0.49323	0.18966	0.10255
	1	25.12889	0.51154	0.19705	0.11037
	2	25.12778	0.50826	0.19590	0.11227
	3	25.12627	0.49835	0.19233	0.11526
	4	25.12506	0.50677	0.19575	0.12212
1	0	25.19705	0.44109	0.17029	0.09435
	1	25.18979	0.45869	0.17748	0.10204
	2	25.19040	0.45064	0.17740	0.10239
	3	25.19129	0.43384	0.16827	0.10341
	4	25.18956	0.43751	0.16977	0.10908
2	0	25.19993	0.44100	0.17512	0.10013
	1	25.19031	0.46046	0.18347	0.10904
	2	25.19176	0.45191	0.18024	0.10978
	3	25.19479	0.43325	0.17347	0.11084
	4	25.19246	0.43756	0.17531	0.11761
3	0	25.15487	0.47223	0.19561	0.11239
	1	25.14069	0.49636	0.20649	0.12319
	2	25.14169	0.49033	0.20435	0.12497
	3	25.14571	0.47379	0.19867	0.12743
	4	25.14115	0.48347	0.20323	0.13806
4	0	25.17702	0.46022	0.20147	0.11679
	1	25.15861	0.48668	0.21388	0.12828
	2	25.16144	0.47867	0.21120	0.13020
	3	25.16926	0.45841	0.20432	0.13246
	4	25.16338	0.46799	0.20936	0.14361

the AMLEs are very simple to use, as they are explicit estimators and also do not need the construction of any special tables.

6.7. Estimation for the Half Logistic Distribution

In this section, we consider the one-parameter half logistic distribution with pdf

$$f(x;\sigma)=\frac{2}{\sigma}\frac{e^{-x/\sigma}}{(1+e^{-x/\sigma})^2},\qquad 0\le x<\infty,\ \sigma>0, \tag{6.7.1}$$

and cdf

$$F(x;\sigma)=\frac{1-e^{-x/\sigma}}{1+e^{-x/\sigma}},\qquad 0\le x<\infty,\ \sigma>0. \tag{6.7.2}$$

The above half logistic distribution has been introduced by Balakrishnan (1985) as a life-time model with increasing hazard rate. The best linear unbiased estimation, the linear unbiased estimation based on optimally selected order statistics, the maximum likelihood estimation, and the approximate maximum likelihood estimation of the scale parameter σ have all been examined by Chan (1989). Similar work for a two-parameter half logistic distribution has been carried out by Balakrishnan and Puthenpura (1986), Balakrishnan and Wong (1990a, b, c), and Wong (1988). The maximum likelihood and the approximate maximum likelihood estimators of the scale parameter σ in (6.7.1) based on Type-I censored samples have been studied by Varadan (1989).

In this section we present the derivation of the AMLE of σ due to Chan (1989). Let

$$X_{r+1:n}\le X_{r+2:n}\le\cdots\le X_{n-s:n} \tag{6.7.3}$$

be a doubly Type-II censored sample available from the half logistic population with its pdf as in (6.7.1). Then the likelihood equation, based on the censored sample in (6.7.3), for σ is obtained by using the relation in (3.10.5) as

$$\begin{aligned}\frac{d\ln L}{d\sigma}&=-\frac{1}{\sigma}\left[A+rZ_{r+1:n}\frac{f(Z_{r+1:n})}{F(Z_{r+1:n})}-\frac{s}{2}Z_{n-s:n}\{1+F(Z_{n-s:n})\}\right.\\&\qquad\left.-\sum_{i=r+1}^{n-s}Z_{i:n}F(Z_{i:n})\right]\\&=0,\end{aligned} \tag{6.7.4}$$

where $A = n - r - s$, $Z_{i:n} = X_{i:n}/\sigma$, $F(z) = (1 - e^{-z})/(1 + e^{-z})$, and $f(z) = 2e^{-z}/(1+e^{-z})^2$. The likelihood equation (6.7.4) does not admit explicit solution. However, by expanding the functions $f(Z_{r+1:n})/F(Z_{r+1:n})$ and $F(Z_{i:n})$ in Taylor series around the points ξ_{r+1} and ξ_i, respectively, where

$$\xi_i = F^{-1}(p_i) = \ln\left(\frac{1+p_i}{1-p_i}\right),$$

with $p_i = 1 - q_i = i/(n+1)$, we may then approximate these functions by

$$\frac{f(Z_{r+1:n})}{F(Z_{r+1:n})} \simeq \gamma - \delta Z_{r+1:n} \tag{6.7.5}$$

and

$$F(Z_{i:n}) \simeq \alpha_i + \beta_i Z_{i:n}, \tag{6.7.6}$$

where

$$\delta = \frac{q_{r+1}(1+p_{r+1}^2)}{p_{r+1}(1+p_{r+1})}, \tag{6.7.7}$$

$$\gamma = \frac{1-p_{r+1}^2}{2p_{r+1}} + \delta\xi_{r+1}, \tag{6.7.8}$$

$$\beta_i = (1-p_i^2)/2, \tag{6.7.9}$$

and

$$\alpha_i = p_i - \beta_i \xi_i. \tag{6.7.10}$$

It is easy to see from Eqs. (6.7.7) and (6.7.9) that $\delta > 0$ and $\beta_i > 0$, respectively. By making use of the linear approximations in (6.7.5) and (6.7.6), we obtain from (6.7.4) the approximate likelihood equation for σ as

$$\begin{aligned}\frac{d\ln L}{d\sigma} &\simeq \frac{d\ln L^*}{d\sigma} \\ &= -\frac{1}{\sigma}\Bigg[A + r\gamma Z_{r+1:n} - \frac{s}{2}(1+\alpha_{n-s})Z_{n-s:n} - \sum_{i=r+1}^{n-s} \alpha_i Z_{i:n} \\ &\quad - r\delta Z_{r+1:n}^2 - \frac{s}{2}\beta_{n-s} Z_{n-s:n}^2 - \sum_{i=r+1}^{n-s} \beta_i Z_{i:n}^2\Bigg] \\ &= 0. \end{aligned} \tag{6.7.11}$$

By solving Eq. (6.7.11), we derive the AMLE of σ as

$$\hat{\sigma} = \frac{-B + (B^2 + 4AC)^{1/2}}{2A}, \tag{6.7.12}$$

where

$$A = n - r - s,$$

$$B = r\gamma X_{r+1:n} - \frac{s}{2}(1 + \alpha_{n-s})X_{n-s:n} - \sum_{i=r+1}^{n-s} \alpha_i X_{i:n},$$

and

$$C = r\delta X_{r+1:n}^2 + \frac{s}{2}\beta_{n-s}X_{n-s:n}^2 + \sum_{i=r+1}^{n-s} \beta_i X_{i:n}^2.$$

It should be mentioned that upon solving Eq. (6.7.11) for σ, we obtain a quadratic equation in σ that has two roots; however, one of them drops out, since $\delta > 0$ and $\beta_i > 0$, and hence $C > 0$.

Chan (1989) has simulated (based on 10,000 Monte Carlo runs) the values of

(1) Unbiasing factor of $\hat{\sigma}$ and
(2) {Variance of the unbiased AMLE of σ}/σ^2,

for all possible choices of censoring and for sample sizes up to 20. In Tables

TABLE 6.7.1
Unbiasing Factor of the AMLE and Variance of the Unbiased AMLE, Unbiasing Factor of the MLE and Variance of the Unbiased MLE, and Variance of the BLUE of σ for the Half Logistic Distribution When $n = 10$

r	s	(1)	(2)	(3)	(4)	(5)
0	0	0.93880	0.06930	1.00378	0.06897	0.07022
0	1	0.97872	0.06855	1.00591	0.07594	0.07702
0	2	0.96466	0.06891	1.01052	0.08501	0.08650
0	3	0.92996	0.06947	1.01691	0.09759	0.09964
1	0	0.80638	0.07392	0.94540	0.06908	0.07023
1	1	0.82828	0.07540	0.94195	0.07608	0.07703
1	2	0.80223	0.07686	0.93838	0.08517	0.08651
1	3	0.75455	0.07843	0.93296	0.09776	0.09966
2	0	0.69879	0.07576	0.89566	0.06953	0.07027
2	1	0.70748	0.07799	0.88695	0.07662	0.07708
2	2	0.67190	0.07963	0.87548	0.08575	0.08657
2	3	0.61346	0.08077	0.85836	0.09846	0.09973
3	0	0.59617	0.07616	0.85367	0.07032	0.07040
3	1	0.59250	0.07853	0.83984	0.07757	0.07724
3	2	0.54748	0.07974	0.82043	0.08673	0.08677
3	3	0.47806	0.08049	0.79101	0.09972	0.10011

TABLE 6.7.2
Unbiasing Factor of the AMLE and Variance of the Unbiased AMLE, Unbiasing Factor of the MLE and Variance of the Unbiased MLE, and Variance of the BLUE of σ for the Half Logistic Distribution When $n = 15$

r	s	(1)	(2)	(3)	(4)	(5)
0	0	0.95824	0.04699	1.00216	0.04677	0.04676
0	1	1.00530	0.04590	1.00278	0.04939	0.04954
0	3	0.99492	0.04598	1.00629	0.05718	0.05745
0	4	0.97025	0.04632	1.00680	0.06246	0.06294
0	6	0.90768	0.04623	1.01465	0.07801	0.07857
1	0	0.86471	0.05039	0.96097	0.04682	0.04676
1	1	0.90111	0.05068	0.95919	0.04943	0.04954
1	3	0.87995	0.05173	0.95559	0.05725	0.05745
1	4	0.85058	0.05252	0.95112	0.06254	0.06294
1	6	0.77437	0.05369	0.94419	0.07808	0.07857
3	0	0.71671	0.05243	0.89112	0.04713	0.04678
3	1	0.73841	0.05348	0.88487	0.04976	0.04956
3	3	0.70083	0.05491	0.86770	0.05767	0.05748
3	4	0.66356	0.05537	0.85343	0.06304	0.06297
3	6	0.56531	0.05548	0.81672	0.07875	0.07862
4	0	0.64790	0.05243	0.86162	0.04750	0.04681
4	1	0.66310	0.05357	0.85320	0.05015	0.04959
4	3	0.61764	0.05474	0.82923	0.05815	0.05753
4	4	0.57643	0.05493	0.80987	0.06362	0.06303
4	6	0.46700	0.05368	0.75679	0.07960	0.07872
6	0	0.51330	0.05154	0.81271	0.04869	0.04703
6	1	0.51582	0.05283	0.80003	0.05141	0.04984
6	3	0.45365	0.05319	0.76195	0.05987	0.05787
6	4	0.40349	0.05241	0.73119	0.06571	0.06343
6	6	0.26719	0.04722	0.63732	0.08239	0.07934

6.7.1–6.7.3, we have presented these values for $n = 10$ and $r, s = 0(1)3$, $n = 15$ and $r, s = 0, 1, 3, 4, 6$, and $n = 20$ and $r, s = 0(2)8$, respectively. For comparison purposes, we have also presented the values of

(3) Unbiasing factor of MLE of σ,
(4) {Variance of the unbiased MLE of σ}$/\sigma^2$, and
(5) {Variance of BLUE of σ}$/\sigma^2$.

The unbiased estimator based on the AMLE of σ in (6.7.12), in addition to being a simple explicit estimator, is observed from Tables 6.7.1–6.7.3 to

TABLE 6.7.3
Unbiasing Factor of the AMLE and Variance of the Unbiased AMLE, Unbiasing Factor of the MLE and Variance of the Unbiased MLE, and Variance of the BLUE of σ for the Half Logistic Distribution when $n = 20$

r	s	(1)	(2)	(3)	(4)	(5)
0	0	0.96910	0.03465	1.00229	0.03455	0.03504
0	2	1.02883	0.03359	1.00318	0.03755	0.03833
0	4	1.01311	0.03397	1.00639	0.04245	0.04300
0	6	0.97406	0.03452	1.00879	0.04900	0.04953
0	8	0.92294	0.03462	1.01183	0.05858	0.05882
2	0	0.83713	0.03831	0.94115	0.03460	0.03505
2	2	0.87696	0.03901	0.93624	0.03760	0.03833
2	4	0.85126	0.04008	0.93091	0.04250	0.04300
2	6	0.80180	0.04113	0.92116	0.04908	0.04953
2	8	0.73646	0.04191	0.90655	0.05869	0.05883
4	0	0.72720	0.03921	0.88950	0.03480	0.03506
4	2	0.75223	0.04048	0.87914	0.03780	0.03835
4	4	0.71824	0.04145	0.86559	0.04275	0.04302
4	6	0.65982	0.04206	0.84380	0.04945	0.04955
4	8	0.58253	0.04211	0.81111	0.05919	0.05885
6	0	0.62305	0.03932	0.84598	0.03518	0.03510
6	2	0.63446	0.04073	0.83026	0.03822	0.03839
6	4	0.59219	0.04138	0.80839	0.04328	0.04308
6	6	0.52464	0.04136	0.77382	0.05019	0.04963
6	8	0.43482	0.04001	0.72051	0.06019	0.05897
8	0	0.52217	0.03881	0.81005	0.03582	0.03522
8	2	0.52020	0.04017	0.78897	0.03891	0.03854
8	4	0.46896	0.04037	0.75831	0.04415	0.04327
8	6	0.39085	0.03932	0.70905	0.05144	0.04988
8	8	0.28545	0.03585	0.62892	0.06206	0.05932

be as efficient as the unbiased estimator based on the MLE of σ and usually more efficient than the best linear unbiased estimator of σ.

Example 6.7.1. Let us consider again the following data (see Example 4.2.2), which represent failure times, in minutes, for a specific type of electrical insulation that was subjected to a continuously increasing voltage stress:

$$12.3, 21.8, 24.4, 28.6, 43.2, 46.9, 70.7, 75.3, 95.5, 98.1, 138.6, 151.9.$$

In Example 4.2.2, by assuming the one-parameter half logistic distribution in (6.7.1) for the above data, we computed the best linear unbiased estimate of the expected failure time to be 66.55 min and its standard error to be 16.09 min. In this case, we obtain the approximate maximum likelihood estimate of σ based on the above given complete sample as

$$\hat{\sigma} = 47.41613 \text{ min.}$$

By using the tables of Chan (1989), we obtain an unbiased estimate of σ as

$$\hat{\sigma}_{\mathrm{U}} = 0.94829\hat{\sigma} = 44.96424 \text{ min,}$$

and its standard error as

$$\text{S.E.}(\hat{\sigma}_{\mathrm{U}}) = 44.96424(0.06451)^{1/2} = 11.42 \text{ min.}$$

Hence, we obtain an unbiased estimate of the expected failure time as

$$\hat{\sigma}_{\mathrm{U}} \ln 4 = 44.96424 \ln 4 = 62.33 \text{ min,}$$

and its standard error as

$$11.42 \ln 4 = 15.83 \text{ min.}$$

Now, suppose the experiment had been stopped as soon as the 11th failure occurred, that is, the largest observation had been censored. Then in Example 4.2.2, by assuming the one-parameter half logistic distribution in (6.7.1) for this censored data, we computed the best linear unbiased estimate of the expected failure time to be 70.01 min and its standard error to be 17.58 min. In this case, we obtain the approximate maximum likelihood estimate of σ as

$$\hat{\sigma} = 49.79094 \text{ min.}$$

By using the tables of Chan (1989), we obtain an unbiased estimate of σ as

$$\hat{\sigma}_{\mathrm{U}} = 0.99304\hat{\sigma} = 49.44 \text{ min,}$$

and its standard error as

$$\text{S.E.}(\hat{\sigma}_{\mathrm{U}}) = 49.44(0.05761)^{1/2} = 11.87 \text{ min.}$$

Hence, we obtain an unbiased estimate of the expected failure time as

$$\hat{\sigma}_{\mathrm{U}} \ln 4 = 49.44 \ln 4 = 68.54 \text{ min}$$

and its standard error as

$$11.87 \ln 4 = 16.46 \text{ min.}$$

It is of interest to mention here that the MLE of σ was computed for the case of the complete sample and the right-censored sample (with the largest observation censored) to be 47.41609 min and 49.79088 min, respectively, and that these values are very close to the corresponding AMLE values reported above.

Chapter 7 Optimal Linear Estimation Based on Selected Order Statistics

7.1. Introduction

In this chapter we consider the problem of optimal linear estimation of the location and scale parameters and quantiles of a population based on selected order statistics. First, we describe in Section 7.2 the work of Bennett (1952) on the optimal asymptotic estimation of the location and scale parameters of a population by using continuous weight functions. We also describe in this section a rather different approach to this problem that has been given by Jung (1955, 1962). Next, in Section 7.3 we discuss the fundamental work of Ogawa (1951, 1962) in determining the optimal spacing of order statistics in large sample sizes for the estimation of the location and scale parameters of a population. In Section 7.4 we describe the simplified linear estimators for the mean and standard deviation of the normal population proposed by Dixon (1957, 1960) in terms of the sample quasi-midrange and quasi-range, respectively. Dixon's estimator of the standard deviation based on the sample range is also explained. Similar work carried out by Raghunandanan and Srinivasan (1970, 1971) for the estimation of the location and scale parameters of the logistic and double exponential populations is also presented in this section. Next, the approximate maximum likelihood estimation method that has been discussed in great detail in Chapter 6 is extended in Section 7.5 to the case in which the estimation is to be based on a few order statistics. By considering a normal population, the AMLEs of the mean and standard deviation based on two symmetric order statistics are derived in this section, and it is also shown that they turn out to be the sample midrange and quasi-range based on the given two symmetric order statistics. Moreover, the AMLE of the standard deviation based on the two extreme order statistics is shown to be almost the same as Dixon's simplified estimator based on the sample range that is described in Section 7.4. In Section 7.6, we discuss the best linear unbiased estimation and the asymptotically best linear unbiased estimation of quantiles of the population distribution. Finally, in Section 7.7 we present a detailed list of references concerning the problems that are discussed in this chapter.

7.2. Bennett's and Jung's Optimal Asymptotic Estimators

In Section 4.4 we considered linear estimators of the form

$$\sum_i c_i X_{i:n} \tag{7.2.1}$$

for the location and scale parameters and presented Lloyd's (1952) derivation of best linear unbiased estimators through least-squares theory. These results were also independently derived by Bennett (1952) in his unpublished Ph.D. thesis. After determining optimal weights for order statistics in linear estimators, Bennett went on to derive optimal asymptotic weights for order statistics by considering general multiply Type-II censored samples. Bennett's (1952) results have also been extended by Chernoff *et al.* (1967).

We shall consider here a doubly censored sample

$$X_{r+1:n} \le X_{r+2:n} \le \cdots \le X_{n-s:n} \tag{7.2.2}$$

and describe Bennett's determination of optimal asymptotic weights. As before, let us denote $Z_{i:n} = (X_{i:n} - \mu)/\sigma$, $\alpha_{i:n} = E(Z_{i:n})$, and $\beta_{i,j:n} = \text{Cov}(Z_{i:n}, Z_{j:n})$. Then, from Section 3.10 we have the covariance of $X_{i:n}$ and $X_{j:n}$ up to order $1/n$ as

$$\begin{aligned}\text{Cov}(X_{i:n}, X_{j:n}) &= \sigma^2 \,\text{Cov}(Z_{i:n}, Z_{j:n}) \\ &= \sigma^2 \beta_{i,j:n} \\ &= \sigma^2 \frac{p_i q_j}{n+2} \frac{1}{f(G_i) f(G_j)}, \qquad i \le j,\end{aligned} \tag{7.2.3}$$

where $p_i = i/(n+1)$, $q_i = 1 - p_i$, and $G_i = F^{-1}(p_i)$ for $r+1 \le i \le n-s$. Then, the inverse of the matrix $\boldsymbol{\beta}$ may be algebraically worked out and is given by (with its (i, j)th element denoted by $(\boldsymbol{\beta}^{-1})_{i,j}$)

$$(\boldsymbol{\beta}^{-1})_{i,i} = nf^2(G_i)\left\{\frac{1}{p_{i+1}-p_i} + \frac{1}{p_i - p_{i-1}}\right\},$$

$$(\boldsymbol{\beta}^{-1})_{i,i-1} = (\boldsymbol{\beta}^{-1})_{i-1,i} = -\frac{nf(G_i)f(G_{i-1})}{p_i - p_{i-1}},$$

$$(\boldsymbol{\beta}^{-1})_{i,j} = 0 \qquad \text{otherwise}, \tag{7.2.4}$$

where we take $p_r = p_{n-s+1} = 0$. From (7.2.4) we get

$$\frac{1}{n}\sum_{i=r+1}^{n-s}\sum_{j=r+1}^{n-s}(\boldsymbol{\beta}^{-1})_{i,j} = \sum_{i=r+1}^{n-s}\frac{\{f(G_{i+1})-f(G_i)\}^2}{p_{i+1}-p_i} + \frac{f^2(G_{r+1})}{p_{r+1}} + \frac{f^2(G_{n-s})}{1-p_{n-s}}. \tag{7.2.5}$$

From Section 3.10 we also have $\alpha_{i:n} = f(G_i) + O(1/n)$; hence, by discarding terms of lower order, we get

$$\frac{1}{n}\sum_{i=r+1}^{n-s}\sum_{j=r+1}^{n-s}\alpha_{i:n}(\boldsymbol{\beta}^{-1})_{i,j} = \sum_{i=r+1}^{n-s}\frac{\{f(G_{i+1})-f(G_i)\}\{G_{i+1}f(G_{i+1})-G_if(G_i)\}}{p_{i+1}-p_i} + \frac{G_{r+1}f^2(G_{r+1})}{p_{r+1}} + \frac{G_{n-s}f^2(G_{n-s})}{1-p_{n-s}}, \tag{7.2.6}$$

and, similarly, we find

$$\frac{1}{n}\sum_{i=r+1}^{n-s}\sum_{j=r+1}^{n-s}\alpha_{i:n}\alpha_{j:n}(\boldsymbol{\beta}^{-1})_{i,j} = \sum_{i=r+1}^{n-s}\frac{\{G_{i+1}f(G_{i+1})-G_if(G_i)\}^2}{p_{i+1}-p_i} + \frac{G_{r+1}^2f^2(G_{r+1})}{p_{r+1}} + \frac{G_{n-s}^2f^2(G_{n-s})}{1-p_{n-s}}. \tag{7.2.7}$$

Let $r/n \to \xi_1$ and $(n-s)/n \to \xi_2$ as $n \to \infty$. Then, we have as $n \to \infty$

$$\frac{1}{n}\sum_{i=r+1}^{n-s}\frac{\{f(G_{i+1})-f(G_i)\}^2}{p_{i+1}-p_i} \to \int_{\xi_1}^{\xi_2}\left\{\frac{df(G(u))}{du}\right\}^2 du = \int_{\xi_1}^{\xi_2}\left\{\frac{f'(G(u))}{f(G(u))}\right\}^2 du \tag{7.2.8}$$

since, as noted in Section 3.10, $G'(u) = 1/f(G(u))$. Let us now use v for $G(u)$ and $\mathfrak{H}(v) = f'(v)/f(v)$. We then obtain from (7.2.5) that

$$\lim_{n\to\infty}\frac{1}{n}\sum_{i=r+1}^{n-s}\sum_{j=r+1}^{n-s}(\boldsymbol{\beta}^{-1})_{i,j} = I_{11} = \int_{\xi_1}^{\xi_2}\mathfrak{H}^2(v)\,du + \frac{f^2(G(\xi_1))}{\xi_1} + \frac{f^2(G(\xi_2))}{1-\xi_2}. \tag{7.2.9}$$

Similarly, from Eqs. (7.2.6) and (7.2.7) we obtain

$$\lim_{n\to\infty}\frac{1}{n}\sum_{i=r+1}^{n-s}\sum_{j=r+1}^{n-s}\alpha_{i:n}(\boldsymbol{\beta}^{-1})_{i,j}=I_{12}$$

$$=\int_{\xi_1}^{\xi_2}\mathfrak{H}(v)\{1+v\mathfrak{H}(v)\}\,du+\frac{G(\xi_1)f^2(G(\xi_1))}{\xi_1}+\frac{G(\xi_2)f^2(G(\xi_2))}{1-\xi_2}\tag{7.2.10}$$

and

$$\lim_{n\to\infty}\frac{1}{n}\sum_{i=r+1}^{n-s}\sum_{j=r+1}^{n-s}\alpha_{i:n}\alpha_{j:n}(\boldsymbol{\beta}^{-1})_{i,j}=I_{22}$$

$$=\int_{\xi_1}^{\xi_2}\{1+v\mathfrak{H}(v)\}^2\,du+\frac{G^2(\xi_1)f^2(G(\xi_1))}{\xi_1}+\frac{G^2(\xi_2)f^2(G(\xi_2))}{1-\xi_2},\tag{7.2.11}$$

since $d/du\,\{vf(v)\}=1+v\mathfrak{H}(v)$. From Eqs. (4.4.5) and (4.4.6) we then obtain the least-squares estimators of μ and σ as

$$\mu^*=\sum_{i=r+1}^{n-s}\gamma_i X_{i:n}\tag{7.2.12}$$

and

$$\sigma^*=\sum_{i=r+1}^{n-s}\delta_i X_{i:n},\tag{7.2.13}$$

where, for $i=r+1,\ r+2,\ldots,n-s$,

$$\gamma_i=\frac{1}{n(I_{11}I_{22}-I_{12}^2)}\left\{I_{22}\sum_{j=r+1}^{n-s}(\boldsymbol{\beta}^{-1})_{i,j}-I_{12}\sum_{j=r+1}^{n-s}\alpha_{j:n}(\boldsymbol{\beta}^{-1})_{i,j}\right\}\tag{7.2.14}$$

and

$$\delta_i=\frac{1}{n(I_{11}I_{22}-I_{12}^2)}\left\{I_{11}\sum_{j=r+1}^{n-s}\alpha_{j:n}(\boldsymbol{\beta}^{-1})_{i,j}-I_{12}\sum_{j=r+1}^{n-s}(\boldsymbol{\beta}^{-1})_{i,j}\right\}.\tag{7.2.15}$$

Now, from (7.2.4) we have for $i=r+2,\ r+3,\ldots,n-s-1$,

$$\frac{1}{n}\sum_{j=r+1}^{n-s}(\boldsymbol{\beta}^{-1})_{i,j}=-\frac{f(G_i)[\{f(G_{i+1})-f(G_i)\}-\{f(G_i)-f(G_{i-1})\}]}{p_{i+1}-p_i},$$

which, asymptotically, gives for $i=r+2,\ r+3,\ldots,n-s-1$,

$$\begin{aligned}\frac{1}{n}\sum_{j=r+1}^{n-s}(\boldsymbol{\beta}^{-1})_{i,j}&\sim -f(G(\xi_i))\,\frac{d^2f(G(\xi_i))}{d\xi_i^2}\,d\xi_i\\&=\phi_1(\xi_i)\,d\xi_i\quad\text{(say)}\\&\sim\phi_1(\xi_i)/n\end{aligned}$$

with $i=[n\xi_i]+1$, $[m]$ indicating the integer part of m. From (7.2.4) we also obtain

$$\frac{1}{n}\sum_{j=r+1}^{n-s}(\boldsymbol{\beta}^{-1})_{r+1,j}=f^2(G_{r+1})\left\{\frac{1}{p_{r+2}-p_{r+1}}+\frac{1}{p_{r+1}}\right\}-\frac{f(G_{r+1})f(G_{r+2})}{p_{r+2}-p_{r+1}},$$

which, asymptotically, gives

$$\frac{1}{n}\sum_{j=r+1}^{n-s}(\boldsymbol{\beta}^{-1})_{r+1,j}\sim\phi_1(\xi_{r+1})\,d\xi_{r+1}+\frac{f^2(G_{r+1})}{p_{r+1}}-f'(G_{r+1})$$

$$\sim\frac{\phi_1(\xi_{r+1})}{n}+\phi^*_{1,r+1},$$

where

$$\phi^*_{1,r+1}=\frac{f^2(G_{r+1})}{p_{r+1}}-f'(G_{r+1}).$$

Similarly, from (7.2.4) we obtain

$$\frac{1}{n}\sum_{j=r+1}^{n-s}(\boldsymbol{\beta}^{-1})_{n-s,j}=f^2(G_{n-s})\left\{\frac{1}{p_{n-s}-p_{n-s-1}}-\frac{1}{p_{n-s}}\right\}-\frac{f(G_{n-s-1})f(G_{n-s})}{p_{n-s}-p_{n-s-1}},$$

which, asymptotically, yields

$$\frac{1}{n}\sum_{j=r+1}^{n-s}(\boldsymbol{\beta}^{-1})_{n-s,j}\sim\phi_1(\xi_{n-s})\,d\xi_{n-s}-\frac{f^2(G_{n-s})}{p_{n-s}}+f'(G_{n-s})$$

$$\sim\frac{\phi_1(\xi_{n-s})}{n}-\phi^*_{1,n-s},$$

where

$$\phi^*_{1,n-s}=\frac{f^2(G_{n-s})}{p_{n-s}}-f'(G_{n-s}).$$

Next, from (7.2.4) we have for $i=r+2,\ r+3,\ldots,n-s-1$,

$$\frac{1}{n}\sum_{j=r+1}^{n-s}\alpha_{j:n}(\boldsymbol{\beta}^{-1})_{i,j}$$

$$=-\frac{f(G_i)[\{G_{i+1}f(G_{i+1})-G_if(G_i)\}-\{G_if(G_i)-G_{i-1}f(G_{i-1})\}]}{p_{i+1}-p_i},$$

which, asymptotically, gives for $i=r+2,\ r+3,\dots,n-s-1$,

$$\frac{1}{n}\sum_{j=r+1}^{n-s}\alpha_{j:n}(\boldsymbol{\beta}^{-1})_{i,j}\sim -f(G(\xi_i))\frac{d^2\{G(\xi_i)f(G(\xi_i))\}}{d\xi_i^2}\,d\xi_i$$

$$=\phi_2(\xi_i)\,d\xi_i \quad \text{(say)}$$

$$\sim \phi_2(\xi_i)/n.$$

From (7.2.4) we also obtain

$$\frac{1}{n}\sum_{j=r+1}^{n-s}\alpha_{j:n}(\boldsymbol{\beta}^{-1})_{r+1,j}$$

$$=G_{r+1}f^2(G_{r+1})\left\{\frac{1}{p_{r+2}-p_{r+1}}+\frac{1}{p_{r+1}}\right\}-\frac{G_{r+2}f(G_{r+1})f(G_{r+2})}{p_{r+2}-p_{r+1}},$$

which, asymptotically, gives

$$\frac{1}{n}\sum_{j=r+1}^{n-s}\alpha_{j:n}(\boldsymbol{\beta}^{-1})_{r+1,j}\sim \phi_2(\xi_{r+1})\,d\xi_{r+1}+\frac{G_{r+1}f^2(G_{r+1})}{p_{r+1}}$$

$$-\{f(G_{r+1})+G_{r+1}f'(G_{r+1})\}$$

$$\sim\frac{\phi_2(\xi_{r+1})}{n}+\phi^*_{2,r+1},$$

where

$$\phi^*_{2,r+1}=\frac{G_{r+1}f^2(G_{r+1})}{p_{r+1}}-\{f(G_{r+1})+G_{r+1}f'(G_{r+1})\}.$$

From (7.2.4) we similarly obtain

$$\frac{1}{n}\sum_{j=r+1}^{n-s}\alpha_{j:n}(\boldsymbol{\beta}^{-1})_{n-s,j}$$

$$=G_{n-s}f^2(G_{n-s})\left\{\frac{1}{p_{n-s}-p_{n-s-1}}-\frac{1}{p_{n-s}}\right\}-\frac{G_{n-s-1}f(G_{n-s-1})f(G_{n-s})}{p_{n-s}-p_{n-s-1}},$$

which yields asymptotically

$$\frac{1}{n}\sum_{j=r+1}^{n-s}\alpha_{j:n}(\boldsymbol{\beta}^{-1})_{n-s,j}\sim \phi_2(\xi_{n-s})\,d\xi_{n-s}-\frac{G_{n-s}f^2(G_{n-s})}{p_{n-s}}$$

$$+\{f(G_{n-s})+G_{n-s}f'(G_{n-s})\}$$

$$\sim\frac{\phi_2(\xi_{n-s})}{n}-\phi^*_{2,n-s},$$

where

$$\phi^*_{2,n-s}=\frac{G_{n-s}f^2(G_{n-s})}{p_{n-s}}-\{f(G_{n-s})+G_{n-s}f'(G_{n-s})\}.$$

In the above calculations, we may note that the functions ϕ_1 and ϕ_2 are of the form (with $v = G(u)$ and $\mathfrak{H}(v) = f'(v)/f(v)$, as before)

$$\phi_1(u) = -f(v)\,\frac{d^2 f(v)}{du^2} = -\mathfrak{H}'(v) \tag{7.2.16}$$

and

$$\phi_2(u) = -f(v)\,\frac{d^2\{v f(v)\}}{du^2} = -\{\mathfrak{H}(v) + v\mathfrak{H}'(v)\}. \tag{7.2.17}$$

Hence, from Eqs. (7.2.14) and (7.2.15), we may determine asymptotically the coefficients γ_i and δ_i (except for the two most extreme coefficients, viz., for $i = r+1$ and $n-s$) by setting $u = p_i = i/(n+1)$ in the following continuous weight functions:

$$\gamma(u) = \frac{\phi_1(u) I_{22} - \phi_2(u) I_{12}}{n(I_{11} I_{22} - I_{12}^2)} \tag{7.2.18}$$

and

$$\delta(u) = \frac{\phi_2(u) I_{11} - \phi_1(u) I_{12}}{n(I_{11} I_{22} - I_{12}^2)}. \tag{7.2.19}$$

For the two extreme cases we may determine asymptotically the coefficients by the formulas

$$\gamma_{r+1} = \gamma(\xi_{r+1}) + \frac{\phi_{1,r+1}^* I_{22} - \phi_{2,r+1}^* I_{12}}{I_{11} I_{22} - I_{12}^2}, \tag{7.2.20}$$

$$\gamma_{n-s} = \gamma(\xi_{n-s}) - \frac{\phi_{1,n-s}^* I_{22} - \phi_{2,n-s}^* I_{12}}{I_{11} I_{22} - I_{12}^2}, \tag{7.2.21}$$

$$\delta_{r+1} = \delta(\xi_{r+1}) + \frac{\phi_{2,r+1}^* I_{11} - \phi_{1,r+1}^* I_{12}}{I_{11} I_{22} - I_{12}^2}, \tag{7.2.22}$$

and

$$\delta_{n-s} = \delta(\xi_{n-s}) - \frac{\phi_{2,n-s}^* I_{11} - \phi_{1,n-s}^* I_{12}}{I_{11} I_{22} - I_{12}^2}. \tag{7.2.23}$$

For the case in which the population distribution is symmetric and the sample is symmetrically Type-II censored, some simplification is possible in all the above formulas. In this case, since $I_{12} = 0$, the continuous weight functions $\gamma(u)$ and $\delta(u)$ in (7.2.18) and (7.2.19), for example, simplify to

$$\gamma(u) = \frac{\phi_1(u)}{n I_{11}} \tag{7.2.24}$$

and

$$\delta(u) = \frac{\phi_2(u)}{nI_{22}}. \tag{7.2.25}$$

In the complete sample case, that is, $r = s = 0$ or $\xi_1 = 1 - \xi_2 = 0$, the last two terms on the RHS of Eqs. (7.2.9) through (7.2.11) all vanish when

$$\lim_{\xi_1 \to 0} \frac{G^2(\xi_1) f^2(G(\xi_1))}{\xi_1} = 0 \quad \text{and} \quad \lim_{\xi_2 \to 1} \frac{G^2(\xi_2) f^2(G(\xi_2))}{1 - \xi_2} = 0. \tag{7.2.26}$$

When the two limiting conditions in (7.2.26) are satisfied, it is easy to show that the inverse of the variance–covariance matrix of the estimators μ^* and σ^* tends to the information matrix, thus establishing the fact that the linear estimators μ^* and σ^* are asymptotically efficient. In order to show this for a population with pdf $f(x; \mu, \sigma)$, we observe with $v = (x - \mu)/\sigma = G(u)$ that, for example,

$$\begin{aligned} E\left\{\frac{\partial \log f(X; \mu, \sigma)}{\partial \mu}\right\}^2 &= E\left\{\frac{\partial \log f(V)}{\partial \mu}\right\}^2 \\ &= \frac{1}{\sigma^2} E\left\{\frac{\partial \log f(V)}{\partial V}\right\}^2 \\ &= \frac{1}{\sigma^2} \int_0^1 \mathfrak{H}^2(v)\, du, \end{aligned}$$

which agrees exactly with the RHS of (7.2.9). Proceeding along these lines, Chernoff *et al.* (1967) have proved that the linear estimators μ^* and σ^* are asymptotically efficient even for the case in which the sample is doubly Type-II censored.

For the purpose of illustration, let us consider the following examples.

Example 7.2.1. Let us consider the normal $N(\mu, \sigma^2)$ population and assume the available sample is complete. In this case, since $f'(v) = -vf(v)$, we have $\mathfrak{H}(v) = -v$ and hence

$$I_{11} = \int_0^1 v^2\, du = E(V^2) = 1,$$

$$I_{12} = 0,$$

and

$$\begin{aligned} I_{22} &= \int_0^1 (1 - v^2)^2\, du = 1 - 2E(V^2) + E(V^4) \\ &= 1 - 2 + 3 = 2. \end{aligned}$$

From (7.2.16) we have

$$\phi_1(u) = 1$$

so that Eq. (7.2.24) gives

$$\gamma(u) = \frac{\phi_1(u)}{nI_{11}} = \frac{1}{n},$$

which yields the estimator μ^* to be exactly the sample mean $\bar{X}$. Also, from (7.2.17) we have

$$\phi_2(u) = 2v$$

so that Eq. (7.2.25) gives

$$\delta(u) = \frac{\phi_2(u)}{nI_{22}} = \frac{v}{n} = \frac{G(u)}{n}.$$

This weight function yields the linear estimator of σ to be

$$\sigma^* = \frac{1}{n}\sum_{i=1}^{n} G(p_i)X_{i:n} = \frac{1}{n}\sum_{i=1}^{n} F^{-1}\left(\frac{i}{n+1}\right) X_{i:n}.$$

Example 7.2.2. Let us now consider the estimation of μ of the normal $N(\mu, \sigma^2)$ population from a symmetrically Type-II censored sample. Then, we have $\xi_1 = 1 - \xi_2 = \xi$, and hence

$$\begin{aligned} I_{11} &= \int_{\xi}^{1-\xi} v^2\, du + \frac{f^2(G(\xi))}{\xi} + \frac{f^2(G(1-\xi))}{\xi} \\ &= \int_{G(\xi)}^{G(1-\xi)} v^2 f(v)\, dv + \frac{2f^2(G(\xi))}{\xi}, \end{aligned}$$

which, by integration by parts, gives

$$I_{11} = 1 - 2\xi + 2G(\xi)f(G(\xi)) + \frac{2f^2(G(\xi))}{\xi}.$$

As we have seen in the last example, we have $\phi_1(u) = 1$. Hence, in this case, the linear estimator of μ is given by

$$\mu^* = \frac{1}{nI_{11}} \sum_{i=r+1}^{n-r} X_{i:n} + a(X_{r+1:n} + X_{n-r:n}),$$

where $r=[n\xi]$, and the additional weight for the two extreme order statistics is obtained from (7.2.20) to be

$$a=\frac{\phi^*_{1,r+1}}{I_{11}}=\frac{\{f^2(G(\xi))/\xi\}-f'(G(\xi))}{I_{11}}$$

$$=\frac{\{f^2(G(\xi))/\xi\}+G(\xi)f(G(\xi))}{I_{11}}.$$

For example, when $\xi=0.05$ we have $I_{11}=0.986$ and $a=0.0437$; when $\xi=0.10$ we have $I_{11}=0.966$ and $a=0.086$. The additional weight given to each of the two extreme order statistics, as pointed out by Chernoff *et al.* (1967), is only slightly less than that suggested by Winsor and also recommended by Tukey (1962). This is of interest, and the asymptotically optimal linear estimator of μ derived above is slightly better than the Winsorized mean. It should also be pointed out here that for $\xi=0.05$ and 0.10, the asymptotically optimal weights determined above are very close (within 1.5%) to the weights of the BLUE of μ based on the symmetrically censored sample for sample size 20 given by Sarhan and Greenberg (1962, p. 248).

Example 7.2.3. Let us consider the logistic population with pdf

$$f(x;\mu,\sigma)=\frac{1}{\sigma}\frac{e^{-(x-\mu)/\sigma}}{\{1+e^{-(x-\mu)/\sigma}\}^2},\qquad -\infty<x<\infty.$$

Let us suppose the available sample from this population is complete. In this case, since $f(v)=F(v)\{1-F(v)\}$, we have $\mathfrak{H}(v)=1-2F(v)=1-2u$, and hence

$$I_{11}=\int_0^1 (1-2u)^2\,du=\tfrac{1}{3}.$$

From Eq. (7.2.16) we have

$$\phi_1(u)=-\mathfrak{H}'(v)=2f(v)=2F(v)\{1-F(v)\}=2u(1-u),$$

so that (7.2.24) gives

$$\gamma(u)=\frac{\phi_1(u)}{nI_{11}}=\frac{6u(1-u)}{n}.$$

This yields the asymptotically optimal linear estimator of μ to be

$$\mu^*=\frac{6}{n}\sum_{i=1}^{n}p_i(1-p_i)X_{i:n}.$$

Next, we get from (7.2.11) that

$$I_{22} = \int_0^1 \{1 + v(1-2u)\}^2 \, du$$

$$= \int_{-\infty}^{\infty} \{1 + v[1-2F(v)]\}^2 f(v) \, dv$$

$$= 1 + \frac{\pi^2}{3} - 2\alpha_{2:2} - 2\alpha_{2:2}^{(2)} + \frac{4}{3}\alpha_{3:3}^{(2)}$$

$$= \frac{1}{9}(\pi^2 + 3).$$

The last equality follows if we use the results that $\alpha_{2:2} = 1$, $\alpha_{2:2}^{(2)} = \pi^2/3$, and $\alpha_{3:3}^{(2)} = 1 + \pi^2/3$ derived from the recurrence relation established in Theorem 3.6.1. From Eq. (7.2.17) we also have

$$\phi_2(u) = -\{\mathfrak{H}(v) + v\mathfrak{H}'(v)\}$$

$$= 2u - 1 + 2u(1-u) \ln\left(\frac{u}{1-u}\right),$$

so that (7.2.25) gives

$$\delta(u) = \frac{\phi_2(u)}{nI_{22}} = \frac{9}{n(\pi^2+3)}\left\{2u - 1 + 2u(1-u) \ln\left(\frac{u}{1-u}\right)\right\}.$$

This yields the asymptotically optimal linear estimator of σ to be

$$\sigma^* = \frac{9}{n(\pi^2+3)} \sum_{i=1}^{n} \left\{2p_i - 1 + 2p_i q_i \ln\left(\frac{p_i}{q_i}\right)\right\} X_{i:n},$$

where $p_i = 1 - q_i = i/(n+1)$.

Example 7.2.4. Let us now consider the Cauchy population with pdf

$$f(x; \mu, \sigma) = \frac{1}{\sigma\pi\left\{1 + \left(\frac{x-\mu}{\sigma}\right)^2\right\}}, \qquad -\infty < x < \infty$$

and cdf

$$F(x; \mu, \sigma) = \frac{1}{2} + \frac{1}{\pi} \tan^{-1}\left(\frac{x-\mu}{\sigma}\right), \qquad -\infty < x < \infty.$$

Suppose the sample available from the above population is complete. In this case, we have $\mathfrak{H}(v) = -2v/(1+v^2)$, and hence

$$
\begin{aligned}
I_{11} &= \int_0^1 \frac{4v^2}{(1+v^2)^2}\,du \\
&= 4\int_0^1 \sin^2 \pi\left(u-\frac{1}{2}\right)\cos^2 \pi\left(u-\frac{1}{2}\right) du \\
&= \frac{4}{\pi} B\left(\frac{3}{2},\frac{3}{2}\right) \\
&= \frac{1}{2}.
\end{aligned}
$$

From Eq. (7.2.16) we have

$$\phi_1(u) = -\mathfrak{H}'(v) = 2(1-v^2)/(1+v^2)^2,$$

so that (7.2.24) gives

$$
\begin{aligned}
\gamma(u) = \frac{\phi_1(u)}{nI_{11}} &= \frac{4}{n}\cos^4 \pi\left(u-\frac{1}{2}\right)\left\{1-\tan^2 \pi\left(u-\frac{1}{2}\right)\right\} \\
&= \frac{4}{n}\cos^2 \pi\left(u-\frac{1}{2}\right)\cos 2\pi\left(u-\frac{1}{2}\right) \\
&= \frac{1}{n}\,\frac{\sin 4\pi(u-\frac{1}{2})}{\tan \pi(u-\frac{1}{2})}.
\end{aligned}
$$

This yields the asymptotically optimal linear estimator of μ to be

$$\mu^* = \frac{1}{n}\sum_{i=1}^{n}\left\{\frac{\sin 4\pi(p_i-\frac{1}{2})}{\tan \pi(p_i-\frac{1}{2})}\right\} X_{i:n}.$$

We next get from (7.2.11) that

$$
\begin{aligned}
I_{22} &= \int_0^1 \left\{\frac{1-v^2}{1+v^2}\right\}^2 du \\
&= \int_0^1 \left\{\cos^2 \pi\left(u-\frac{1}{2}\right) - \sin^2 \pi\left(u-\frac{1}{2}\right)\right\}^2 du \\
&= \frac{2}{\pi}\left\{B\left(\frac{1}{2},\frac{5}{2}\right) - B\left(\frac{3}{2},\frac{3}{2}\right)\right\} \\
&= \frac{1}{2}.
\end{aligned}
$$

From Eq. (7.2.17) we have

$$\phi_2(u) = -\{\mathfrak{H}(v) + v\mathfrak{H}'(v)\} = 4v/(1+v^2)^2,$$

so that (7.2.25) gives

$$\delta(u) = \frac{\phi_2(u)}{nI_{22}} = \frac{8}{n} \tan \pi\left(u - \frac{1}{2}\right) \cos^4 \pi\left(u - \frac{1}{2}\right).$$

This yields the asymptotically optimal linear estimator of σ to be

$$\sigma^* = \frac{8}{n} \sum_{i=1}^{n} \tan \pi\left(p_i - \frac{1}{2}\right) \cos^4 \pi\left(p_i - \frac{1}{2}\right) X_{i:n}.$$

When either μ or σ is known, some simplification is possible in the form of the asymptotically optimal linear estimators. Let us first consider the case in which the scale parameter σ is known. In this case, $X_{i:n}$ may be replaced by $X_{i:n} - \sigma\alpha_{i:n}$, leaving $\boldsymbol{\beta}^{-1}$ unchanged so that the asymptotically optimal linear estimator of μ can be written as

$$\mu^* = \sum_{i=r+1}^{n-s} \gamma_i^* (X_{i:n} - \sigma\alpha_{i:n}), \tag{7.2.27}$$

where, for $i = r+1, r+2, \ldots, n-s$,

$$\gamma_i^* = \frac{1}{nI_{11}} \sum_{j=r+1}^{n-s} (\boldsymbol{\beta}^{-1})_{i,j}; \tag{7.2.28}$$

as before, we may determine asymptotically the coefficients γ_i^* (except for the two most extreme cases, viz., $i = r+1$ and $n-s$) by setting $u = p_i = i/(n+1)$ in the continuous weight function

$$\gamma^*(u) = \frac{\phi_1(u)}{nI_{11}}, \tag{7.2.29}$$

and for the two extreme cases the coefficients are given by the formulas

$$\gamma_{r+1}^* = \gamma^*(\xi_{r+1}) + \frac{\phi_{1,r+1}^*}{I_{11}} \tag{7.2.30}$$

and

$$\gamma_{n-s}^* = \gamma^*(\xi_{n-s}) - \frac{\phi_{1,n-s}^*}{I_{11}}, \tag{7.2.31}$$

where $\phi_1(u)$, $\phi_{1,r+1}^*$ and $\phi_{1,n-s}^*$ are as defined earlier. The variance of the estimator μ^* in (7.2.27) is $\sigma^2/(nI_{11})$. Also, if the exact value of $\alpha_{i:n}$ is not available, $\alpha_{i:n}$ in (7.2.27) may be replaced by $G(p_i)$.

Similarly, let us now consider the case in which the location parameter μ is known. In this case, $X_{i:n}$ may be replaced by $X_{i:n}-\mu$, leaving $\boldsymbol{\beta}^{-1}$ unchanged once again so that the asymptotically optimal linear estimator of σ can be written as

$$\sigma^* = \sum_{i=r+1}^{n-s} \delta_i^*(X_{i:n}-\mu), \tag{7.2.32}$$

where, for $i=r+1,\ r+2,\ldots,n-s$,

$$\delta_i^* = \frac{1}{nI_{22}} \sum_{j=r+1}^{n-s} \alpha_{j:n}(\boldsymbol{\beta}^{-1})_{i,j}; \tag{7.2.33}$$

we may determine asymptotically the coefficients δ_i^* (except for the two most extreme cases) by setting $u=p_i$ in the following continuous weight function:

$$\delta^*(u) = \frac{\phi_2(u)}{nI_{22}}. \tag{7.2.34}$$

The coefficients for the two extreme cases are given by

$$\delta_{r+1}^* = \delta^*(\xi_{r+1}) + \frac{\phi_{2,r+1}^*}{I_{22}} \tag{7.2.35}$$

and

$$\delta_{n-s}^* = \delta^*(\xi_{n-s}) - \frac{\phi_{2,n-s}^*}{I_{22}}, \tag{7.2.36}$$

where $\phi_2(u)$, $\phi_{2,r+1}^*$, and $\phi_{2,n-s}^*$ are as defined earlier. The variance of the estimator σ^* in (7.2.32) is $\sigma^2/(nI_{22})$.

The main advantage of this method of estimation is that it just requires the tabulation of the auxiliary functions $\gamma(u)$ and $\delta(u)$, instead of extensive tables of coefficients for different choices of censoring. Another advantage of this method is that even for small sample sizes, the loss in efficiency of these estimators relative to the BLUEs is only moderate in most cases. This method of estimation has been adopted by several authors, including Johns and Lieberman (1966), D'agostino (1971), Bain (1972), Engelhardt and Bain (1973, 1974), and D'agostino and Lee (1975, 1976), for a wide variety of populations.

A different method of optimal asymptotic estimation of the location and scale parameters of a population due to Jung (1955, 1962) will now be discussed. Some properties of the estimators derived by this approach have been studied by Chan (1971, 1974). Jung started with linear estimators of the form

$$T(h)=\frac{1}{n}\sum_{i=1}^{n} h(p_i)X_{i:n}, \tag{7.2.37}$$

where, as earlier, $p_i = i/(n+1)$, and $h(u)$ is a continuous differentiable function defined in the interval (0, 1). Under the assumption that $h(u)$ and its first four derivatives exist and are bounded in (0, 1), by using Taylor series expansions Jung (1955) has shown that

$$E(T(h)) = M(h) + O\left(\frac{1}{n}\right) \tag{7.2.38}$$

and

$$\operatorname{Var}(T(h)) = \frac{\sigma^2}{n} V(h) + O\left(\frac{1}{n^2}\right), \tag{7.2.39}$$

where

$$M(h) = \int_{-\infty}^{\infty} (\mu + \sigma x) h(F(x)) f(x)\, dx \tag{7.2.40}$$

and

$$V(h) = \int_{-\infty}^{\infty}\int_{-\infty}^{\infty} K(x, y) h(F(x)) h(F(y))\, dy\, dx, \tag{7.2.41}$$

with

$$K(x, y) = \begin{bmatrix} F(x)\{1-F(y)\}, & \text{for } x<y \\ \{1-F(x)\}F(y), & \text{for } x>y \end{bmatrix}; \tag{7.2.42}$$

the remainder terms $O(1/n)$ and $O(1/n^2)$ depend on the upper bounds of the derivatives of $h(u)$. We now wish to choose the function $h(u)$ such that $\eta^* = T(h)$ provides an estimate of the parameter $\eta = \ell_1\mu + \ell_2\sigma$, where ℓ_1 and ℓ_2 are arbitrary constants. Furthermore, among all linear estimators $T(h)$ that are unbiased for η (that is, $M(h) = \eta$), we wish to find the estimator $\eta^* = T(h_0)$ that is asymptotically best (that is, for which $V(h)$ is a minimum).

Let us now denote $H(x) = h(F(x))$. Then we wish to minimize

$$V(h) = \int_{-\infty}^{\infty}\int_{-\infty}^{\infty} K(x, y) H(x) H(y)\, dy\, dx \tag{7.2.43}$$

with respect to $H(x)$, subject to the two conditions

$$\int_{-\infty}^{\infty} H(x)f(x)\,dx = \ell_1 \tag{7.2.44}$$

and

$$\int_{-\infty}^{\infty} xH(x)f(x)\,dx = \ell_2. \tag{7.2.45}$$

By calculus of variation, Jung (1955) derived the optimal choice of $H(x)$ to be

$$H_0(x) = k_1 H_1(x) + k_2 H_2(x), \tag{7.2.46}$$

where $H_1(x)$ and $H_2(x)$ are the unique solutions of the integral equations

$$\int_{-\infty}^{\infty} K(x, y)H_1(y)\,dy = f(x) \tag{7.2.47}$$

and

$$\int_{-\infty}^{\infty} K(x, y)H_2(y)\,dy = xf(x); \tag{7.2.48}$$

k_1 and k_2 in (7.2.46) are to be determined from Eqs. (7.2.44) and (7.2.45). From the special form of $K(x, y)$, the solutions $H_1(x)$ and $H_2(x)$ may be obtained by differentiating Eqs. (7.2.47) and (7.2.48).

Let us now denote

$$\kappa_1(x) = -\frac{f'(x)}{f(x)}, \qquad \kappa_2(x) = -x\frac{f'(x)}{f(x)} - 1,$$

$$D_{ij} = \int_{-\infty}^{\infty} \kappa_i(x)\kappa_j(x)f(x)\,dx, \qquad i, j = 1, 2,$$

$$\mathbf{D} = \begin{bmatrix} D_{11} & D_{12} \\ D_{12} & D_{22} \end{bmatrix},$$

$$\mathbf{k} = \begin{pmatrix} k_1 \\ k_2 \end{pmatrix} \quad \text{and} \quad \boldsymbol{\ell} = \begin{pmatrix} \ell_1 \\ \ell_2 \end{pmatrix}.$$

Then the asymptotically best linear estimator $\eta^* = T(h_0)$ of $\eta = \ell_1\mu + \ell_2\sigma$ is given by

$$\eta^* = T(h_0) = \frac{1}{n}\sum_{i=1}^{n} h_0(p_i)X_{i:n}, \tag{7.2.49}$$

with

$$h_0(F(x)) = H_0(x) = -k_1\kappa_1'(x) - k_2\kappa_2'(x), \tag{7.2.50}$$

where

$$\mathbf{k} = \mathbf{D}^{-1}\boldsymbol{\ell}. \tag{7.2.51}$$

The variance of the estimator η^* in (7.2.49) is given by

$$\mathrm{Var}(\eta^*) = \frac{\sigma^2}{n}\boldsymbol{\ell}'\mathbf{D}^{-1}\boldsymbol{\ell} + O\left(\frac{1}{n^2}\right). \tag{7.2.52}$$

By showing that

$$D_{11} = E\left\{\frac{\partial \ln f(X;\mu,\sigma)}{\partial\mu}\right\}^2,$$

$$D_{12} = E\left\{\frac{\partial \ln f(X;\mu,\sigma)}{\partial\mu}\,\frac{\partial \ln f(X;\mu,\sigma)}{\partial\sigma}\right\},$$

and

$$D_{22} = E\left\{\frac{\partial \ln f(X;\mu,\sigma)}{\partial\sigma}\right\}^2,$$

we may easily establish that the estimators μ^* and σ^* are asymptotically jointly efficient, and also that each of them is efficient when the other is taken to be a nuisance parameter.

Example 7.2.5. Let us consider the Student's t distribution with m degrees of freedom ($m \geq 3$) with pdf

$$f(x) = \frac{1}{\sqrt{m\pi}}\,\frac{\Gamma\left(\frac{m+1}{2}\right)}{\Gamma\left(\frac{m}{2}\right)}\left(1+\frac{x^2}{m}\right)^{-(m+1)/2}.$$

In this case, we obtain

$$\kappa_1(x) = -\frac{f'(x)}{f(x)} = \frac{m+1}{m}\,\frac{x}{1+\frac{x^2}{m}},$$

$$\kappa_2(x) = -x\frac{f'(x)}{f(x)} - 1 = \frac{m+1}{m}\,\frac{x^2}{1+\frac{x^2}{m}} - 1,$$

$$H_1(x) = h_1(F(x)) = -\kappa_1'(x) = \frac{m+1}{m} \frac{1 - \frac{x^2}{m}}{\left(1 + \frac{x^2}{m}\right)^2},$$

and

$$H_2(x) = h_2(F(x)) = -\kappa_2'(x) = \frac{m+1}{m} \frac{2x}{\left(1 + \frac{x^2}{m}\right)^2}.$$

Next, we find

$$D_{11} = \int_{-\infty}^{\infty} \left\{ \frac{m+1}{m} \frac{x}{1 + \frac{x^2}{m}} \right\}^2 f(x)\, dx = \frac{m+1}{m+3},$$

$$D_{12} = \int_{-\infty}^{\infty} \left\{ \frac{m+1}{m} \frac{x}{1 + \frac{x^2}{m}} \right\}^2 \left\{ \frac{m+1}{m} \frac{x^2}{1 + \frac{x^2}{m}} - 1 \right\} f(x)\, dx$$

$$= 0 \quad \text{(since the integral is an odd function of } x\text{)},$$

and

$$D_{22} = \int_{-\infty}^{\infty} \left\{ \frac{m+1}{m} \frac{x^2}{1 + \frac{x^2}{m}} - 1 \right\}^2 f(x)\, dx = \frac{2m}{m+3}.$$

So, for the estimation of μ we choose $\ell_1 = 1$ and $\ell_2 = 0$ so that we get

$$\mathbf{k} = \begin{bmatrix} \frac{m+1}{m+3} & 0 \\ 0 & \frac{2m}{m+3} \end{bmatrix}^{-1} \begin{bmatrix} 1 \\ 0 \end{bmatrix} = \begin{bmatrix} \frac{m+3}{m+1} \\ 0 \end{bmatrix},$$

and the asymptotically best linear estimator of μ to be

$$\mu^* = T(h_0),$$

where

$$h_0(F(x)) = H_0(x) = \frac{m+3}{m} \frac{1 - \frac{x^2}{m}}{\left(1 + \frac{x^2}{m}\right)^2}.$$

The variance of this estimator is obtained from (7.2.52) to be

$$\operatorname{Var}(\mu^*) = \frac{\sigma^2}{n}\frac{m+3}{m+1} + O\left(\frac{1}{n^2}\right).$$

Similarly, for the estimation of σ we choose $\ell_1 = 0$ and $\ell_2 = 1$ so that we get

$$\mathbf{k} = \begin{bmatrix} \dfrac{m+1}{m+3} & 0 \\ 0 & \dfrac{2m}{m+3} \end{bmatrix}^{-1} \begin{bmatrix} 0 \\ 1 \end{bmatrix} = \begin{bmatrix} 0 \\ \dfrac{m+3}{2m} \end{bmatrix}$$

and the asymptotically best linear estimator of σ to be

$$\sigma^* = T(h_0),$$

where

$$h_0(F(x)) = H_0(x) = \frac{(m+1)(m+3)}{m^2}\frac{x}{\left(1+\dfrac{x^2}{m}\right)^2}.$$

The variance of this estimator is obtained from (7.2.52) to be

$$\operatorname{Var}(\sigma^*) = \frac{\sigma^2}{n}\frac{m+3}{2m} + O\left(\frac{1}{n^2}\right).$$

It is of interest to mention here that for the case of the normal distribution, by letting $m \to \infty$ we derive the weight functions for the estimators μ^* and σ^* to be 1 and $F^{-1}(u) = G(u)$ (the inverse of the standard normal cdf), respectively. These are exactly the same as the weight functions derived in Example 7.2.1. But the weight function $F^{-1}(u)$ of the estimator σ^* is unbounded in (0, 1) and hence Jung's method is not applicable in this case. However, as already seen in Example 7.2.1, Bennett's method does yield the asymptotically optimal linear estimator of σ to be one whose weight function is $F^{-1}(u)$ and whose variance is $\sigma^2/(2n)$.

7.3. Ogawa's Optimal Estimators Based on Selected Order Statistics

We discussed the best linear unbiased estimation of the location and scale parameters based on complete and doubly Type-II censored samples in great detail in Chapter 4. The drawback is that the coefficients in these estimators are functions of the means, variances, and covariances of order statistics that for many population distributions, as mentioned earlier in Chapter 2, can be computed numerically only for small sample sizes. However, as seen in the last section, the BLUEs in the case of large sample sizes may be approximated by the asymptotically optimal linear estimators, which are in terms of the pdf and cdf of the population. It has been observed that both the BLUEs and the asymptotically optimal linear estimators are asymptotically most efficient among estimators that are based on the same number of order statistics.

In this section we describe the best linear unbiased estimation and the asymptotically best linear estimation methods based on k optimally selected order statistics which yield the largest efficiency among the linear estimators based on all $\binom{n}{k}$ choices of subsets of k order statistics from a sample of size n. It was Mosteller (1946) who first advocated the use of only a few order statistics instead of all, especially when the sample size is large, for estimation purposes, based on the consideration that the resulting loss in efficiency is often offset by the saving of effort in collecting data and computing the estimates and by their robustness features. Considerable attention has been paid to this problem in the last 30 years or so. While a brief history of linear estimation based on selected order statistics can be found in Harter (1971), a more recent expository review article is due to Chan and Cheng (1988), and most of the details presented in this section may be found in their article.

Let

$$X_{n_1:n} \le X_{n_2:n} \le \cdots \le X_{n_k:n} \tag{7.3.1}$$

be the k selected order statistics from a sample of size n. Let us now denote

$$\mathbf{X} = (X_{n_1:n} \quad X_{n_2:n} \quad \cdots \quad X_{n_k:n})',$$

$$\mathbf{1} = (1 \quad 1 \quad \cdots \quad 1)'_{k\times 1},$$

$$\mathbf{Z} = (\mathbf{X} - \mu\mathbf{1})/\sigma,$$

$$\boldsymbol{\alpha} = (\alpha_{n_1:n} \quad \alpha_{n_2:n} \quad \cdots \quad \alpha_{n_k:n})',$$

and

$$\boldsymbol{\beta} = \begin{bmatrix} \beta_{n_1,n_1:n} & \beta_{n_1,n_2:n} & \cdots & \beta_{n_1,n_k:n} \\ \beta_{n_1,n_2:n} & \beta_{n_2,n_2:n} & \cdots & \beta_{n_2,n_k:n} \\ \vdots & \vdots & \cdots & \vdots \\ \beta_{n_1,n_k:n} & \beta_{n_2,n_k:n} & \cdots & \beta_{n_k,n_k:n} \end{bmatrix}.$$

Then, by applying the generalized Gauss–Markov theorem and proceeding exactly on the same lines as in Section 4.4, we may derive the BLUEs of μ and σ. Thus, when the scale parameter σ is known, we may obtain the BLUE of μ to be

$$\mu^* = \frac{\mathbf{1}'\boldsymbol{\beta}^{-1}\mathbf{X}}{\mathbf{1}'\boldsymbol{\beta}^{-1}\mathbf{1}} - \sigma \frac{\mathbf{1}'\boldsymbol{\beta}^{-1}\alpha}{\mathbf{1}'\boldsymbol{\beta}^{-1}\mathbf{1}}, \tag{7.3.2}$$

and the variance of this estimator is given by

$$\mathrm{Var}(\mu^*) = \sigma^2/(\mathbf{1}'\boldsymbol{\beta}^{-1}\mathbf{1}). \tag{7.3.3}$$

Similarly, when the location parameter μ is known, we may obtain the BLUE of σ to be

$$\sigma^* = \frac{\boldsymbol{\alpha}'\boldsymbol{\beta}^{-1}\mathbf{X}}{\boldsymbol{\alpha}'\boldsymbol{\beta}^{-1}\boldsymbol{\alpha}} - \mu \frac{\boldsymbol{\alpha}'\boldsymbol{\beta}^{-1}\mathbf{1}}{\boldsymbol{\alpha}'\boldsymbol{\beta}^{-1}\boldsymbol{\alpha}}, \tag{7.3.4}$$

and the variance of this estimator is given by

$$\mathrm{Var}(\sigma^*) = \sigma^2/(\boldsymbol{\alpha}'\boldsymbol{\beta}^{-1}\boldsymbol{\alpha}). \tag{7.3.5}$$

The estimators μ^* and σ^* in (7.3.2) and (7.3.4) are based on the k optimally selected order statistics that give the minimum $\mathrm{Var}(\mu^*)$ and $\mathrm{Var}(\sigma^*)$ in (7.3.3) and (7.3.5), respectively, among the estimators based on all possible $\binom{n}{k}$ choices of subsets of k order statistics from a sample of size n.

When the parameters μ and σ are both unknown, the BLUEs of μ and σ are similarly derived to be

$$\mu^* = \left\{ \frac{\boldsymbol{\alpha}'\boldsymbol{\beta}^{-1}\boldsymbol{\alpha}\mathbf{1}'\boldsymbol{\beta}^{-1} - \boldsymbol{\alpha}'\boldsymbol{\beta}^{-1}\mathbf{1}\boldsymbol{\alpha}'\boldsymbol{\beta}^{-1}}{(\boldsymbol{\alpha}'\boldsymbol{\beta}^{-1}\boldsymbol{\alpha})(\mathbf{1}'\boldsymbol{\beta}^{-1}\mathbf{1}) - (\boldsymbol{\alpha}'\boldsymbol{\beta}^{-1}\mathbf{1})^2} \right\} \mathbf{X}, \tag{7.3.6}$$

and

$$\sigma^* = \left\{ \frac{\mathbf{1}'\boldsymbol{\beta}^{-1}\mathbf{1}\boldsymbol{\alpha}'\boldsymbol{\beta}^{-1} - \mathbf{1}'\boldsymbol{\beta}^{-1}\boldsymbol{\alpha}\mathbf{1}'\boldsymbol{\beta}^{-1}}{(\boldsymbol{\alpha}'\boldsymbol{\beta}^{-1}\boldsymbol{\alpha})(\mathbf{1}'\boldsymbol{\beta}^{-1}\mathbf{1}) - (\boldsymbol{\alpha}'\boldsymbol{\beta}^{-1}\mathbf{1})^2} \right\} \mathbf{X}, \tag{7.3.7}$$

and the variance and covariance of these estimators are given by

$$\mathrm{Var}(\mu^*) = \sigma^2 \left\{ \frac{\boldsymbol{\alpha}'\boldsymbol{\beta}^{-1}\boldsymbol{\alpha}}{(\boldsymbol{\alpha}'\boldsymbol{\beta}^{-1}\boldsymbol{\alpha})(\mathbf{1}'\boldsymbol{\beta}^{-1}\mathbf{1}) - (\boldsymbol{\alpha}'\boldsymbol{\beta}^{-1}\mathbf{1})^2} \right\}, \tag{7.3.8}$$

$$\mathrm{Var}(\sigma^*) = \sigma^2 \left\{ \frac{\mathbf{1}'\boldsymbol{\beta}^{-1}\mathbf{1}}{(\boldsymbol{\alpha}'\boldsymbol{\beta}^{-1}\boldsymbol{\alpha})(\mathbf{1}'\boldsymbol{\beta}^{-1}\mathbf{1}) - (\boldsymbol{\alpha}'\boldsymbol{\beta}^{-1}\mathbf{1})^2} \right\}, \tag{7.3.9}$$

and

$$\mathrm{Cov}(\mu^*, \sigma^*) = -\sigma^2 \left\{ \frac{\boldsymbol{\alpha}'\boldsymbol{\beta}^{-1}\mathbf{1}}{(\boldsymbol{\alpha}'\boldsymbol{\beta}^{-1}\boldsymbol{\alpha})(\mathbf{1}'\boldsymbol{\beta}^{-1}\mathbf{1}) - (\boldsymbol{\alpha}'\boldsymbol{\beta}^{-1}\mathbf{1})^2} \right\}. \tag{7.3.10}$$

In this case the estimators μ^* and σ^* in (7.3.6) and (7.3.7) are based on the k optimally selected order statistics that give the minimum value of the generalized variance, viz., $\mathrm{Var}(\mu^*)\,\mathrm{Var}(\sigma^*) - \{\mathrm{Cov}(\mu^*, \sigma^*)\}^2$, among the estimators based on all possible $\binom{n}{k}$ choices of subsets of k order statistics from a sample of size n. We may note here that for a symmetric distribution with symmetrically chosen ranks, that is, $n_{k-i+1} = n - n_i + 1$ for $i = 1, 2, \ldots, [k/2]+1$, we have $\boldsymbol{\alpha}'\boldsymbol{\beta}^{-1}\mathbf{1} = 0$, and hence the estimators μ^* and σ^* in (7.3.6) and (7.3.7) (adjusted for bias) reduce to the estimators in (7.3.2) and (7.3.4), respectively.

The ranks of the optimally selected order statistics will be called 'optimal ranks' from now on and denoted by $(n_1^0\ n_2^0\ \cdots\ n_k^0)$. Craig (1943) derived the BLUE σ^* in (7.3.4) for the rectangular and polynomial distributions. Godwin (1949b) derived the BLUE σ^* in (7.3.4) based on the complete sample for several distributions. The significance of optimal selection of k order statistics has been well illustrated by Chan and Cheng (1988) with numerical results through which they have shown that high efficiency may be achieved by the k-optimum BLUE even when k/n is small.

For a few distributions, the BLUEs of μ and σ based on complete samples may even involve only extreme order statistics. For example, for the exponential distribution when σ is known, the estimator μ^* in (7.3.2) is

$$\mu^* = X_{1:n} - \sigma/n;$$

for the symmetric rectangular distribution, the estimators μ^* and σ^* in (7.3.6) and (7.3.7) may be shown to be

$$\mu^* = \frac{1}{2}(X_{1:n} + X_{n:n})$$

and

$$\sigma^* = \frac{n+1}{n-1}(X_{n:n} - X_{1:n}).$$

Kulldorff (1963a) and Kulldorff and Vännman (1973) have respectively shown that for distributions such as exponential and Pareto, $\mathrm{Var}(\sigma^*)$ in (7.3.5) can be expressed explicitly as

$$\sigma^2 \Big/ \left\{ \sum_{i=1}^{k} h(n_{i-1}, n_i) \right\}, \qquad 0 = n_0 < n_1 < \cdots < n_k \le n.$$

Then the following dynamic programming procedure converts the complex k-variate discrete optimization problem to a one-variate problem repeated k times. Let

$$H_{j-1}(n_j) = \max_{1 \le n_1 < \cdots < n_j} \sum_{i=1}^{j-1} h(n_{i-1}, n_i), \qquad j = 2, 3, \ldots, k+1, \, n_{k+1} = n.$$

Then

$$H_j(n_{j+1}) = \max_{j \le n_j < n_{j+1}} \{H_{j-1}(n_j) + h(n_j, n_{j+1})\}$$

and, hence,

$$\min\{\text{Var}(\sigma^*)\} = \sigma^2 / H_k(n).$$

Kulldorff (1963b) also established that if $(n_1^0 \; n_2^0 \; \cdots \; n_k^0)$ are the optimal ranks for the estimator σ^* in (7.3.4), then $(1 \; n_1^0+1 \; n_2^0+1 \; \cdots \; n_k^0+1)$ are the optimal ranks for the estimators μ^* and σ^* in (7.3.6) and (7.3.7) based on $k+1$ order statistics selected from a sample of size $n+1$. Determining the optimal ranks is the problem of maximizing a function of k discrete variables subject to the constraint

$$1 \le n_1 \le n_2 \le \cdots \le n_k \le n.$$

If $\binom{n}{k}$ is not large, a rather straightforward method is to compare the variances or the generalized variances of all $\binom{n}{k}$ possible estimators and choose the optimal estimator to be the one with the smallest variance. The advantage of this direct method, as mentioned by Chan and Cheng (1982), is that the same computer program can be used for any distribution. It should be pointed out here that for symmetric distributions, the optimal ranks are not necessarily symmetric about $(n+1)/2$ for odd n; for example, for the normal distribution when $n=7$ and $k=3$, the optimal ranks for μ^* in (7.3.2) are $(1 \; 3 \; 6)$; similarly, for the logistic distribution when $n=11$ and $k=2$, the optimal ranks for the estimators μ^* and σ^* in (7.3.6) and (7.3.7) are given by $(2 \; 9)$. However, for even values of n, numerical results reveal that the k optimal ranks are always symmetric.

In the following we develop the asymptotically best linear estimation of μ and σ based on k optimally selected order statistics. Let us first define a spacing (ξ_i) to be a fixed number, k, of proportions

$$0 < \xi_1 < \xi_2 < \cdots < \xi_k < 1.$$

Let us also use $\xi_0 = 0$ and $\xi_{k+1} = 1$. Let us determine the ranks of the k order statistics in (7.3.1) by $n_i = [n\xi_i] + 1$. $X_{n_i:n}$ is termed as the sample quantile of order ξ_i. Then, as $n \to \infty$, we have from Section 3.11 that

$$\alpha_{n_i:n} = F^{-1}(\xi_i) + O\left(\frac{1}{n}\right) \tag{7.3.11}$$

and

$$\beta_{n_i,n_j:n} = \frac{\xi_i(1-\xi_j)}{nf(F^{-1}(\xi_i))f(F^{-1}(\xi_j))} + O\left(\frac{1}{n^2}\right). \tag{7.3.12}$$

By replacing $\boldsymbol{\alpha}$ and $\boldsymbol{\beta}$ in the formulas of the BLUEs in Eqs. (7.3.2) through (7.3.10) by their approximate expressions in (7.3.11) and (7.3.12), respectively, we may obtain the corresponding asymptotically best linear estimators $\tilde{\mu}^*$ and $\tilde{\sigma}^*$ and their asymptotic variances and covariance. These estimators are asymptotically unbiased and are asymptotically equal to the BLUEs, as they have the same limiting normal distribution; see Miké (1971). For a given spacing, their asymptotic variance or asymptotic generalized variance is equal to the Cramer–Rao lower bound based on the same set of order statistics, and hence they are asymptotically efficient. Chan and Cheng (1988) have demonstrated that the asymptotic efficiencies of these asymptotically best linear estimators are quite high for a variety of populations, even when only a few order statistics are used. Following the pioneering work of Mosteller (1946), Yamanouchi (1949) derived the asymptotically best linear estimators of μ and σ for the normal case in which one of the parameters is known. A couple of years later, Ogawa (1951, 1952) carried out a detailed systematic study on the linear estimation based on selected order statistics by applying the generalized Gauss–Markov theorem directly to the joint asymptotic normal distribution of $\mathbf{X}$.

Now, by making use of the approximation of the covariance $\beta_{n_i,n_j:n}$ in (7.3.12) and the inverse of the variance–covariance matrix $\boldsymbol{\beta}$ obtained from this approximation as given in (7.2.4), we work out the asymptotic variance of $\tilde{\mu}^*$, when σ is known, to be

$$\mathrm{Var}(\tilde{\mu}^*) = \sigma^2/(nK_{11}) \tag{7.3.13}$$

and the asymptotic variance of $\tilde{\sigma}^*$, when μ is known, to be

$$\mathrm{Var}(\tilde{\sigma}^*) = \sigma^2/(nK_{22}); \tag{7.3.14}$$

similarly, when both parameters are unknown, the asymptotic variances and covariance of $\tilde{\mu}^*$ and $\tilde{\sigma}^*$ are obtained to be

$$\mathrm{Var}(\tilde{\mu}^*) = \frac{\sigma^2}{n}\frac{K_{22}}{K_{11}K_{22} - K_{12}^2}, \tag{7.3.15}$$

$$\mathrm{Var}(\tilde{\sigma}^*) = \frac{\sigma^2}{n} \frac{K_{11}}{K_{11}K_{22} - K_{12}^2}, \tag{7.3.16}$$

and

$$\mathrm{Cov}(\tilde{\mu}^*, \tilde{\sigma}^*) = -\frac{\sigma^2}{n} \frac{K_{12}}{K_{11}K_{22} - K_{12}^2}, \tag{7.3.17}$$

where

$$K_{11} = \sum_{i=1}^{k+1} \frac{\{f(F^{-1}(\xi_i)) - f(F^{-1}(\xi_{i-1}))\}^2}{\xi_i - \xi_{i-1}}, \tag{7.3.18}$$

$$K_{12} = \sum_{i=1}^{k+1} \frac{\{f(F^{-1}(\xi_i)) - f(F^{-1}(\xi_{i-1}))\}\{F^{-1}(\xi_i)f(F^{-1}(\xi_i)) - F^{-1}(\xi_{i-1})f(F^{-1}(\xi_{i-1}))\}}{\xi_i - \xi_{i-1}}, \tag{7.3.19}$$

and

$$K_{22} = \sum_{i=1}^{k+1} \frac{\{F^{-1}(\xi_i)f(F^{-1}(\xi_i)) - F^{-1}(\xi_{i-1})f(F^{-1}(\xi_{i-1}))\}^2}{\xi_i - \xi_{i-1}}, \tag{7.3.20}$$

with

$$\begin{aligned} f(F^{-1}(\xi_0)) &= F^{-1}(\xi_0)f(F^{-1}(\xi_0)) = f(F^{-1}(\xi_{k+1})) \\ &= F^{-1}(\xi_{k+1})f(F^{-1}(\xi_{k+1})) = 0. \end{aligned}$$

Mathematically, therefore, finding an optimum spacing becomes the problem of maximizing the functions K_{11}, K_{22}, or $(K_{11}K_{22} - K_{12}^2)$ of the k continuous variables $(\xi_1 \ \xi_2 \ \cdots \ \xi_k)$ subject to the constraint that $0 < \xi_1 < \xi_2 < \cdots < \xi_k < 1$.

Several methods have been used in literature for finding optimum spacings for individual distributions. We refrain from going into a description of these methods here. Interested readers may refer to the article by Chan and Cheng (1988) for a review on these methods and the references cited in it. Chan and Cheng (1988) have also presented a brief discussion on the adequacy of using the asymptotically best linear estimators as an approximation to the BLUEs in small samples based on a comparison of the efficiencies of these estimators.

We now mention some of the early examples from literature in order to demonstrate the usefulness of the linear estimation methods based on selected order statistics. Benson (1949), in order to estimate the standard deviation of the distribution of single thread strength in textiles, used the estimator $\tilde{\sigma}^*$ derived by minimizing the variance in (7.3.14) based on the normal distribution with $k=2$ selected order statistics; Benson observed this estimator to be robust against biases. The sample of $n=70$ observations were recorded in this case by a Moscrop automatic tester as points on a chart. As an accurate conversion of the 70 points to numerical values was time-consuming, and also an estimate based on inaccurate conversion would usually cause gross accumulated error, sample quantiles were used for estimation since they could be easily picked up from the chart. For estimating the average age at death for infants in the United States in 1955, Greenberg and Sarhan (1958) used a one-parameter exponential distribution and based the estimation on five optimal spacings from a sample of size $n=106{,}903$. In Table 7.3.1 we have given for the one-parameter exponential distribution the optimal spacings (for $1 \leq k \leq 25$), the corresponding coefficients, and the relative efficiencies of the estimator $\tilde{\sigma}^*$. From this table we see that for $k=5$, the optimal spacings are given by $\xi_1^0 = 0.393054$, $\xi_2^0 = 0.667045$, $\xi_3^0 = 0.843356$, $\xi_4^0 = 0.943378$, and $\xi_5^0 = 0.988495$; hence, the lower 39% of data was not required for the estimation of the parameter σ. This, as rightly pointed out by Greenberg and Sarhan (1958), is particularly useful in countries where ages of early deaths are not properly recorded.

For the purpose of illustration, let us consider the one-parameter exponential population with pdf

$$f(x;\sigma) = \frac{1}{\sigma} e^{-x/\sigma}, \qquad x > 0, \sigma > 0. \tag{7.3.21}$$

In this case we have

$$u_i = F^{-1}(\xi_i) = -\ln(1-\xi_i), \qquad i = 1, 2, \ldots, k;$$

by noting that σ^2/n is the Cramer–Rao lower bound in this case, we get the relative efficiency of the estimator $\tilde{\sigma}^*$ from (7.3.14) to be

$$\begin{aligned} K_{22} &= \sum_{i=1}^{k+1} \frac{\{u_i f(u_i) - u_{i-1} f(u_{i-1})\}^2}{\xi_i - \xi_{i-1}} \\ &= \sum_{i=1}^{k+1} \frac{\{u_i e^{-u_i} - u_{i-1} e^{-u_{i-1}}\}^2}{e^{-u_{i-1}} - e^{-u_i}}. \end{aligned} \tag{7.3.22}$$

The asymptotically best linear estimator of σ is then given by

$$\tilde{\sigma}^* = \frac{1}{K_{22}} \sum_{i=1}^{k} a_i X_{n_i:n} = \sum_{i=1}^{n} b_i X_{n_i:n}, \tag{7.3.23}$$

TABLE 7.3.1

Optimum Spacings, the Corresponding Coefficients, and Relative Efficiencies for $\tilde{\sigma}^*$ of the One-Parameter Exponential Distribution

	$k=1$, $K_{22}=0.647610$			$k=2$, $K_{22}=0.820263$			$k=3$, $K_{22}=0.891048$		
i	u_i^0	ξ_i^0	b_i	u_i^0	ξ_i^0	b_i	u_i^0	ξ_i^0	b_i
1	1.593624	0.796812	0.627501	1.017578	0.638531	0.523191	0.754033	0.529535	0.447688
2	—	—	—	2.611202	0.926554	0.179080	1.771611	0.829941	0.226589
3	—	—	—	—	—	—	3.365235	0.965446	0.077558

	$k=4$, $K_{22}=0.926911$			$k=5$, $K_{22}=0.947573$			$k=6$, $K_{22}=0.960561$		
i	u_i^0	ξ_i^0	b_i	u_i^0	ξ_i^0	b_i	u_i^0	ξ_i^0	b_i
1	0.600431	0.451425	0.390703	0.499316	0.393054	0.346332	0.427564	0.347904	0.310880
2	1.354464	0.741914	0.236088	1.099747	0.667045	0.231965	0.926880	0.604213	0.222788
3	2.372042	0.906710	0.119492	1.853780	0.843356	0.140168	1.527311	0.782881	0.149218
4	3.965666	0.981045	0.040900	2.871358	0.943378	0.070944	2.281344	0.897853	0.090167
5	—	—	—	4.464982	0.988495	0.024283	3.298922	0.963077	0.045637
6	—	—	—	—	—	—	4.892546	0.992498	0.015621

TABLE 7.3.1 (contd.)

	$k=7$, $K_{22}=0.969255$			$k=8$, $K_{22}=0.975360$			$k=9$, $K_{22}=0.979811$		
i	u_i^0	ξ_i^0	b_i	u_i^0	ξ_i^0	b_i	u_i^0	ξ_i^0	b_i
1	0.373938	0.311980	0.281942	0.332324	0.282745	0.257890	0.299072	0.258494	0.237596
2	0.801502	0.551345	0.211973	0.706262	0.506515	0.200958	0.631396	0.468151	0.190358
3	1.300818	0.727691	0.151908	1.133826	0.678200	0.151087	1.005334	0.634078	0.148335
4	1.901249	0.850618	0.101744	1.633142	0.804685	0.108275	1.432898	0.761384	0.111523
5	2.655282	0.929721	0.061480	2.233573	0.892855	0.072520	1.932214	0.855173	0.079922
6	3.672860	0.974596	0.031117	2.987606	0.949592	0.043821	2.532645	0.920551	0.053529
7	5.266484	0.994838	0.010651	4.005184	0.981779	0.022179	3.286678	0.962622	0.032346
8	—	—	—	5.598808	0.996298	0.007592	4.304256	0.986489	0.016371
9	—	—	—	—	—	—	5.897880	0.997255	0.005604

	$k=10$, $K_{22}=0.983156$			$k=11$, $K_{22}=0.985733$			$k=12$, $K_{22}=0.987761$		
i	u_i^0	ξ_i^0	b_i	u_i^0	ξ_i^0	b_i	u_i^0	ξ_i^0	b_i
1	0.271892	0.238063	0.220246	0.249236	0.220604	0.205246	0.230049	0.205505	0.192139
2	0.570964	0.435019	0.180417	0.521128	0.406150	0.171210	0.479285	0.380774	0.162732
3	0.903288	0.594765	0.144547	0.820200	0.559656	0.140249	0.751177	0.528189	0.135747
4	1.277226	0.721190	0.112637	1.152524	0.684161	0.112365	1.050249	0.650149	0.111198
5	1.704790	0.818189	0.084684	1.526462	0.782697	0.087560	1.382573	0.749068	0.089090
6	2.204106	0.889651	0.060688	1.954026	0.858298	0.065830	1.756511	0.827354	0.069423
7	2.804537	0.939465	0.040647	2.453342	0.913994	0.047176	2.184075	0.887418	0.052194
8	3.558570	0.971520	0.024562	3.053773	0.952819	0.031597	2.683391	0.931669	0.037404
9	4.576148	0.989706	0.012431	3.807806	0.977803	0.019093	3.283822	0.962515	0.025052
10	6.169772	0.997908	0.004255	4.825384	0.991976	0.009664	4.037855	0.982365	0.015138
11	—	—	—	6.419008	0.998370	0.003308	5.055433	0.993625	0.007662
12	—	—	—	—	—	—	6.649057	0.998705	0.002623

TABLE 7.3.1 (contd.)

	$k=13$, $K_{22}=0.989385$			$k=14$, $K_{22}=0.990706$			$k=15$, $K_{22}=0.991795$		
i	u_i^0	ξ_i^0	b_i	u_i^0	ξ_i^0	b_i	u_i^0	ξ_i^0	b_i
1	0.213623	0.192347	0.180594	0.199463	0.180829	0.170379	0.187012	0.170566	0.161269
2	0.443672	0.358324	0.154926	0.413086	0.338395	0.147740	0.386475	0.320552	0.141162
3	0.692908	0.499880	0.131215	0.643135	0.474358	0.126742	0.600098	0.451242	0.122406
4	0.964800	0.618941	0.109456	0.892371	0.590317	0.107344	0.830147	0.564015	0.105009
5	1.263872	0.717442	0.089662	1.164263	0.687847	0.089544	1.079383	0.660195	0.088937
6	1.596196	0.797334	0.071836	1.463335	0.768537	0.073351	1.351275	0.741090	0.074189
7	1.970134	0.860562	0.055977	1.795659	0.833982	0.058767	1.650347	0.808017	0.060773
8	2.397698	0.909073	0.042086	2.169597	0.885776	0.045794	1.982671	0.862299	0.048690
9	2.897014	0.944812	0.030160	2.597161	0.925515	0.034429	2.356609	0.905259	0.037941
10	3.497445	0.969725	0.020200	3.096477	0.954792	0.024673	2.784173	0.938220	0.028526
11	4.251478	0.985757	0.012206	3.696908	0.975200	0.016526	3.283489	0.962503	0.020442
12	5.269056	0.994852	0.006178	4.450941	0.988332	0.009986	3.883920	0.979430	0.013692
13	6.862680	0.998954	0.002115	5.468519	0.995783	0.005054	4.637953	0.990323	0.008273
14	—	—	—	7.062143	0.999143	0.001730	5.655531	0.996502	0.004187
15	—	—	—	—	—	—	7.249155	0.999289	0.001433

TABLE 7.3.1 (contd.)

	$k=16$, $K_{22}=0.992703$			$k=17$, $K_{22}=0.993468$			$k=18$, $K_{22}=0.994119$		
i	u_i^0	ξ_i^0	b_i	u_i^0	ξ_i^0	b_i	u_i^0	ξ_i^0	b_i
1	0.175879	0.161281	0.153017	0.166230	0.153149	0.145576	0.157695	0.145890	0.138954
2	0.362891	0.304338	0.135136	0.342109	0.289729	0.129482	0.323925	0.276695	0.124257
3	0.562354	0.430134	0.118287	0.529121	0.410877	0.114352	0.499804	0.393350	0.110520
4	0.775977	0.539746	0.102570	0.728584	0.517408	0.100095	0.686816	0.496824	0.097604
5	1.006026	0.634331	0.087992	0.942207	0.610233	0.086795	0.886279	0.587813	0.085436
6	1.255262	0.714999	0.074525	1.172256	0.690332	0.074459	1.099902	0.667096	0.074084
7	1.527154	0.782847	0.062167	1.421492	0.758646	0.063063	1.329951	0.735510	0.063555
8	1.826226	0.838980	0.050925	1.693384	0.816104	0.052606	1.579187	0.793857	0.053827
9	2.158550	0.884508	0.040800	1.992456	0.863640	0.043092	1.851079	0.842932	0.044902
10	2.532488	0.920539	0.031793	2.324780	0.902195	0.034525	2.150151	0.883533	0.036782
11	2.960052	0.948184	0.023903	2.698718	0.932708	0.026903	2.482475	0.916464	0.029469
12	3.459368	0.968550	0.017130	3.126282	0.956119	0.020227	2.856413	0.942525	0.022963
13	4.059799	0.982747	0.011473	3.625598	0.973367	0.014495	3.283977	0.962521	0.017265
14	4.813832	0.991883	0.006933	4.226029	0.985390	0.009709	3.783293	0.977252	0.012372
15	5.831410	0.997066	0.003509	4.980062	0.993126	0.005867	4.383724	0.987521	0.008287
16	7.425034	0.999404	0.001201	5.997640	0.997515	0.002969	5.137757	0.994129	0.005007
17	—	—	—	7.591264	0.999495	0.001016	6.155335	0.997878	0.002534
18	—	—	—	—	—	—	7.748959	0.999569	0.000867

TABLE 7.3.1 (contd.)

	$k = 19$, $K_{22} = 0.994677$			$k = 20$, $K_{22} = 0.995159$			$k = 21$, $K_{22} = 0.995579$		
i	u_i^0	ξ_i^0	b_i	u_i^0	ξ_i^0	b_i	u_i^0	ξ_i^0	b_i
1	0.149531	0.138888	0.132805	0.142344	0.132677	0.127039	0.135781	0.126966	0.121813
2	0.307226	0.264516	0.119588	0.291875	0.253138	0.115129	0.278125	0.242798	0.110862
3	0.473456	0.377154	0.106939	0.449570	0.362098	0.103671	0.427656	0.347964	0.100469
4	0.649335	0.477607	0.095117	0.615800	0.459791	0.092705	0.585351	0.443090	0.090471
5	0.836347	0.566710	0.084001	0.791679	0.546917	0.082457	0.751581	0.528380	0.080901
6	1.035810	0.645061	0.073529	0.978691	0.624197	0.072821	0.927460	0.604443	0.071957
7	1.249433	0.713333	0.063759	1.178154	0.692154	0.063742	1.114472	0.671912	0.063549
8	1.479482	0.772244	0.054697	1.391777	0.751367	0.055273	1.313935	0.731240	0.055626
9	1.728718	0.822488	0.046326	1.621826	0.802462	0.047417	1.527558	0.782935	0.048234
10	2.000610	0.864747	0.038644	1.871062	0.846040	0.040160	1.757607	0.827543	0.041379
11	2.299682	0.899709	0.031655	2.142954	0.882692	0.033500	2.006843	0.865588	0.035046
12	2.632006	0.928066	0.025362	2.442026	0.913016	0.027442	2.278735	0.897586	0.029234
13	3.005944	0.950508	0.019763	2.774350	0.937610	0.021986	2.577807	0.924060	0.023948
14	3.433508	0.967726	0.014858	3.148288	0.957074	0.017132	2.910131	0.945531	0.019187
15	3.932824	0.980412	0.010648	3.575852	0.972008	0.012881	3.284069	0.962525	0.014951
16	4.533255	0.989254	0.007132	4.075168	0.983011	0.009231	3.711633	0.975562	0.011241
17	5.287288	0.994945	0.004309	4.675599	0.990680	0.006183	4.210949	0.985168	0.008055
18	6.304866	0.998173	0.002181	5.429632	0.995615	0.003736	4.811380	0.991863	0.005395
19	7.898490	0.999629	0.000747	6.447210	0.998415	0.001891	5.565413	0.996172	0.003260
20	—	—	—	8.040834	0.999678	0.000647	6.582991	0.998616	0.001650
21	—	—	—	—	—	—	8.176615	0.999719	0.000565

TABLE 7.3.1 (contd.)

	$k=22$ $K_{22}=0.995946$			$k=23$ $K_{22}=0.996270$			$k=24$ $K_{22}=0.996556$		
i	u_i^0	ξ_i^0	b_i	u_i^0	ξ_i^0	b_i	u_i^0	ξ_i^0	b_i
1	0.130313	0.122179	0.117160	0.124844	0.117365	0.112923	0.119375	0.112525	0.108645
2	0.266094	0.233633	0.106890	0.255157	0.225205	0.103376	0.244219	0.216684	0.100188
3	0.408438	0.335312	0.097281	0.390938	0.323578	0.094313	0.374532	0.312389	0.091717
4	0.557969	0.427630	0.088162	0.533282	0.413324	0.085836	0.510313	0.399692	0.083677
5	0.715664	0.511133	0.079387	0.682813	0.494806	0.077789	0.652657	0.479339	0.076156
6	0.881894	0.586002	0.070990	0.840508	0.568509	0.070048	0.802188	0.551653	0.069016
7	1.057773	0.652772	0.063142	1.006738	0.634591	0.062638	0.959883	0.617062	0.062147
8	1.244785	0.711997	0.055764	1.182617	0.693524	0.055714	1.126113	0.675709	0.055574
9	1.444248	0.764077	0.048811	1.369629	0.745799	0.049203	1.301992	0.728011	0.049430
10	1.657871	0.809456	0.042326	1.569092	0.791766	0.043068	1.489004	0.774403	0.043654
11	1.887920	0.848614	0.036310	1.782715	0.831819	0.037346	1.688467	0.815197	0.038211
12	2.137156	0.882010	0.030753	2.012764	0.866381	0.032038	1.902090	0.850744	0.033134
13	2.409048	0.910099	0.025653	2.262000	0.895858	0.027135	2.132139	0.881417	0.028425
14	2.708120	0.933338	0.021014	2.533892	0.920650	0.022635	2.381375	0.907577	0.024074
15	3.040444	0.952186	0.016836	2.832964	0.941162	0.018542	2.653267	0.929579	0.020082
16	3.414382	0.967103	0.013119	3.165288	0.957798	0.014855	2.952339	0.947783	0.016451
17	3.841946	0.978548	0.009864	3.539226	0.970964	0.011576	3.284663	0.962547	0.013180
18	4.341262	0.986980	0.007069	3.966790	0.981066	0.008703	3.658601	0.974231	0.010270
19	4.941693	0.992858	0.004734	4.466106	0.988508	0.006237	4.086165	0.983196	0.007722
20	5.695726	0.996640	0.002861	5.066537	0.993696	0.004177	4.585481	0.989801	0.005534
21	6.713304	0.998785	0.001448	5.820570	0.997034	0.002524	5.185912	0.994405	0.003706
22	8.306928	0.999753	0.000496	6.838148	0.998928	0.001278	5.939945	0.997368	0.002240
23	—	—	—	8.431772	0.999782	0.000437	6.957523	0.999049	0.001134
24	—	—	—	—	—	—	8.551147	0.999807	0.000388

TABLE 7.3.1 (contd.)

	$k = 25$, $K_{22} = 0.996810$		
i	u_i^0	ξ_i^0	b_i
1	0.115000	0.108634	0.104715
2	0.234375	0.208935	0.096817
3	0.359219	0.301779	0.089281
4	0.489532	0.387087	0.081733
5	0.625313	0.464906	0.074567
6	0.767657	0.535901	0.067865
7	0.917188	0.600359	0.061503
8	1.074883	0.658662	0.055382
9	1.241113	0.710938	0.049524
10	1.416992	0.757558	0.044049
11	1.604004	0.798910	0.038902
12	1.803467	0.835273	0.034052
13	2.017090	0.866958	0.029527
14	2.247139	0.894299	0.025330
15	2.496375	0.917617	0.021454
16	2.768267	0.937229	0.017896
17	3.067339	0.953455	0.014660
18	3.399663	0.966615	0.011745
19	3.773601	0.977031	0.009152
20	4.201165	0.985022	0.006881
21	4.700481	0.990909	0.004931
22	5.300912	0.995013	0.003303
23	6.054945	0.997654	0.001996
24	7.072523	0.999152	0.001010
25	8.666147	0.999828	0.000346

where

$$a_i = e^{-u_i}\left\{\frac{u_i\, e^{-u_i} - u_{i-1}\, e^{-u_{i-1}}}{e^{-u_{i-1}} - e^{-u_i}} - \frac{u_{i+1}\, e^{-u_{i+1}} - u_i\, e^{-u_i}}{e^{-u_i} - e^{-u_{i+1}}}\right\}, \tag{7.3.24}$$

$$b_i = a_i / K_{22}, \tag{7.3.25}$$

and

$$n_i = [n\xi_i] + 1 \quad \text{for } i = 1, 2, \ldots, k.$$

The required optimal spacing is the set $(\xi_1^0\ \xi_2^0\ \cdots\ \xi_k^0)$, or equivalently $(u_1^0\ u_2^0\ \cdots\ u_k^0)$, which gives the maximum value of the relative efficiency K_{22} in (7.3.22). We may determine the optimal spacings step by step as follows.

k = 1. From (7.3.22) we get in this case

$$K_{22} = \frac{u_1^2}{e^{u_1} - 1}, \tag{7.3.26}$$

which has only one maximum $K_{22} = 0.647610$ at $u_1^0 = 1.593624$ and the corresponding value of $\xi_1^0 = 0.796812$.

k = 2. By setting $u_2 = u_1 + v$, we obtain from (7.3.22) that

$$K_{22} = e^{-u_1}\left\{\frac{u_1^2}{e^{u_1} - 1} + u_1^2 + \frac{v^2}{e^{v} - 1}\right\} \tag{7.3.27}$$

which gets maximized at the point ($u_1^0 = 1.017578$, $v^0 = 1.593624$), or ($u_1^0 = 1.017578$, $u_2^0 = 2.611202$), yielding ($\xi_1^0 = 0.638531$, $\xi_2^0 = 0.926554$), with the maximum value of K_{22} being 0.820263.

k = 3. By setting $u_2 = u_1 + v$ and $u_3 = u_2 + w = u_1 + v + w$, we obtain from (7.3.22) that

$$K_{22} = e^{-u_1}\left\{\frac{u_1^2}{e^{u_1} - 1} + u_1^2 + e^{-v}\left(\frac{v^2}{e^{v} - 1} + v^2 + \frac{w^2}{e^{w} - 1}\right)\right\}, \tag{7.3.28}$$

which gets maximized at the point ($u_1^0 = 0.754033$, $v^0 = 1.017578$, $w^0 = 1.593624$) or ($u_1^0 = 0.754033$, $u_2^0 = 1.771611$, $u_3^0 = 3.365235$), yielding ($\xi_1^0 = 0.529535$, $\xi_2^0 = 0.829941$, $\xi_3^0 = 0.965446$), with the maximum value of K_{22} being 0.891048.

In a similar way, the k-optimal spacings have been calculated for $k = 1(1)25$ and are presented in Table 7.3.1 along with the corresponding coefficients and the relative efficiencies of $\tilde{\sigma}^*$. It may be noted from this table that there is an increasing rate in relative efficiencies against the number of sample quantiles that have been selected. However, we observe that after $k = 10$ the gain in relative efficiency is not appreciable.

We shall present the following example, given by Ogawa (1962), in order to illustrate the use of the above-developed estimator.

Example 7.3.1. The data on the time intervals in days between explosions in mines involving more than 10 men killed from December 6, 1875, to May 29, 1951, has been given by Maguire *et al.* (1952). This data of size 109, with observations arranged in increasing order of magnitude, is presented in Table 7.3.2.

Suppose we want to estimate σ based on $k = 10$ optimally selected order statistics. From Table 7.3.1, we find

$$n_1^0 = 26, \quad n_2^0 = 48, \quad n_3^0 = 65, \quad n_4^0 = 79, \quad n_5^0 = 90,$$

$$n_6^0 = 97, \quad n_7^0 = 103, \quad n_8^0 = 106, \quad n_9^0 = 108, \quad n_{10}^0 = 109,$$

and

$$K_{22} = 0.983156.$$

We then obtain

$$\tilde{\sigma}^* = \sum_{i=1}^{10} b_i X_{n_i^0:n} = 242.1849.$$

When we compare the above estimate to the classical estimate of 241 based on the entire sample, we find a very close agreement.

7.4. Dixon's Simplified Linear Estimators

Dixon (1957) suggested some simple linear estimators for the mean and standard deviation of a normal population that are highly efficient. These estimators are in terms of quasi-midrange and quasi-range, respectively. Similar work has been carried out by Raghunandanan and Srinivasan (1970, 1971) for the logistic and double exponential distributions.

Let us denote the ith quasi-midrange of the sample by

$$V_i = \frac{1}{2}(X_{i:n} + X_{n-i+1:n}), \tag{7.4.1}$$

TABLE 7.3.2
Time Intervals in Days between Explosions in Mines Involving More Than 10 Men Killed from 6 December 1875 to 29 May 1951

Order	Observation	Order	Observation	Order	Observation
1	1	36	72	73	255
2	4	37	72	74	271
3	4	38	75	75	275
4	7	39	78	76	275
5	11	40	78	77	275
6	13	41	81	78	286
7	15	42	93	79	291
8	15	43	96	80	312
9	17	44	99	81	312
10	18	45	108	82	315
11	19	46	113	83	326
12	19	47	114	84	326
13	20	48	120	85	329
14	20	49	120	86	330
15	22	50	123	87	336
16	23	51	124	88	338
17	28	52	129	89	345
18	29	53	131	90	348
19	31	54	137	91	354
20	32	55	145	92	361
21	36	56	151	93	364
22	37	57	156	94	369
23	47	58	171	95	378
24	48	59	176	96	390
25	49	60	182	97	457
26	50	61	188	98	467
27	54	62	189	99	498
28	54	63	195	100	517
29	55	64	203	101	566
30	58	65	208	102	644
31	59	66	215	103	745
32	59	67	217	104	871
33	61	68	217	105	1205
34	61	69	217	106	1312
35	66	70	224	107	1357
		71	228	108	1613
		72	233	109	1630

and the ith quasi-range of the sample by

$$W_i = X_{n-i+1:n} - X_{i:n}. \tag{7.4.2}$$

For $i = 1$, W_1 is simply the sample range. Dixon (1957) defined an estimator of μ as that V_i with the smallest variance. In Table 7.4.1 we have given these estimators for n up to 20 and the values of their variances and the corresponding efficiency relative to the sample mean. Next, Dixon (1957) proposed an estimator of σ as the one with the smallest variance among estimators of the form

$$C \sum_{i=1}^{[n/2]} \delta_i W_i, \tag{7.4.3}$$

where each δ_i can take on the value 0 or 1 and C is a constant determined by making use of the expected values of normal order statistics so as to make the estimator in (7.4.3) unbiased for σ. In Table 7.4.2 we have given

TABLE 7.4.1
Estimator in (7.4.1) for the Mean of the Normal Population and Values of Variance and Efficiency Relative to the Sample Mean

n	i	(Variance)/σ^2	Eff
2	1	0.500	1.000
3	1	0.362	0.920
4	2	0.298	0.838
5	2	0.231	0.867
6	2	0.193	0.865
7	2	0.168	0.849
8	3	0.149	0.837
9	3	0.132	0.843
10	3	0.119	0.840
11	3	0.109	0.832
12	4	0.100	0.831
13	4	0.0924	0.833
14	4	0.0860	0.830
15	4	0.0808	0.825
16	5	0.0756	0.827
17	5	0.0711	0.827
18	5	0.0673	0.825
19	6	0.0640	0.823
20	6	0.0607	0.824
∞	0.27	—	0.810

TABLE 7.4.2
Estimator in (7.4.3) for the Standard Deviation of the Normal Population and Values of Variance and Relative Efficiencies

n	Estimator	(Variance)/σ^2	Eff_1	Eff_2
2	$0.8862\ W_1$	0.571	1.000	1.000
3	$0.5908\ W_1$	0.275	1.000	0.992
4	$0.4857\ W_1$	0.183	0.986	0.975
5	$0.4299\ W_1$	0.138	0.966	0.955
6	$0.2619\ (W_1+W_2)$	0.109	0.968	0.957
7	$0.2370\ (W_1+W_2)$	0.0895	0.978	0.967
8	$0.2197\ (W_1+W_2)$	0.0761	0.980	0.970
9	$0.2068\ (W_1+W_2)$	0.0664	0.979	0.968
10	$0.1968\ (W_1+W_2)$	0.0591	0.974	0.964
11	$0.1608\ (W_1+W_2+W_4)$	0.0529	0.977	0.967
12	$0.1524\ (W_1+W_2+W_4)$	0.0478	0.981	0.972
13	$0.1456\ (W_1+W_2+W_4)$	0.0436	0.984	0.975
14	$0.1399\ (W_1+W_2+W_4)$	0.0401	0.985	0.977
15	$0.1352\ (W_1+W_2+W_4)$	0.0372	0.985	0.977
16	$0.1311\ (W_1+W_2+W_4)$	0.0347	0.983	0.975
17	$0.1050\ (W_1+W_2+W_3+W_5)$	0.0325	0.985	0.978
18	$0.1020\ (W_1+W_2+W_3+W_5)$	0.0305	0.986	0.978
19	$0.09939\ (W_1+W_2+W_3+W_5)$	0.0288	0.986	0.979
20	$0.10446\ (W_1+W_2+W_4+W_6)$	0.0272	0.987	0.980

these estimators for n up to 20 and the values of their variances. Also given in this table are the values of the efficiency of these estimators relative to the BLUE of σ based on the entire sample of size n (values of Eff_1) and the efficiency relative to the unbiased sample standard deviation (values of Eff_2). From this table we observe that the loss in efficiency is negligible. Dixon (1957) also made use of the sample range and proposed a simple estimator of σ as

$$C^* W_1 = C^*(X_{n:n} - X_{1:n}), \tag{7.4.4}$$

where the constant C^* is determined as to make the estimator in (7.4.4) unbiased for σ. In Table 7.4.3 we have given the values of C^*, variances of the estimator in (7.4.4) and its efficiency relative to the BLUE of σ based on the entire sample of size n (values of Eff_1) and the efficiency relative to the unbiased sample standard deviation (values of Eff_2), for sample sizes up to 20. We note from this table that the estimator in (7.4.4) is quite efficient even though it is a very simple estimator based on just the two extreme order statistics; however, it is considerably less efficient than the estimator

TABLE 7.4.3
Estimator in (7.4.4) for the Standard Deviation of the Normal Population and Values of Variance and Relative Efficiencies

n	C^*	(Variance)$/\sigma^2$	Eff_1	Eff_2
2	0.886	0.571	1.000	1.000
3	0.591	0.275	1.000	0.992
4	0.486	0.183	0.984	0.975
5	0.430	0.138	0.966	0.955
6	0.395	0.112	0.944	0.933
7	0.370	0.0949	0.922	0.911
8	0.351	0.0829	0.900	0.890
9	0.337	0.0740	0.878	0.869
10	0.325	0.0671	0.858	0.850
11	0.315	0.0616	0.839	0.831
12	0.307	0.0571	0.821	0.814
13	0.300	0.0533	0.805	0.797
14	0.294	0.0502	0.787	0.781
15	0.288	0.0474	0.772	0.766
16	0.283	0.0451	0.756	0.751
17	0.279	0.0430	0.744	0.738
18	0.275	0.0412	0.731	0.725
19	0.271	0.0395	0.719	0.712
20	0.268	0.0381	0.703	0.700

in (7.4.3). For example, when $n = 20$, the estimator in (7.4.3) is 98% efficient relative to the unbiased sample standard deviation, while the estimator in (7.4.4) has an efficiency of only 70%.

Raghunandanan and Srinivasan (1970) carried out similar work for the logistic population with pdf

$$f(x; \mu, \sigma) = \frac{\pi}{\sigma\sqrt{3}} \frac{e^{-\pi(x-\mu)/\sigma\sqrt{3}}}{\{1 + e^{-\pi(x-\mu)/\sigma\sqrt{3}}\}^2}, \qquad -\infty < x < \infty. \tag{7.4.5}$$

They have given an estimator for μ of the form in (7.4.1). These estimators are given in Table 7.4.4 for sample sizes up to 20. The variance of this estimator and its efficiency relative to the BLUE based on the entire sample are also presented in this table. We note that this simple estimator of μ is quite efficient in this case and has its efficiency close to 90% almost in all cases. Raghunandanan and Srinivasan (1970) have also proposed an estimator for σ of the form in (7.4.3). These estimators are presented in Table 7.4.5 for sample sizes up to 20. Also given in this table are the values

TABLE 7.4.4
Estimator in (7.4.1) for the Mean of the Logistic Population and Values of Variance and Efficiency Relative to the BLUE

n	i	(Variance)/σ^2	Eff
2	1	0.5000	1.000
3	2	0.3921	0.831
4	2	0.2599	0.927
5	2	0.2071	0.922
6	3	0.1795	0.881
7	3	0.1482	0.910
8	3	0.1294	0.908
9	4	0.1171	0.889
10	4	0.1035	0.903
11	4	0.0940	0.905
12	5	0.0870	0.892
13	5	0.0795	0.899
14	5	0.0738	0.899
15	6	0.0693	0.892
16	6	0.0645	0.898
17	6	0.0607	0.897
18	7	0.0576	0.892
19	7	0.0542	0.897
20	7	0.0516	0.895

Produced with kind permission of Biometrika Trustees.

of the variance of this estimator and its efficiency relative to the BLUE based on the entire sample. We note that this estimator of σ is remarkably efficient and has an efficiency never less than 97%. Similar estimators have been given by the above authors for the symmetrically Type-II censored sample case as well.

By considering a double exponential population with pdf

$$f(x;\,\mu,\sigma)=\frac{1}{2\sigma}\,e^{-|(x-\mu)/\sigma|}, \qquad -\infty<x<\infty \tag{7.4.6}$$

and making use of the moments of order statistics from (7.4.6) tabulated by Govindarajulu (1966) for sample sizes 20 and less, Raghunandanan and Srinivasan (1971) have constructed estimators for μ and σ of the form (7.4.1) and (7.4.3), respectively. These estimators, together with their variance and efficiency relative to the corresponding BLUE based on the entire

TABLE 7.4.5
Estimator in (7.4.3) for the Standard Deviation of the Logistic Population and Values of Variance and Efficiency Relative to the BLUE

n	Estimator	(Variance)/σ^2	Eff
2	$0.9069\ W_1$	0.6449	1.000
3	$0.6046\ W_1$	0.3333	1.000
4	$0.3887\ (W_1+W_2)$	0.2324	0.970
5	$0.3109\ (W_1+W_2)$	0.1719	0.991
6	$0.2694\ (W_1+W_2)$	0.1376	0.995
7	$0.2101\ (W_1+W_2+W_3)$	0.1159	0.988
8	$0.1879\ (W_1+W_2+W_3)$	0.0989	0.995
9	$0.1724\ (W_1+W_2+W_3)$	0.0865	0.997
10	$0.1608\ (W_1+W_2+W_3)$	0.0773	0.993
11	$0.1346\ (W_1+W_2+W_3+W_4)$	0.0694	0.997
12	$0.1266\ (W_1+W_2+W_3+W_4)$	0.0631	0.998
13	$0.1154\ (W_1+W_2+W_3+W_4+W_6)$	0.0580	0.996
14	$0.1049\ (W_1+W_2+W_3+W_4+W_5)$	0.0535	0.997
15	$0.1215\ (W_1+W_2+W_4+W_5)$	0.0506	0.980
16	$0.0922\ (W_1+W_2+W_3+W_4+W_5+W_7)$	0.0464	0.998
17	$0.08808\ (W_1+W_2+W_3+W_4+W_5+W_7)$	0.0434	0.999
18	$0.084572\ (W_1+W_2+W_3+W_4+W_5+W_7)$	0.0409	0.999
19	$0.076906\ (W_1+W_2+W_3+W_4+W_5+W_6+W_8)$	0.0386	0.999
20	$0.074073\ (W_1+W_2+W_3+W_4+W_5+W_6+W_8)$	0.0366	1.000

Produced with kind permission of Biometrika Trustees.

sample, are presented in Tables 7.4.6 and 7.4.7. From Table 7.4.6 we observe that this simple estimator of μ is quite efficient and has efficiency never less than 90%. From Table 7.4.7 we note that the estimator in (7.4.3) for σ is remarkably efficient and has an efficiency always more than 99%. It should be mentioned here that Eff = 1 does not imply that the estimator in (7.4.3) coincides with the BLUE of σ; it simply means that the best coefficients are close to the value of C in (7.4.3). Raghunandanan and Srinivasan (1971) have also given similar estimators for μ and σ for the case in which the available sample is symmetrically Type-II censored.

7.5. Balakrishnan's Approximate Maximum Likelihood Estimation

In this section we extend the approximate maximum likelihood estimation method discussed in great detail in Chapter 6 to the case in which the estimation is to be based on a few order statistics. For the purpose of

TABLE 7.4.6
Estimator in (7.4.1) for the Mean of the Double Exponential Population and Values of Variance and Efficiency Relative to the BLUE

n	i	(Variance)/σ^2	Eff
2	1	1.000000	1.000
3	2	0.638890	0.923
4	2	0.420135	0.989
5	3	0.351180	0.902
6	3	0.260905	0.977
7	3	0.225805	0.940
8	4	0.187310	0.968
9	4	0.164795	0.959
10	5	0.145225	0.963
11	5	0.129605	0.967
12	6	0.118125	0.960
13	6	0.106670	0.970
14	7	0.099285	0.959
15	7	0.090540	0.972
16	7	0.085190	0.960
17	8	0.078575	0.972
18	8	0.074175	0.967
19	9	0.069350	0.973
20	9	0.065670	0.970

Reproduced from Raghunandanan and Srinivasan (1971), Table 1, p. 689, with permission of the Technometrics Management Committee.

illustration, we consider here two symmetric order statistics from a sample of size n from the normal $N(\mu, \sigma^2)$ distribution and derive the approximate maximum likelihood estimators (AMLEs) of μ and σ. Quite interestingly, the AMLEs of μ and σ turn out to be the sample midrange and quasi-range based on the given two symmetric order statistics, respectively. Furthermore, the estimator of σ based on the two extreme order statistics becomes almost the same as Dixon's simplified estimator based on the sample range that has been discussed in the last section. As in Chapter 6, we derive approximate expressions for the variances of the estimators and show that the two estimators are jointly highly efficient as compared to the BLUEs of μ and σ based on two order statistics.

To fix the ideas, let $X_{r+1:n}$ and $X_{n-r:n}$ be two symmetric order statistics available from a random sample of size n from a normal $N(\mu, \sigma^2)$ population. Further, let $Z_{i:n} = (X_{i:n} - \mu)/\sigma$ for $i = r+1,\ n-r$. Then the likelihood

TABLE 7.4.7
Estimator in (7.4.3) for σ of the Double Exponential Population and Values of Variance and Efficiency Relative to the BLUE

n	Estimator	(Variance)/σ^2	Eff
2	$0.666667\ W_1$	0.7778	1.000
3	$0.444444\ W_1$	0.4321	1.000
4	$0.289157\ (W_1+W_2)$	0.300624	0.993
5	$0.231325\ (W_1+W_2)$	0.229000	1.000
6	$0.183486\ (W_1+W_2+W_3)$	0.186515	0.996
7	$0.157274\ (W_1+W_2+W_3)$	0.156500	1.000
8	$0.134254\ (W_1+W_2+W_3+W_4)$	0.135438	0.997
9	$0.119337\ (W_1+W_2+W_3+W_4)$	0.119000	1.000
10	$0.108696\ (W_1+W_2+W_3+W_4)$	0.106392	0.998
11	$0.096208\ (W_1+W_2+W_3+W_4+W_5)$	0.096025	0.999
12	$0.088919\ (W_1+W_2+W_3+W_4+W_5)$	0.087615	0.999
13	$0.080615\ (W_1+W_2+W_3+W_4+W_5+W_6)$	0.080508	0.999
14	$0.075306\ (W_1+W_2+W_3+W_4+W_5+W_6)$	0.074500	1.000
15	$0.069383\ (W_1+W_2+W_3+W_4+W_5+W_6+W_7)$	0.069315	0.999
16	$0.065343\ (W_1+W_2+W_3+W_4+W_5+W_6+W_7)$	0.064810	0.999
17	$0.060904\ (W_1+W_2+W_3+W_4+W_5+W_6+W_7+W_8)$	0.060859	0.999
18	$0.057727\ (W_1+W_2+W_3+W_4+W_5+W_6+W_7+W_8)$	0.057356	0.999
19	$0.054276\ (W_1+W_2+W_3+W_4+W_5+W_6+W_7+W_8+W_9)$	0.054244	0.999
20	$0.051710\ (W_1+W_2+W_3+W_4+W_5+W_6+W_7+W_8+W_9)$	0.051447	0.999

Reproduced from Raghunandanan and Srinivasan (1971), Table 2, p. 690, with permission of the Technometrics Management Committee.

function based on the two symmetric order statistics is given by

$$L=\frac{n!}{(r!)^2(n-2r-2)!\sigma^2}\{F(Z_{r+1:n})\}^r\{F(Z_{n-r:n})-F(Z_{r+1:n})\}^{n-2r-2}$$
$$\times\{1-F(Z_{n-r:n})\}^r f(Z_{r+1:n})f(Z_{n-r:n}), \qquad (7.5.1)$$

where f and F denote the density and cumulative distribution functions of the standard normal population, respectively. Equation (7.5.1) yields the log-likelihood function to be

$$\ln L=\text{Const}-2\ln\sigma+r\ln F(Z_{r+1:n})+r\ln\{1-F(Z_{n-r:n})\}$$
$$+(n-2r-2)\ln\{F(Z_{n-r:n})-F(Z_{r+1:n})\}-\frac{1}{2}Z^2_{r+1:n}-\frac{1}{2}Z^2_{n-r:n}. \qquad (7.5.2)$$

From Eq. (7.5.2) we obtain the likelihood equations for μ and σ to be

$$\frac{\partial \ln L}{\partial \mu} = -\frac{1}{\sigma}[rg_1(Z_{r+1:n}) - rg_2(Z_{n-r:n}) + (n-2r-2)h_1(Z_{r+1:n}, Z_{n-r:n}) - Z_{r+1:n} - Z_{n-r:n}] = 0 \tag{7.5.3}$$

and

$$\frac{\partial \ln L}{\partial \sigma} = -\frac{2}{\sigma} - \frac{1}{\sigma}[rZ_{r+1:n}g_1(Z_{r+1:n}) - rZ_{n-r:n}g_2(Z_{n-r:n}) + (n-2r-2)h_2(Z_{r+1:n}, Z_{n-r:n}) - Z^2_{r+1:n} - Z^2_{n-r:n}] = 0, \tag{7.5.4}$$

where

$$g_1(Z_{r+1:n}) = f(Z_{r+1:n})/F(Z_{r+1:n}), \tag{7.5.5}$$

$$g_2(Z_{n-r:n}) = f(Z_{n-r:n})/\{1 - F(Z_{n-r:n})\}, \tag{7.5.6}$$

$$h_1(Z_{r+1:n}, Z_{n-r:n}) = \{f(Z_{n-r:n}) - f(Z_{r+1:n})\}/\{F(Z_{n-r:n}) - F(Z_{r+1:n})\}, \tag{7.5.7}$$

and

$$h_2(Z_{r+1:n}, Z_{n-r:n}) = \frac{Z_{n-r:n}f(Z_{n-r:n}) - Z_{r+1:n}f(Z_{r+1:n})}{F(Z_{n-r:n}) - F(Z_{r+1:n})}. \tag{7.5.8}$$

The likelihood equations (7.5.3) and (7.5.4) do not admit explicit estimators for μ and σ. However, by following the steps of Section 6.3, we may approximate the functions $g_1(Z_{r+1:n})$ and $g_2(Z_{n-r:n})$ in (7.5.5) and (7.5.6), respectively, by the linear functions

$$g_1(Z_{r+1:n}) \simeq \alpha - \beta Z_{r+1:n} \tag{7.5.9}$$

and

$$g_2(Z_{n-r:n}) \simeq \alpha + \beta Z_{n-r:n}, \tag{7.5.10}$$

where

$$p_{r+1} = 1 - q_{r+1} = (r+1)/(n+1),$$

$$\xi_{r+1} = F^{-1}(p_{r+1}),$$

$$\alpha = f(\xi_{r+1})\{1 + \xi^2_{r+1} + \xi_{r+1}f(\xi_{r+1})/p_{r+1}\}/p_{r+1}, \tag{7.5.11}$$

and

$$\beta = f(\xi_{r+1})\{f(\xi_{r+1}) + p_{r+1}\xi_{r+1}\}/p^2_{r+1}. \tag{7.5.12}$$

It has also been shown in Section 6.3 that $\beta > 0$.

Following the lines of Section 6.3, let us now consider the function

$$h_1^*(Z_{r+1:n}, Z_{n-r:n}) = f(Z_{n-r:n})/\{F(Z_{n-r:n}) - F(Z_{r+1:n})\}.$$

By expanding this function in a bivariate Taylor series around $(\xi_{r+1}, -\xi_{r+1})$ (see Arnold and Balakrishnan (1989) and David (1981)), we may then approximate it by

$$h_1^*(Z_{r+1:n}, Z_{n-r:n}) \simeq \gamma_0 + \gamma_1 Z_{r+1:n} - \gamma_2 Z_{n-r:n}, \tag{7.5.13}$$

where

$$\gamma_1 = \left\{\frac{f(\xi_{r+1})}{1-2p_{r+1}}\right\}^2 \geq 0,$$

$$\gamma_2 = \frac{f(\xi_{r+1})}{(1-2p_{r+1})^2}\{f(\xi_{r+1}) - \xi_{r+1}(1-2p_{r+1})\} \geq 0,$$

and

$$\gamma_0 = \frac{f(\xi_{r+1})}{1-2p_{r+1}} - (\gamma_1 + \gamma_2)\xi_{r+1} \geq 0.$$

Proceeding similarly, we also obtain the linear approximation

$$\begin{aligned} h_2^*(Z_{r+1:n}, Z_{n-r:n}) &= f(Z_{r+1:n})/\{F(Z_{n-r:n}) - F(Z_{r+1:n})\} \\ &\simeq \gamma_0 + \gamma_2 Z_{r+1:n} - \gamma_1 Z_{n-r:n}. \end{aligned} \tag{7.5.14}$$

By making use of the linear approximations in (7.5.13) and (7.5.14), we may approximate the functions $h_1(Z_{r+1:n}, Z_{n-r:n})$ and $h_2(Z_{r+1:n}, Z_{n-r:n})$ in Eqs. (7.5.7) and (7.5.8) by

$$h_1(Z_{r+1:n}, Z_{n-r:n}) \simeq -\delta(Z_{r+1:n} + Z_{n-r:n}) \tag{7.5.15}$$

and

$$\begin{aligned} h_2(Z_{r+1:n}, Z_{n-r:n}) \simeq{}& \gamma_0(Z_{n-r:n} - Z_{r+1:n}) + 2\gamma_1 Z_{r+1:n} Z_{n-r:n} \\ & - \gamma_2(Z_{r+1:n}^2 + Z_{n-r:n}^2), \end{aligned} \tag{7.5.16}$$

where $\delta = \gamma_2 - \gamma_1 \geq 0$. Upon using the approximations in (7.5.9), (7.5.10), (7.5.15), and (7.5.16), we obtain the approximate likelihood equations for μ and σ from Eqs. (7.5.3) and (7.5.4) to be

$$\begin{aligned} \frac{\partial \ln L}{\partial \mu} &\simeq \frac{\partial \ln L^*}{\partial \mu} \\ &= -\frac{1}{\sigma}[r(\alpha - \beta Z_{r+1:n}) - r(\alpha + \beta Z_{n-r:n}) - Z_{r+1:n} - Z_{n-r:n} \\ &\quad -(n-2r-2)\delta(Z_{r+1:n} + Z_{n-r:n})] \end{aligned} \tag{7.5.17}$$

and

$$\frac{\partial \ln L}{\partial \sigma} \simeq \frac{\partial \ln L^*}{\partial \sigma}$$

$$= -\frac{2}{\sigma} - \frac{1}{\sigma}[rZ_{r+1:n}(\alpha - \beta Z_{r+1:n}) - rZ_{n-r:n}(\alpha + \beta Z_{n-r:n})$$

$$+(n-2r-2)\{\gamma_0(Z_{n-r:n} - Z_{r+1:n}) + 2\gamma_1 Z_{r+1:n} Z_{n-r:n}$$

$$-\gamma_2(Z_{r+1:n}^2 + Z_{n-r:n}^2)\} - Z_{r+1:n}^2 - Z_{n-r:n}^2]$$

$$= 0. \quad (7.5.18)$$

From Eq. (7.5.17) we derive the AMLE of μ to be

$$\hat{\mu} = \frac{1}{2}(X_{r+1:n} + X_{n-r:n}), \quad (7.5.19)$$

which is simply the sample midrange based on the two given symmetric order statistics. Now, upon using the estimator of μ from (7.5.19) in Eq. (7.5.18) and solving for σ, we obtain the following quadratic equation:

$$2\sigma^2 + B(X_{n-r:n} - X_{r+1:n})\sigma - C(X_{n-r:n} - X_{r+1:n})^2 = 0, \quad (7.5.20)$$

where

$$B = (n-2r-2)\gamma_0 - r\alpha$$

and

$$C = (n-2r-2)\gamma_1 + \frac{1}{2}\{1 + r\beta + (n-2r-2)\delta\}.$$

From Eq. (7.5.20) we derive the AMLE of σ as

$$\hat{\sigma} = (X_{n-r:n} - X_{r+1:n})\left\{\frac{-B + (B^2 + 8C)^{1/2}}{4}\right\}, \quad (7.5.21)$$

which is simply a constant multiple of the quasi-range based on the given two symmetric order statistics. It should be mentioned here that upon solving Eq. (7.5.20) for σ, we get two roots, one of which drops out since $C \geq 0$.

Remark 1. It is easy to see that the estimator $\hat{\mu}$ is unbiased for μ. Further, since the estimator $\hat{\sigma}$ in (7.5.21) is an approximate solution of the likelihood equation, it is asymptotically unbiased for σ. It is observed that the estimator $\hat{\sigma}$ has only a small bias even for small sample sizes. The estimator $\hat{\sigma}$ in

(7.5.21), for the case in which the two extreme observations are given, is just a constant multiple of the sample range. In this case, as mentioned in the last section, Dixon (1957) has considered the estimator $\tilde{\sigma} = C^*(X_{n:n} - X_{1:n})$ and has tabulated the values of C^* (the unbiasing factor for the sample range) and $\text{Var}(\tilde{\sigma})$ for sample sizes up to 20; see Table 7.4.3.

For example, we have the following values:

$n = 10$:

$$\text{Unbiasing factor of } \hat{\sigma} = 0.317731$$

$$\text{Unbiasing factor of } \tilde{\sigma} = 0.324938$$

$$(\text{Variance of } \hat{\sigma})/\sigma^2 = 0.064143$$

$$(\text{Variance of } \tilde{\sigma})/\sigma^2 = 0.067077$$

$n = 20$:

$$\text{Unbiasing factor of } \hat{\sigma} = 0.264845$$

$$\text{Unbiasing factor of } \tilde{\sigma} = 0.267741$$

$$(\text{Variance of } \hat{\sigma})/\sigma^2 = 0.038892$$

$$(\text{Variance of } \tilde{\sigma})/\sigma^2 = 0.038064$$

$n = 30$:

$$\text{Unbiasing factor of } \hat{\sigma} = 0.239684$$

$$\text{Unbiasing factor of } \tilde{\sigma} = 0.244767$$

$$(\text{Variance of } \hat{\sigma})/\sigma^2 = 0.029950$$

$$(\text{Variance of } \tilde{\sigma})/\sigma^2 = 0.028744.$$

From the above values it is quite apparent that the AMLE of σ in (7.5.21) based on the two extreme order statistics is nearly the same as Dixon's simplified estimator based on the sample range. Furthermore, the estimator $\hat{\sigma}$ in (7.5.21) shows that the estimator based on the sample quasi-range is approximately the MLE of σ.

From (7.5.17) and (7.5.18) we derive

$$E\left(-\frac{\partial^2 \ln L}{\partial \mu^2}\right) \simeq E\left(-\frac{\partial^2 \ln L^*}{\partial \mu^2}\right) = m/\sigma^2 \tag{7.5.22}$$

and

$$E\left(-\frac{\partial^2 \ln L}{\partial \sigma^2}\right) \simeq E\left(-\frac{\partial^2 \ln L^*}{\partial \sigma^2}\right) = D/\sigma^2, \tag{7.5.23}$$

where

$$m = 2\{1 + r\beta + (n - 2r - 2)\delta\} \tag{7.5.24}$$

and

$$D = 6\{1 + r\beta + (n - 2r - 2)\gamma_2\}E(Z^2_{n-r:n}) \\ -6(n - 2r - 2)\gamma_1 E(Z_{r+1:n}Z_{n-r:n}) - 4BE(Z_{n-r:n}) - 2. \tag{7.5.25}$$

From these expressions we obtain

$$\text{Var}(\hat{\mu}) \simeq \sigma^2/m \tag{7.5.26}$$

and

$$\text{Var}(\hat{\sigma}) \simeq \sigma^2/D. \tag{7.5.27}$$

It is important to mention here that the estimators $\hat{\mu}$ and $\hat{\sigma}$ are uncorrelated, as they are based on two symmetric order statistics. It may also be noted from the fact that $E(-\partial^2 \ln L/\partial\mu\, \partial\sigma) = 0$. From (7.5.26) and (7.5.27) we may compute the variances of the two estimators by using the extensive tables of means and product moments of normal order statistics prepared by Harter (1961) and Tietjen *et al.* (1977).

We may also use the formulas for $\text{Var}(\hat{\mu})$ and $\text{Var}(\hat{\sigma})$ in (7.5.26) and (7.5.27), respectively, in order to select the two optimal symmetric order statistics for a given sample size. The individual optimal estimators of μ and σ can be based on that value of r for which the variance is minimum, and the joint estimation of μ and σ can be based on that value of r for which the sum of the variances, for example, is minimum. The individual optimal estimator of μ was determined by this process for various sample sizes; the results for sample sizes 10, 20, and 30 are given below. Also given are the values of the variance of Dixon's estimator $\tilde{\mu}$ of μ, based on two selected order statistics; see Section 7.4 for details.

n	r	r'	Exact $\text{Var}(\hat{\mu})/\sigma^2$	Approximate $\text{Var}(\hat{\mu})/\sigma^2$	Exact $\text{Var}(\tilde{\mu})/\sigma^2$
10	2	2	0.117263	0.119018	0.117263
20	5	5	0.060234	0.060681	0.060234
30	7	—	0.040476	0.040756	—

In the above table, r' denotes the optimal choice of r for Dixon's (1957) estimator of μ. We see that for $n = 10$ and 20, the approximate maximum likelihood estimation approach yields exactly the same estimator of μ as was given by Dixon (1957).

The joint optimal estimators of μ and σ were also determined by this approximate maximum likelihood estimation approach for various sample sizes; the results for sample sizes 10, 20, and 30 are presented in the following table. Also given in this table are the values of r', $\text{Var}(\mu^*)/\sigma^2$, and $\text{Var}(\sigma^*)/\sigma^2$ taken from the tables of Chan and Chan (1973), where μ^* and σ^* are the BLUEs of μ and σ based on the two optimally selected symmetric order statistics, viz., $X_{r'+1:n}$ and $X_{n-r':n}$.

	$n=10$		$n=20$	
	BLUE	AMLE	BLUE	AMLE
Value of r'	1	—	2	—
Value of r	—	1	—	2
Coefficients for μ	0.5000	0.5000	0.5000	0.5000
	0.5000	0.5000	0.5000	0.5000
(Exact Variance)/σ^2	0.128948	0.128948	0.070900	0.070900
(Approximate Variance)/σ^2	—	0.123407	—	0.068132
Coefficients for σ	−0.4993	−0.4626	−0.4421	−0.4269
	+0.4993	+0.4626	+0.4421	+0.4269
(Exact Variance)/σ^2	0.085337	0.073264	0.040586	0.037847
(Approximate Variance)/σ^2	—	0.054630	—	0.031039

	$n=30$	
	BLUE	AMLE
Value of r'	—	—
Value of r	—	4
Coefficients for μ	—	0.5000
	—	0.5000
(Exact Variance)/σ^2	—	0.045043
(Approximate Variance)/σ^2	—	0.044086
Coefficients for σ	—	−0.4733
	—	+0.4733
(Exact Variance)/σ^2	—	0.027798
(Approximate Variance)/σ^2	—	0.024488

Remark 2. It is quite clear from the above table that the two approaches yield identical optimal estimators for μ for sample sizes 10 and 20. It is also clear that the optimal estimators for μ and σ based on the AMLE approach are based on exactly the same two symmetric order statistics as those of the optimal BLUEs of μ and σ determined by Chan and Chan (1973). Furthermore, we observe that the coefficients of the two order

statistics in the optimal AMLE of σ are almost the same as those of the optimal BLUE of σ.

This strongly suggests the possible use of the AMLE approach for the optimal estimation of μ and σ based on k selected order statistics. It should be remarked here that the optimal estimation by AMLE approach described in this section, because of its simplicity and computational ease, may possibly be adopted for values of k greater than 2 and larger sample sizes, but this warrants further investigation.

7.6. Estimation of Population Quantiles

In Chapter 4 and also in Sections 2 and 3 of this chapter, we paid a great deal of attention to the linear estimation of the location and scale parameters of a population. We focus our attention in this section on the estimation of quantiles of the population distribution, and we discuss the best linear unbiased estimation and also the asymptotically best linear unbiased estimation of population quantiles.

Order statistics arise in a natural way while one estimates the quantile function defined by

$$G(u) = F^{-1}(u) = \inf\{x\colon F(x) \geq u\}. \tag{7.6.1}$$

Based on a random sample of size n and the corresponding order statistics $X_{1:n} \leq X_{2:n} \leq \cdots \leq X_{n:n}$ obtained from a population with distribution function $F(x)$, an estimate of F is given by the empirical distribution function $F_n(x)$ defined by

$$F_n(x) = \begin{cases} 0, & \text{for } x < X_{1:n}, \\ (i-1)/n, & \text{for } X_{i-1:n} \leq x < X_{i:n}, \\ 1, & \text{for } x \geq X_{n:n}. \end{cases} \tag{7.6.2}$$

It is well known that $F_n(x) \to F(x)$ in probability. So, a natural estimate of the quantile function $G(u)$ in (7.6.1) is given by the sample quantile function $G_n(u)$ defined by

$$\begin{aligned} G_n(u) = F_n^{-1}(u) &= \inf\{x\colon F_n(x) \geq u\} \\ &= X_{i:n}, \qquad \text{for } \frac{i-1}{n} < u \leq \frac{i}{n}. \end{aligned} \tag{7.6.3}$$

The above sample quantile function has been studied in detail by Csörgo and Revesz (1981). They have established several properties of the function $G_n(u)$ that are similar to the known properties of the empirical distribution function $F_n(x)$. It should be noted that $G_n(u)$ gives a nonparametric estimate of $G(u)$ that is based on a single order statistic, regardless of the form of the population distribution $F(x)$. A modified estimate of $G(u)$ has been proposed by Parzen (1979) that uses adjacent order statistics and is given by

$$\tilde{G}_n(u) = n\left(\frac{i}{n} - u\right) X_{i-1:n} + n\left(u - \frac{i-1}{n}\right) X_{i:n}, \qquad \text{for } \frac{i-1}{n} < u \le \frac{i}{n}. \tag{7.6.4}$$

But, when the form of the population distribution $F(x)$ is known (with the location and scale parameters being unknown), more efficient estimators of the quantile function $G(u)$ than the ones in (7.6.3) and (7.6.4) may be constructed by the use of a few selected order statistics through the BLUE or the ABLUE approach.

Let $X_{1:n} \le X_{2:n} \le \cdots \le X_{n:n}$ be the order statistics obtained from a population with cdf $F(x; \mu, \sigma)$. We assume that the parameters μ and σ are unknown while the form of F is known. Then, $Z_{i:n} = (X_{i:n} - \mu)/\sigma$, $1 \le i \le n$, are order statistics from the standardized population with cdf $F(z)$. Now, with $G(u; \mu, \sigma)$ and $G(u)$ denoting the quantile functions of the original and the standardized populations, respectively, we observe the relation

$$G(u; \mu, \sigma) = \mu + \sigma G(u). \tag{7.6.5}$$

For a specified $u \in (0, 1)$, we are interested in estimating the function $G(u; \mu, \sigma)$ by using a few selected order statistics.

Let $n_1, n_2, \ldots, n_k$ (where $1 \le n_1 < n_2 < \cdots < n_k \le n$) denote the ranks of the k selected order statistics from a sample of size n. Further, let us use the notations

$$\mathbf{X} = (X_{n_1:n} \quad X_{n_2:n} \quad \cdots \quad X_{n_k:n})',$$

$$\boldsymbol{\alpha} = (\alpha_{n_1:n} \quad \alpha_{n_2:n} \quad \cdots \quad \alpha_{n_k:n})',$$

$$\mathbf{1} = (1 \quad 1 \quad \cdots \quad 1)'_{k \times 1},$$

and

$$\boldsymbol{\beta} = \begin{bmatrix} \beta_{n_1,n_1:n} & \beta_{n_1,n_2:n} & \cdots & \beta_{n_1,n_k:n} \\ \beta_{n_1,n_2:n} & \beta_{n_2,n_2:n} & \cdots & \beta_{n_2,n_k:n} \\ \vdots & \vdots & \cdots & \vdots \\ \beta_{n_1,n_k:n} & \beta_{n_2,n_k:n} & \cdots & \beta_{n_k,n_k:n} \end{bmatrix}.$$

Then the BLUEs of μ and σ based on the k selected order statistics are given by

$$\mu^* = -\boldsymbol{\alpha}'\boldsymbol{\Delta}\mathbf{X} \tag{7.6.6}$$

and

$$\sigma^* = \mathbf{1}'\boldsymbol{\Delta}\mathbf{X}, \tag{7.6.7}$$

where $\boldsymbol{\Delta}$ is a skew-symmetric matrix of order k given by

$$\boldsymbol{\Delta} = \left\{\frac{\boldsymbol{\beta}^{-1}(\mathbf{1}\boldsymbol{\alpha}' - \boldsymbol{\alpha}\mathbf{1}')\boldsymbol{\beta}^{-1}}{(\boldsymbol{\alpha}'\boldsymbol{\beta}^{-1}\boldsymbol{\alpha})(\mathbf{1}'\boldsymbol{\beta}^{-1}\mathbf{1}) - (\boldsymbol{\alpha}'\boldsymbol{\beta}^{-1}\mathbf{1})^2}\right\}. \tag{7.6.8}$$

Also, the variances and covariance of these estimators are given by

$$\mathrm{Var}(\mu^*) = \sigma^2\left\{\frac{\boldsymbol{\alpha}'\boldsymbol{\beta}^{-1}\boldsymbol{\alpha}}{(\boldsymbol{\alpha}'\boldsymbol{\beta}^{-1}\boldsymbol{\alpha})(\mathbf{1}'\boldsymbol{\beta}^{-1}\mathbf{1}) - (\boldsymbol{\alpha}'\boldsymbol{\beta}^{-1}\mathbf{1})^2}\right\}, \tag{7.6.9}$$

$$\mathrm{Var}(\sigma^*) = \sigma^2\left\{\frac{\mathbf{1}'\boldsymbol{\beta}^{-1}\mathbf{1}}{(\boldsymbol{\alpha}'\boldsymbol{\beta}^{-1}\boldsymbol{\alpha})(\mathbf{1}'\boldsymbol{\beta}^{-1}\mathbf{1}) - (\boldsymbol{\alpha}'\boldsymbol{\beta}^{-1}\mathbf{1})^2}\right\}, \tag{7.6.10}$$

and

$$\mathrm{Cov}(\mu^*, \sigma^*) = -\sigma^2\left\{\frac{\boldsymbol{\alpha}'\boldsymbol{\beta}^{-1}\mathbf{1}}{(\boldsymbol{\alpha}'\boldsymbol{\beta}^{-1}\boldsymbol{\alpha})(\mathbf{1}'\boldsymbol{\beta}^{-1}\mathbf{1}) - (\boldsymbol{\alpha}'\boldsymbol{\beta}^{-1}\mathbf{1})^2}\right\}. \tag{7.6.11}$$

Because of the relation in (7.6.5), we therefore obtain the BLUE of the quantile function $G(u;\,\mu,\sigma)$ based on the k selected order statistics to be

$$G^*(u;\,\mu,\sigma) = \mu^* + \sigma^* G(u) = \{G(u)\mathbf{1}' - \boldsymbol{\alpha}'\}\boldsymbol{\Delta}\mathbf{X}, \tag{7.6.12}$$

where $G(u)$ is the known quantile function of the standardized population. Furthermore, the variance of this estimator is derived by using Eqs. (7.6.9) through (7.6.11) to be

$$\mathrm{Var}(G^*(u;\,\mu,\sigma)) = \sigma^2\left\{\frac{\boldsymbol{\alpha}'\boldsymbol{\beta}^{-1}\boldsymbol{\alpha} + G^2(u)\mathbf{1}'\boldsymbol{\beta}^{-1}\mathbf{1} - 2G(u)\boldsymbol{\alpha}'\boldsymbol{\beta}^{-1}\mathbf{1}}{(\boldsymbol{\alpha}'\boldsymbol{\beta}^{-1}\boldsymbol{\alpha})(\mathbf{1}'\boldsymbol{\beta}^{-1}\mathbf{1}) - (\boldsymbol{\alpha}'\boldsymbol{\beta}^{-1}\mathbf{1})^2}\right\}. \tag{7.6.13}$$

In the special case when the population distribution $F(z)$ is symmetric about zero and the ranks of the selected order statistics are symmetrically placed, that is, $n_i + n_{k-i+1} = n+1$, we have $\boldsymbol{\alpha}'\boldsymbol{\beta}^{-1}\mathbf{1} = 0$ and, consequently, the above formulas reduce to

$$\mu^* = \frac{\mathbf{1}'\boldsymbol{\beta}^{-1}}{\mathbf{1}'\boldsymbol{\beta}^{-1}\mathbf{1}}\mathbf{X}, \tag{7.6.14}$$

$$\sigma^* = \frac{\boldsymbol{\alpha}'\boldsymbol{\beta}^{-1}}{\boldsymbol{\alpha}'\boldsymbol{\beta}^{-1}\boldsymbol{\alpha}}\mathbf{X}, \tag{7.6.15}$$

$$\mathrm{Var}(\mu^*) = \sigma^2/(\mathbf{1}'\boldsymbol{\beta}^{-1}\mathbf{1}), \tag{7.6.16}$$

$$\mathrm{Var}(\sigma^*) = \sigma^2/(\boldsymbol{\alpha}'\boldsymbol{\beta}^{-1}\boldsymbol{\alpha}), \tag{7.6.17}$$

$$G^*(u;\,\mu,\sigma) = \left\{\frac{\mathbf{1}'\boldsymbol{\beta}^{-1}}{\mathbf{1}'\boldsymbol{\beta}^{-1}\mathbf{1}} + G(u)\,\frac{\boldsymbol{\alpha}'\boldsymbol{\beta}^{-1}}{\boldsymbol{\alpha}'\boldsymbol{\beta}^{-1}\boldsymbol{\alpha}}\right\}\mathbf{X}, \tag{7.6.18}$$

and

$$\operatorname{Var}(G^*(u;\mu,\sigma)) = \sigma^2\left\{\frac{1}{\mathbf{1}'\boldsymbol{\beta}^{-1}\mathbf{1}} + \frac{G^2(u)}{\boldsymbol{\alpha}'\boldsymbol{\beta}^{-1}\boldsymbol{\alpha}}\right\}. \tag{7.6.19}$$

In either case, the optimal ranks for the k selected order statistics are those values of $n_1, n_2, \ldots, n_k$ which minimize the value of $\operatorname{Var}(G^*(u;\mu,\sigma))$ computed from either (7.6.13) or (7.6.19). This optimal choice of $(n_1, n_2, \ldots, n_k)$ is usually found by searching through all $\binom{n}{k}$ possible collection of ranks, since analytic solutions are almost always difficult to obtain. For example, the optimal spacings have been determined for the exponential distribution by Umbach *et al.* (1981a) and Ali *et al.* (1982), and for the Pareto distribution by Ali *et al.* (1981a).

For the derivation of the ABLUE of quantiles, let us first of all call a k-tuple $(p_1, p_2, \ldots, p_k)$ as a spacing if $0<p_1<p_2<\cdots p_k<1$. Then, with $0<p_1<p_2<\cdots<p_k<1$ fixed, $n_i=[np_i]+1$ ($[m]$ indicating the greatest integer contained in m), $G_i=G(p_i)=F^{-1}(p_i)$, $\mathbf{1}=(1\ 1\ \cdots\ 1)'_{k\times 1}$, $\mathbf{G}=(G_1\ G_2\ \cdots\ G_k)$, and $\mathbf{X}=(X_{n_1:n}\ X_{n_2:n}\ \cdots\ X_{n_k:n})'$, Mosteller (1946) has shown that the asymptotic distribution of $\mathbf{X}$ is a k-variate normal distribution with mean $\mu\mathbf{1}+\sigma\mathbf{G}$ and covariance matrix $(\sigma^2/n)\tilde{\boldsymbol{\beta}}$, where

$$\tilde{\boldsymbol{\beta}}_{ij} = \frac{p_i(1-p_j)}{f(G_i)f(G_j)}, \qquad 1\le i\le j\le k; \tag{7.6.20}$$

see Section 3.11 for details. Now by following the same lines as those used in the derivation of the BLUEs, we derive the ABLUEs of μ and σ through the Gauss–Markov theorem as

$$\tilde{\mu}^* = -\mathbf{G}'\tilde{\boldsymbol{\Delta}}\mathbf{X} \tag{7.6.21}$$

and

$$\tilde{\sigma}^* = \mathbf{1}'\tilde{\boldsymbol{\Delta}}\mathbf{X}, \tag{7.6.22}$$

where $\tilde{\boldsymbol{\Delta}}$ is a skew-symmetric matrix of order k given by

$$\tilde{\boldsymbol{\Delta}} = \left\{\frac{\tilde{\boldsymbol{\beta}}^{-1}(\mathbf{1}\mathbf{G}'-\mathbf{G}\mathbf{1}')\tilde{\boldsymbol{\beta}}^{-1}}{(\mathbf{G}'\tilde{\boldsymbol{\beta}}^{-1}\mathbf{G})(\mathbf{1}'\tilde{\boldsymbol{\beta}}^{-1}\mathbf{1})-(\mathbf{G}'\tilde{\boldsymbol{\beta}}^{-1}\mathbf{1})^2}\right\}. \tag{7.6.23}$$

Also, the variances and covariance of these estimators are given by

$$\operatorname{Var}(\tilde{\mu}^*) = \frac{\sigma^2}{n}\left\{\frac{\mathbf{G}'\tilde{\boldsymbol{\beta}}^{-1}\mathbf{G}}{(\mathbf{G}'\tilde{\boldsymbol{\beta}}^{-1}\mathbf{G})(\mathbf{1}'\tilde{\boldsymbol{\beta}}^{-1}\mathbf{1})-(\mathbf{G}'\tilde{\boldsymbol{\beta}}^{-1}\mathbf{1})^2}\right\}, \tag{7.6.24}$$

$$\operatorname{Var}(\tilde{\sigma}^*) = \frac{\sigma^2}{n}\left\{\frac{\mathbf{1}'\tilde{\boldsymbol{\beta}}^{-1}\mathbf{1}}{(\mathbf{G}'\tilde{\boldsymbol{\beta}}^{-1}\mathbf{G})(\mathbf{1}'\tilde{\boldsymbol{\beta}}^{-1}\mathbf{1})-(\mathbf{G}'\tilde{\boldsymbol{\beta}}^{-1}\mathbf{1})^2}\right\}, \tag{7.6.25}$$

and

$$\operatorname{Cov}(\tilde{\mu}^*, \tilde{\sigma}^*) = -\frac{\sigma^2}{n}\left\{\frac{\mathbf{G}'\tilde{\boldsymbol{\beta}}^{-1}\mathbf{1}}{(\mathbf{G}'\tilde{\boldsymbol{\beta}}^{-1}\mathbf{G})(\mathbf{1}'\tilde{\boldsymbol{\beta}}^{-1}\mathbf{1}) - (\mathbf{G}'\tilde{\boldsymbol{\beta}}^{-1}\mathbf{1})^2}\right\}. \tag{7.6.26}$$

Because of the relation in (7.6.5), we derive the ABLUE of the quantile function $G(u; \mu, \sigma)$ based on the k selected order statistics $X_{n_1:n}$, $X_{n_2:n}, \ldots, X_{n_k:n}$ to be

$$\tilde{G}^*(u; \mu, \sigma) = \tilde{\mu}^* + \tilde{\sigma}^* G(u) = \{G(u)\mathbf{1}' - \mathbf{G}'\}\tilde{\boldsymbol{\Delta}}\mathbf{X}, \tag{7.6.27}$$

where $G(u)$ is the known quantile function of the standardized population. Also, the variance of this estimator is obtained by using Eqs. (7.6.24) through (7.6.26) to be

$$\operatorname{Var}(\tilde{G}^*(u; \mu, \sigma)) = \frac{\sigma^2}{n}\left\{\frac{\mathbf{G}'\tilde{\boldsymbol{\beta}}^{-1}\mathbf{G} + G^2(u)\mathbf{1}'\tilde{\boldsymbol{\beta}}^{-1}\mathbf{1} - 2G(u)\mathbf{G}'\tilde{\boldsymbol{\beta}}^{-1}\mathbf{1}}{(\mathbf{G}'\tilde{\boldsymbol{\beta}}^{-1}\mathbf{G})(\mathbf{1}'\tilde{\boldsymbol{\beta}}^{-1}\mathbf{1}) - (\mathbf{G}'\tilde{\boldsymbol{\beta}}^{-1}\mathbf{1})^2}\right\}. \tag{7.6.28}$$

As before, some simplification is possible in the above formulas in the special case when the population distribution $F(z)$ is symmetric about zero and the ranks of the k selected order statistics are symmetrically placed, that is, $n_i + n_{k-i+1} = n+1$. In this case, since $\mathbf{G}'\tilde{\boldsymbol{\beta}}^{-1}\mathbf{1} = 0$, we obtain

$$\tilde{\mu}^* = \frac{\mathbf{1}'\tilde{\boldsymbol{\beta}}^{-1}}{\mathbf{1}'\tilde{\boldsymbol{\beta}}^{-1}\mathbf{1}}\mathbf{X}, \tag{7.6.29}$$

$$\tilde{\sigma}^* = \frac{\mathbf{G}'\tilde{\boldsymbol{\beta}}^{-1}}{\mathbf{G}'\tilde{\boldsymbol{\beta}}^{-1}\mathbf{G}}\mathbf{X}, \tag{7.6.30}$$

$$\operatorname{Var}(\tilde{\mu}^*) = \frac{\sigma^2}{n(\mathbf{1}'\boldsymbol{\beta}^{-1}\mathbf{1})}, \tag{7.6.31}$$

$$\operatorname{Var}(\tilde{\sigma}^*) = \frac{\sigma^2}{n(\mathbf{G}'\boldsymbol{\beta}^{-1}\mathbf{G})}, \tag{7.6.32}$$

$$\tilde{G}^*(u; \mu, \sigma) = \left\{\frac{\mathbf{1}'\tilde{\boldsymbol{\beta}}^{-1}}{\mathbf{1}'\tilde{\boldsymbol{\beta}}^{-1}\mathbf{1}} + G(u)\frac{\mathbf{G}'\tilde{\boldsymbol{\beta}}^{-1}}{\mathbf{G}'\tilde{\boldsymbol{\beta}}^{-1}\mathbf{G}}\right\}\mathbf{X}, \tag{7.6.33}$$

and

$$\operatorname{Var}(\tilde{G}^*(u; \mu, \sigma)) = \frac{\sigma^2}{n}\left\{\frac{1}{\mathbf{1}'\tilde{\boldsymbol{\beta}}^{-1}\mathbf{1}} + \frac{G^2(u)}{\mathbf{G}'\tilde{\boldsymbol{\beta}}^{-1}\mathbf{G}}\right\}. \tag{7.6.34}$$

Now, the matrix $\tilde{\boldsymbol{\beta}}^{-1}$ may be explicitly written down as follows:

$$(\tilde{\boldsymbol{\beta}}^{-1})_{i,i} = f^2(G_i)\left\{\frac{1}{p_{i+1} - p_i} + \frac{1}{p_i - p_{i-1}}\right\}, \qquad 1 \le i \le k, \tag{7.6.35}$$

$$(\tilde{\boldsymbol{\beta}}^{-1})_{i,i-1} = (\tilde{\boldsymbol{\beta}}^{-1})_{i-1,i} = -f(G_{i-1})f(G_i)/(p_i - p_{i-1}), \quad 2 \le i \le k, \tag{7.6.36}$$

and

$$(\tilde{\boldsymbol{\beta}}^{-1})_{i,j} = 0 \quad \text{for } |i-j| \ge 2, \tag{7.6.37}$$

where $p_0 = 0$ and $p_{k+1} = 1$. By making use of the formulas in Eqs. (7.6.35) through (7.6.37), we may write

$$\begin{aligned} K_1 &= \mathbf{1}'\tilde{\boldsymbol{\beta}}^{-1}\mathbf{1} \\ &= \sum_{i=1}^{k+1} \frac{\{f(G_i) - f(G_{i-1})\}^2}{p_i - p_{i-1}}, \end{aligned} \tag{7.6.38}$$

$$\begin{aligned} K_2 &= \mathbf{1}'\tilde{\boldsymbol{\beta}}^{-1}\mathbf{1} \\ &= \sum_{i=1}^{k+1} \frac{\{G_i f(G_i) - G_{i-1} f(G_{i-1})\}^2}{p_i - p_{i-1}}, \end{aligned} \tag{7.6.39}$$

and

$$\begin{aligned} K_3 &= \mathbf{G}'\tilde{\boldsymbol{\beta}}^{-1}\mathbf{1} \\ &= \sum_{i=1}^{k+1} \frac{\{G_i f(G_i) - G_{i-1} f(G_{i-1})\}\{f(G_i) - f(G_{i-1})\}}{p_i - p_{i-1}}, \end{aligned} \tag{7.6.40}$$

where $f(G_0) = f(G_{k+1}) = 0$ and $G_0 f(G_0) = G_{k+1} f(G_{k+1}) = 0$. By using the formulas in (7.6.38) through (7.6.40), we may express the ABLUEs of μ and σ in Eqs. (7.6.21) and (7.6.22), respectively, as

$$\tilde{\mu}^* = (K_2 Y_1 - K_3 Y_2)/(K_1 K_2 - K_3^2) \tag{7.6.41}$$

and

$$\tilde{\sigma}^* = (K_1 Y_2 - K_3 Y_1)/(K_1 K_2 - K_3^2), \tag{7.6.42}$$

where

$$Y_1 = \sum_{i=1}^{k} a_i X_{n_i:n} \tag{7.6.43}$$

and

$$Y_2 = \sum_{i=1}^{k} b_i X_{n_i:n}, \tag{7.6.44}$$

with

$$a_i = f(G_i)\left\{\frac{f(G_i) - f(G_{i-1})}{p_i - p_{i-1}} - \frac{f(G_{i+1}) - f(G_i)}{p_{i+1} - p_i}\right\} \tag{7.6.45}$$

and

$$b_i = f(G_i)\left\{\frac{G_i f(G_i) - G_{i-1} f(G_{i-1})}{p_i - p_{i-1}} - \frac{G_{i+1} f(G_{i+1}) - G_i f(G_i)}{p_{i+1} - p_i}\right\} \quad (7.6.46)$$

for $i = 1, 2, \ldots, k$. As a result, we obtain the ABLUE of the quantile function $G(u; \mu, \sigma)$ based on the k selected order statistics $X_{n_1:n}, X_{n_2:n}, \ldots, X_{n_k:n}$ from (7.6.27) to be

$$\tilde{G}^*(u; \mu, \sigma) = \frac{1}{K_1 K_2 - K_3^2}\{(K_2 - K_3 G(u)) Y_1 - (K_3 - K_1 G(u)) Y_2\}, \quad (7.6.47)$$

and the variance of this estimator is obtained from (7.6.28) to be

$$\operatorname{Var}(\tilde{G}^*(u; \mu, \sigma)) = \frac{\sigma^2}{n}\left\{\frac{K_2 + K_1 G^2(u) - 2K_3 G(u)}{K_1 K_2 - K_3^2}\right\}. \quad (7.6.48)$$

In the special case in which the population distribution is symmetric and the spacing is also symmetric, we see from (7.6.40) that $K_3 = 0$, so that the ABLUE in (7.6.47) becomes

$$\tilde{G}^*(u; \mu, \sigma) = \frac{Y_1}{K_1} + \frac{Y_2}{K_2} G(u), \quad (7.6.49)$$

and the variance in (7.6.48) becomes

$$\operatorname{Var}(\tilde{G}^*(u; \mu, \sigma)) = \frac{\sigma^2}{n}\left\{\frac{1}{K_1} + \frac{1}{K_2} G^2(u)\right\}. \quad (7.6.50)$$

Hence, by choosing $(p_1, p_2, \ldots, p_k)$ that minimizes the variance in (7.6.48) or (7.6.50), we may select the optimal order statistics for estimating the quantile function $G(u; \mu, \sigma)$. This has been carried out, for example, by Kubat and Epstein (1980) for the normal and Gumbel distributions with $k = 2$ and 3.

7.7. Details of Other Related Work

Literature on the problems of estimation dealt with in this chapter is rather extensive. For the sake of completeness, however, we list in this section several important articles that have appeared on these topics. For the convenience of the reader we present two sets of references, the first one giving those dealing with the estimation of the location and scale parameters based on selected order statistics, and the second set giving those dealing with the estimation of population quantiles:

Adatia and Chan (1981), Balmer *et al.* (1974), Beyer *et al.* (1976), Birnbaum and Laska (1967), Birnbaum *et al.* (1971), Birnbaum and Miké (1970), Bloch (1966), Bofinger (1975), Cane (1974), Chan (1969, 1970), Chan and Chan (1973), Chan *et al.* (1971, 1972, 1973a, b, 1974), Chan and Cheng (1971, 1972, 1973, 1974, 1982, 1988), Chan and Kabir (1969), Chan and Mead (1971a, b), Chan (1989), Cheng (1975, 1983), Chernoff (1971), Dalenius (1950), Dyer (1973), Dyer and Whisenand (1973b), Eisenberger (1968), Eisenberger and Posner (1965), Fama and Roll (1971), Gupta and Gnanadesikan (1966), Hammersley and Morton (1954), Harter (1961b, 1971), Hassanein (1968, 1969a, b, 1971, 1972, 1977), Higuchi (1956), Johns and Lieberman (1966), Jung (1955), Kabir and Ahsanullah (1979), Kaminsky (1972, 1973, 1974), Kulldorff (1961, 1963a, b, 1964, 1973), Kulldorff and Vännman (1973), Mann and Fertig (1977), Miké (1971), Miyamoto (1972), Ogawa (1976), Saleh (1966, 1967), Saleh and Ali (1966), Saleh *et al.* (1984, 1985), Saleh and Sen (1985), Särndal (1962, 1964), Shelnutt (1966), Shelnutt *et al.* (1973), Siddiqui and Raghunandanan (1967), Tischendorf (1955), Ukita (1955), Weiss (1963), Wong (1988), and Yamanouchi (1949).

Ali *et al.* (1981a, b, 1982, 1983, 1985), Breiman *et al.* (1979), Cohen *et al.* (1985), Erto and Guida (1985), Eubank (1981a, b, 1986), Harrell and Davis (1982), Hassanein *et al.* (1984, 1985, 1986a, b), Hosking and Wallis (1987), Kaigh and Lachenbruch (1982), Kappenman (1987), Keating (1983), Koutrouvelis (1981), Kubat and Epstein (1980), Likeš (1985), Oppenlander (1986), Oppenlander *et al.* (1988), Parzen (1978, 1979), Reiss (1980), Robertson (1977), Rukhin and Strawderman (1982), Saleh (1981), Saleh *et al.* (1982, 1983), Saleh and Hassanein (1986), Umbach *et al.* (1981a, b, 1984), Walsh (1957), and Weissman (1978, 1980).

Chapter 8 Cohen–Whitten Estimators: Using First Order Statistics

8.1. Preliminary Remarks

This chapter provides an opportunity to present important applications of the first order statistics in estimating the threshold parameter of a positively skewed distribution. This first order statistic is, of course, an upper bound on the threshold parameter, and it contains more information about this parameter than any other sample observation, perhaps more than all other sample observations combined. In complete samples, the first order statistic is the smallest of n complete observations. In right-censored samples, it is the smallest of N observations of which n are complete while c are censored, and $N = n + c$. This chapter will be limited to consideration of complete samples only.

Modifications of the traditional moment and maximum likelihood estimators which incorporate the first order statistic for estimating parameters of the three-parameter Weibull, lognormal, inverse Gaussian, and gamma distributions have been given by Cohen (1951, 1975), Cohen *et al.* (1984, 1985), Cohen and Whitten (1980, 1981, 1982a, b, 1985, 1986, 1988), Cohen and Norgaard (1977), Chan *et al.* (1984), and Whitten *et al.* (1988), and perhaps by others. In most instances, these estimators are easy to calculate, and they avoid most of the regularity problems that are often prevalent with maximum likelihood estimators and sometimes with moment estimators. Modified moment estimators, which are featured in this chapter, are, like moment estimators, unbiased with respect to population means and variances. Although they might not be optimal in a strict sense, various simulation studies have shown them to be near optimal as demonstrated by comparisons of estimate variances and/or mean square errors.

Estimating equations that employ the first order statistic are $E(X_{1:N}) = x_{1:n}$ for the Weibull and the exponential distributions. For the inverse Gaussian and the gamma distributions, they are $E[F(X_{1:n})] = F(x_{1:n})$. For the lognormal distribution, the equation is $E[\ln(X_{1:n} - \gamma)] = \ln(X_{1:n} - \gamma)$, where γ is the threshold parameter. Expected values of the first order statistic in the Weibull and exponential distributions result in explicit functions of sample data that can be readily evaluated. For the inverse Gaussian and gamma distributions, $E[F(X_{1:n})] = 1/(n+1)$ for complete samples, while $E[F(X_{1:n})] = 1/(N+1)$ for censored samples. For the lognormal distribution, $E[\ln(X_{1:n} - \gamma)]$ can be expressed in terms of expected values of the first order statistic in the standard normal distribution (0, 1) and existing tables, cf. Harter (1970b), facilitate their evaluation. Although for theoretical reasons, it might have been desirable also to employ $E[X_{1:n}]$ in estimating parameters of the lognormal, inverse Gaussian, and gamma distribution, computational complications dictated otherwise and led to the choices given above.

Consideration in this chapter is limited to modified moment estimators, MME, for parameters of the Weibull, exponential, lognormal, inverse Gaussian, gamma, and related distributions.

8.2. The Weibull Distribution

Let X denote a random variable that is Weibull (γ, δ, β) with probability density function, pdf,

$$
\begin{aligned}
f(x; \gamma, \delta, \beta) &= \frac{\delta}{\beta^{\delta}} (x-\gamma)^{\delta-1} \\
&\quad \times \exp\{-[(x-\gamma)/\beta]^{\delta}\}, \qquad \gamma < x < \infty, \delta > 0, \beta > 0, \\
&= 0 \quad \text{otherwise.}
\end{aligned} \tag{8.2.1}
$$

The corresponding cumulative distribution function, cdf, is

$$
F(x; \gamma, \delta, \beta) = 1 - \exp\{-[(x-\gamma)/\beta]^{\delta}\}. \tag{8.2.2}
$$

In the notation employed here, γ is the threshold parameter, δ is the shape parameter, and β is the scale parameter. The expected value, variance, and third standard moment of the Weibull distribution are

$$\begin{aligned} E(X) &= \gamma + \beta\Gamma_1, \\ V(X) &= \beta^2[\Gamma_2 - \Gamma_1^2], \\ \alpha_3(X) &= [\Gamma_3 - 3\Gamma_2\Gamma_1 + 2\Gamma_1^3]/[\Gamma_2 - \Gamma_1^2]^{3/2}, \end{aligned} \tag{8.2.3}$$

where

$$\Gamma_k = \Gamma(1 + k/\delta), \qquad k = 1, 2, \ldots, \tag{8.2.4}$$

and $\Gamma(\;)$ is the gamma function

$$\Gamma(z) = \int_0^\infty t^{z-1}\, e^{-t}\, dt. \tag{8.2.5}$$

It follows from the pdf of the first order statistic given in Chapter 2 that the pdf of the first order statistic in a sample of size n from a Weibull distribution (γ, δ, β) is

$$\begin{aligned} f_{1:n}(x_1; \gamma, \delta, \beta') &= \frac{\delta}{(\beta')^\delta}(x_1 - \gamma)^{\delta-1} \exp\{-[(x_1 - \gamma)/\beta']^\delta\}, \\ &\qquad \gamma < x_1 < \infty,\ \delta > 0,\ \beta' > 0, \\ &= 0 \quad \text{elsewhere}, \end{aligned} \tag{8.2.6}$$

where

$$\beta' = \beta/n^{1/\delta}. \tag{8.2.7}$$

Thus it can be stated that if X is Weibull (γ, δ, β), then $X_{1:n}$ is Weibull (γ, δ, β'). Consequently $E(X_1)$ and $V(X_1)$ follow from (8.2.3) with β replaced by β'. Note that in (8.2.6) and elsewhere in this chapter that $X_{1:n}$ and $x_{1:n}$ are abbreviated to X_1 and x_1, respectively.

A. *Modified Moment Estimates*

Estimating equations for the MME are $E(X) = \bar{x}$, $V(X) = s^2$, and $E(X_1) = x_1$, where $\bar{x} = \sum_1^n x_i/n$, $s^2 = \sum_1^n (x_i - \bar{x})^2/(n-1)$, n is the sample size, and x_1 is the smallest sample observation. The estimating equations subsequently become

$$\begin{aligned} \gamma + \beta\Gamma_1 &= \bar{x}, \\ \beta^2[\Gamma_2 - \Gamma_1^2] &= s^2, \\ \gamma + (\beta/n^{1/\delta})\Gamma_1 &= x_1. \end{aligned} \tag{8.2.8}$$

Following a few simple algebraic manipulations, the three equations of

(8.2.8) are reduced to

$$
\begin{aligned}
W(n, \hat{\delta}) &= \frac{s^2}{(\bar{x} - x_1)^2}, \\
\hat{\gamma} &= (n^{1/\hat{\delta}} - \bar{x})/(n^{1/\hat{\delta}} - 1), \qquad (8.2.9) \\
\hat{\beta} &= n^{1/\hat{\delta}}(\bar{x} - x_1)/(n^{1/\hat{\delta}} - 1)\hat{\Gamma}_1,
\end{aligned}
$$

where

$$
W(n, \delta) = \frac{\Gamma_2 - \Gamma_1^2}{[(1 - n^{-1/\delta})\Gamma_1]^2}. \qquad (8.2.10)
$$

With $\bar{x}$, s^2, and x_1 available from the sample data, the first equation of (8.2.9) can be solved for $\hat{\delta}$. Subsequently $\hat{\gamma}$ and $\hat{\beta}$ follow from the last two equations of (8.2.9).

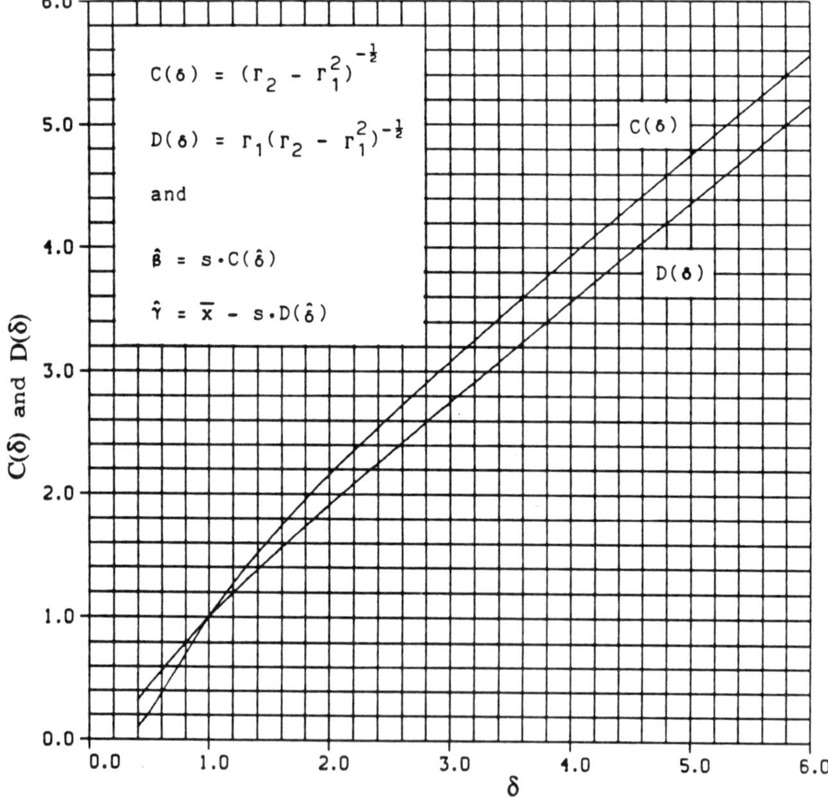

FIGURE 8.2.1. Graphs of $C(\delta)$ and $D(\delta)$ for the Weibull distribution. Reproduced from Cohen *et al.* (1984), Figure 2, page 163, with permission of the American Society for Quality Control.

Equivalent expressions for $\hat{\gamma}$ and $\hat{\beta}$ that are sometimes more convenient for routine calculations are

$$\hat{\beta} = s \cdot C(\hat{\delta}),$$
$$\hat{\gamma} = \bar{x} - s \cdot D(\hat{\delta}), \tag{8.2.11}$$

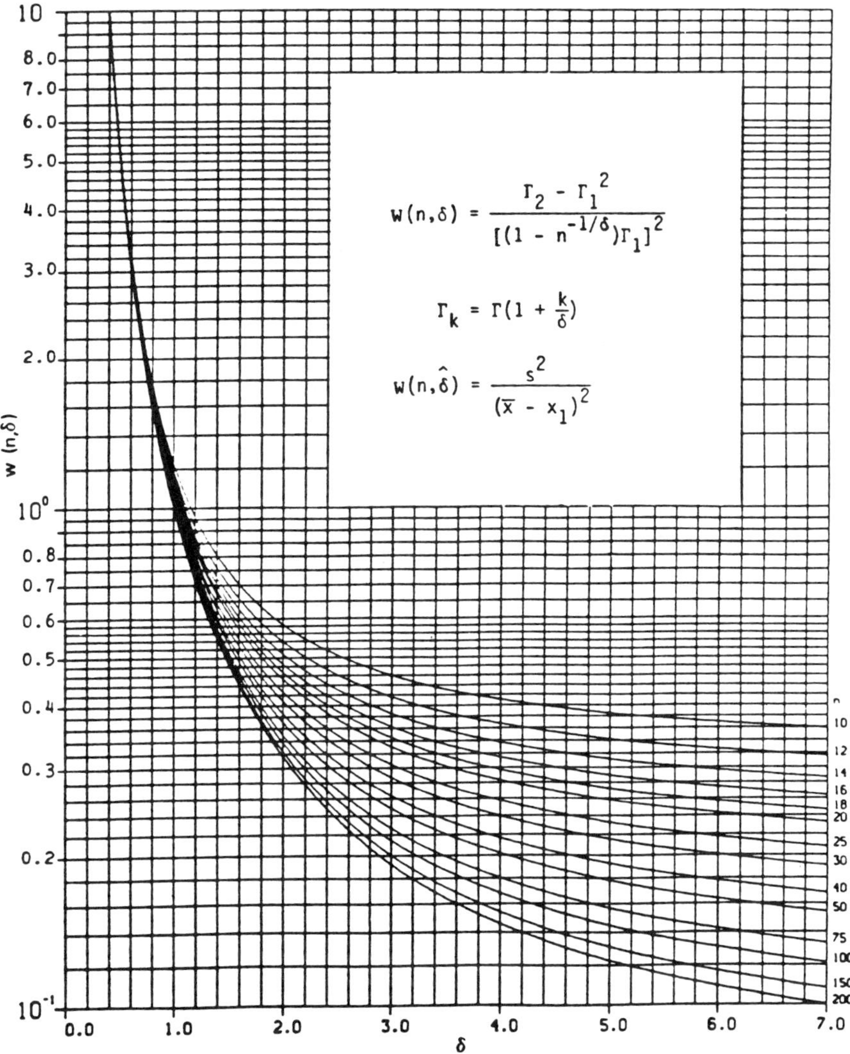

FIGURE 8.2.2. Graphs of $W(n, \hat{\delta}) = s^2/(\bar{x} - x_1)^2$ for the Weibull distribution. Reproduced from Cohen *et al.* (1984), Figure 1, page 162, with permission of the American Society for Quality Control.

where

$$C(\delta) = [\Gamma_2 - \Gamma_1^2]^{-1/2},$$
$$D(\delta) = \Gamma_1 \cdot [\Gamma_2 - \Gamma_1^2]^{-1/2}. \tag{8.2.12}$$

In order to facilitate the calculation of estimates $\hat{\gamma}$, $\hat{\delta}$, and $\hat{\beta}$ from sample data, tables of $W(n, \delta)$, $C(\delta)$, and $D(\delta)$ have been provided by Cohen *et al.* (1984) and Cohen and Whitten (1988). In addition, graphs of these functions, which enable quick approximate calculations to be made, were also given by Cohen and Whitten. These graphs are reproduced here as Figs. 8.2.1 and 8.2.2, respectively.

8.3. The Lognormal Distribution

The lognormal distribution has received attention from numerous writers. Among these are Aitchison and Brown (1957), Crow and Shimizu (1988), Johnson and Kotz (1970), and many others. This distribution derives its name from the relationship that exists between random variables X and $Y = \ln(X - \gamma)$. If Y is distributed normally (μ, σ^2), then X is lognormal (γ, μ, σ^2) with pdf

$$f(x; \gamma, \mu, \sigma^2) = \frac{1}{\sigma\sqrt{2\pi}(x-\gamma)} \exp\left\{-\frac{1}{2\sigma^2}[\ln(x-\gamma) - \mu]^2\right\},$$
$$\gamma < x < \infty, \ \sigma > 0,$$
$$= 0 \quad \text{elsewhere.} \tag{8.3.1}$$

The cdf may be expressed as

$$F(x; \gamma, \mu, \sigma^2) = \Phi\left[\frac{\ln(x-\gamma) - \mu}{\sigma}\right], \tag{8.3.2}$$

where $\Phi(\ )$ is the cumulative standard normal distribution. In the lognormal distribution, γ is the threshold parameter, μ is the scale parameter, and σ^2 is the shape parameter, although μ and σ^2 are the mean and variance respectively of the normal distribution. It is sometimes more convenient to employ

$$\beta = \exp(\mu) \tag{8.3.3}$$

as the lognormal scale parameter and

$$\omega = \exp(\sigma^2) \tag{8.3.4}$$

as the shape parameter. In order to permit more meaningful comparisons with other skewed distributions, it is often desirable to employ α_3, the third standard moment, as the primary shape parameter.

A. *Some Fundamentals*

The expected value, median, mode, variance, coefficient of variation, and Pearson's betas (β_1, β_2) as given by Yuan (1933) are

$$E(Z) = \gamma + \beta\sqrt{\omega},$$

$$\text{Me}(X) = \gamma + \beta, \qquad \text{Mo}(X) = \gamma + \frac{\beta}{\omega},$$

$$V(X) = \beta^2\omega(\omega - 1), \qquad v(X) = \sqrt{\omega - 1}, \tag{8.3.5}$$

$$\beta_1 = \alpha_3^2 = (\omega + 2)^2(\omega - 1),$$

$$\beta_2 = \alpha_4 = \omega^4 + 2\omega^3 + 3\omega^2 - 3,$$

where $E(\cdot)$ is the expected value symbol, $v(\cdot)$ is the coefficient of variation ($v(X) = \sqrt{V(X)}/[E(X) - \gamma]$), and α_3 and α_4 are the third and fourth standard moments.

If we make the standardizing transformation

$$Z = \frac{X - E(X)}{\sqrt{V(X)}}, \tag{8.3.6}$$

the pdf of the standard lognormal distribution, with mean zero, unit variance, and shape parameter ω, becomes

$$g(z; 0, 1, \omega) = \frac{\sqrt{\omega - 1}}{\sqrt{2\pi \ln \omega}[1 + z\sqrt{\omega - 1}]} \exp\left\{\frac{-1}{2(\ln \omega)}\right\}[\ln\{\sqrt{\omega} + z\sqrt{\omega(\omega - 1)}\}]^2,$$

$$-(\omega - 1)^{-1/2} < z, \ \omega > 1,$$

$$= 0 \quad \text{otherwise.} \tag{8.3.7}$$

The pdf of (8.3.7) can also be expressed in terms of the standard normal pdf $\phi(\cdot)$ as

$$g(z; 0, 1, \omega) = \frac{1}{\sqrt{\ln \omega}}\left(\frac{\sqrt{\omega - 1}}{1 + z\sqrt{\omega - 1}}\right)\phi\left[\frac{\ln\{\sqrt{\omega} + z\sqrt{\omega(\omega - 1)}\}}{\sqrt{\ln \omega}}\right]. \tag{8.3.8}$$

The standard cdf becomes

$$G(z; 0, 1, \omega) = \Phi\left[\frac{\ln[\sqrt{\omega} + z\sqrt{\omega(\omega - 1)}]}{\sqrt{\ln \omega}}\right], \tag{8.3.9}$$

where $\Phi(\ )$ is the cdf of the standard normal distribution (0, 1).

B. *Parameter Estimation*

Although moment and maximum likelihood estimators are the preferred estimators for most distributions, regularity problems are encountered when they are employed to estimate parameters of the three-parameter lognormal distribution. Hill (1963) has shown that global maximum likelihood estimators lead to inadmissible estimates, and Heyde (1963) demonstrated that the lognormal distribution is not uniquely determined by its moments. However, local maximum likelihood estimators, for which partial derivatives of the likelihood function with respect to distribution parameters are set equal to zero, seem to yield acceptable estimates in most instances. A more complete account of moment and local maximum likelihood estimators for lognormal parameters can be found in the volume by Cohen and Whitten (1988).

In an effort to avoid regularity problems, Cohen and Whitten (1980) and later Cohen *et al.* (1985) proposed modified moment estimators, MME, which employ the first order statistic. Estimating equations based on a random sample of size n are

$$E(X)=\bar{x}, \qquad V(X)=s^2, \quad \text{and} \quad E[\ln(X_1-\gamma)]=\ln(x_1-\gamma). \tag{8.3.10}$$

The third equation of (8.3.10) can be reduced to

$$\gamma+\beta\exp[\sqrt{\ln\omega}\,E(Z_{1:n})]=x_1, \tag{8.3.11}$$

where $E(Z_{1:n})$ is the expected value of the first order statistic in a random sample of size n from a standard normal distribution (0, 1). Values of $E(Z_{1:n})$ can be obtained from Harter's tables (1970a, b). Selected values from this source are reproduced here as Table 8.3.1. Linear interpolation in this table will usually provide sufficient accuracy for most practical applications. Appropriate substitutions from (8.3.5) into (8.3.10) together with (8.3.11) enable us to write the estimating equations as

$$\begin{gathered}\hat{\gamma}+\hat{\beta}\sqrt{\hat{\omega}}=\bar{x}, \qquad \hat{\beta}^2\hat{\omega}(\hat{\omega}-1)=s^2,\\ \hat{\gamma}+\hat{\beta}\exp[\sqrt{\ln\hat{\omega}}\,E(Z_{1:n})]=x_1.\end{gathered} \tag{8.3.12}$$

After a few simple algebraic manipulations, the three equations of (8.3.12) are reduced to

$$\begin{gathered}J(n,\hat{\omega})=\frac{s^2}{(\bar{x}-x_1)^2},\\ \hat{\gamma}=\bar{x}-s/\sqrt{\hat{\omega}-1},\\ \hat{\beta}=s/\sqrt{\hat{\omega}(\hat{\omega}-1)},\end{gathered} \tag{8.3.13}$$

where

$$J(n,\omega)=\frac{\omega(\omega-1)}{[\sqrt{\omega}-\exp\{\sqrt{\ln\omega}\,E(Z_{1:n})\}]^2}. \tag{8.3.14}$$

TABLE 8.3.1
Expected Values of the First Order Statistics from the Standard Normal Distribution (0, 1)

n	$E(Z_{1:n})$	n	$E(Z_{1:n})$	n	$E(Z_{1:n})$
5	-1.16296	36	-2.11812	100	-2.50759
10	-1.53875	38	-2.14009	125	-2.58634
12	-1.62923	40	-2.16078	150	-2.64925
14	-1.70338	45	-2.20772	175	-2.70148
16	-1.75699	50	-2.24907	200	-2.74604
18	-1.82003	55	-2.28598	225	-2.78485
20	-1.86748	60	-2.31928	250	-2.81918
22	-1.90969	65	-2.34958	280	-2.85572
24	-1.94767	70	-2.37736	300	-2.87777
26	-1.98216	75	-2.40299	315	-2.89327
28	-2.01371	80	-2.42677	350	-2.92651
30	-2.04276	85	-2.44894	375	-2.94810
32	-2.06967	90	-2.46970	400	-2.96818
34	-2.09471	95	-2.48920	1000[a]	-3.09053

Source: Extracted from Harter's (1961) tables.
[a] Entry approximated as $E(Z_{1:1000}) = \Phi^{-1}(1/1001)$.

The first equation of (8.3.13) can be solved for $\hat{\omega}$, and estimates $\hat{\gamma}$ and $\hat{\beta}$ then follow from the second and third equations. Estimates $\hat{\mu}$ and $\hat{\sigma}$ follow as

$$\hat{\mu} = \ln \hat{\beta} \quad \text{and} \quad \hat{\sigma} = \sqrt{\ln \hat{\omega}}. \tag{8.3.15}$$

In order to facilitate the solution of the first equation of (8.3.13), tables of $J(n, \sigma)$ have been provided by Cohen *et al.* (1985) and by Cohen and Whitten (1988) with n and σ as the arguments. Of course $\omega = \exp \sigma^2$. Estimates of σ can be obtained by inverse interpolation in these tables. In addition to the tables, graphs of $J(n, \sigma)$, reproduced here as Fig. 8.3.1, were also prepared. Estimates that are sufficiently accurate for many practical applications can be read directly from these graphs.

8.4. The Inverse Gaussian Distribution

The inverse Gaussian (IG) distribution has been studied extensively by numerous investigators, including Schrödinger (1915), Smoluchowsky (1915), Wald (1944), Tweedie (1956, 1957a, b), Wasan (1968), Wasan and Roy (1969), Chhikara and Folks (1974, 1988), Folks and Chhikara (1978), Padgett and Wei (1979), Cheng and Amin (1981), Chan *et al.* (1983, 1984),

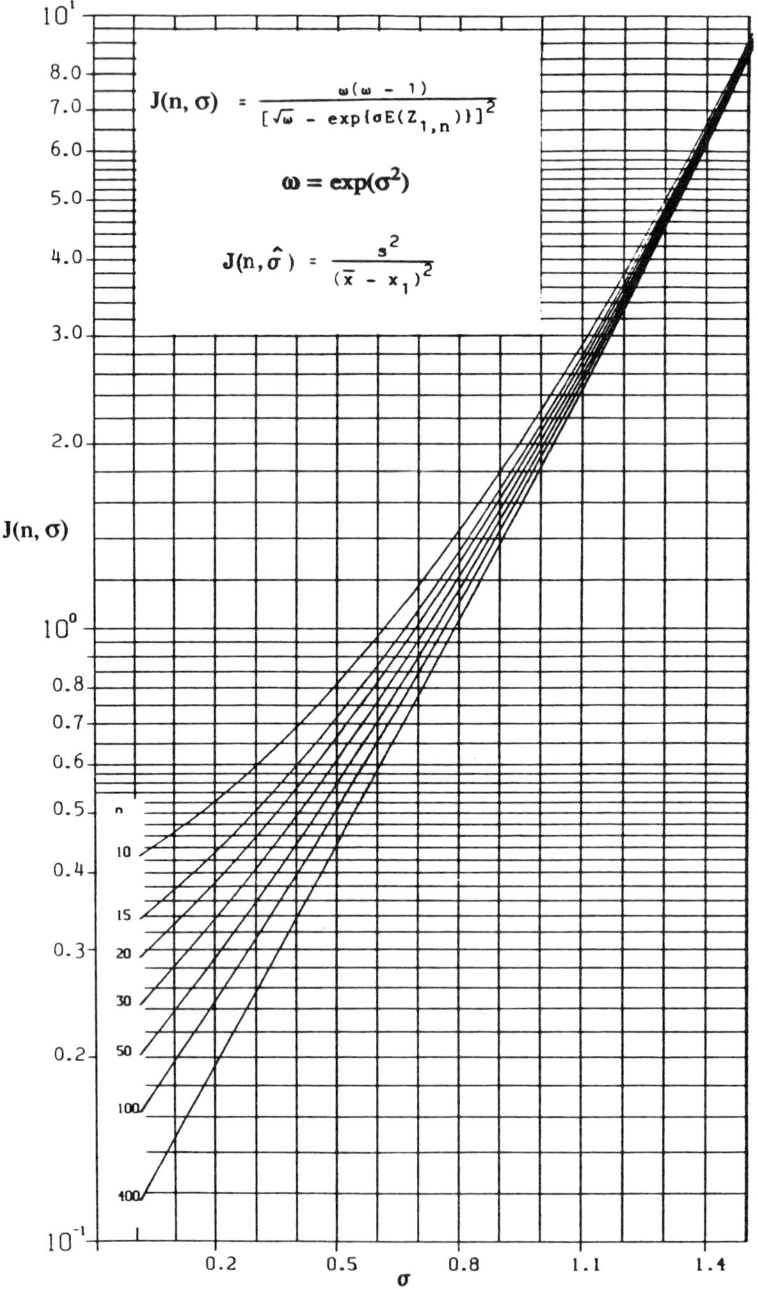

FIGURE 8.3.1. Graphs of $[\omega(\omega-1)]/[\sqrt{\omega}-\exp\{\sigma E(Z_{1:n})\}]^2$ for the lognormal distribution. Reproduced from Cohen *et al.* (1985), Figure 1, page 96, with permission of the American Society for Quality Control.

and Cohen and Whitten (1985). An excellent expository account of this distribution is given by Johnson and Kotz (1970).

The IG distribution provides a useful alternative to the Weibull, lognormal, gamma, and other positively skewed distributions as a model in various lifespan and reliability investigations. It is free from the regularity problems that have previously been mentioned in connection with the Weibull and the lognormal distributions.

This section is concerned with modified moment estimators (MME) which employ the first two sample moments plus the first order statistic. Although moment and maximum likelihood estimators are considered to be traditional estimators for parameters of the IG distribution, the MME enjoy distinct advantages whenever the threshold parameter must be estimated.

A. *Some Fundamentals*

In the parameterization of Chan *et al.* (1983), the pdf of the three-parameter IG distribution is

$$f(x;\gamma,\mu,\sigma)=\frac{1}{\sigma\sqrt{2\pi}}\left(\frac{\mu}{(x-\gamma)}\right)^{3/2}\exp\left\{-\frac{1}{2}\left(\frac{\mu}{(x-\gamma)}\right)\left[\frac{(x-\gamma)-\mu}{\sigma}\right]^2\right\},$$
$$\gamma<x<\infty,\ \mu>0,\ \sigma>0,$$
$$=0 \quad \text{elsewhere.} \tag{8.4.1}$$

The expected value (mean), variance, third standard moment (a measure of skewness), and fourth standard moment are

$$E(X)=\gamma+\mu, \qquad V(X)=\sigma^2,$$
$$\alpha_3(X)=\sqrt{\beta_1}=3\sigma/\mu, \tag{8.4.2}$$
$$\alpha_4(X)=\beta_2=3+(5/3)\beta_1.$$

In this notation, γ is the threshold parameter, $\gamma+\mu$ is the mean, σ^2 is the variance, $\alpha_3=3\sigma/\mu$ is the shape parameter, and α_4 is the fourth standard moment, while β_1 and β_2 are Pearson's betas.

The pdf of the standardized IG distribution $(0, 1, \alpha_3)$ as derived by Chan *et al.* (1983), where $Z=[X-E(X)]/\sigma$, is

$$g(z;0,1,\alpha_3)=\frac{1}{\sqrt{2\pi}}\left(\frac{1}{3+\alpha_3 z}\right)^{3/2}$$
$$\times\exp\left\{-\frac{z^2}{2}\left(\frac{3}{3+\alpha_3 z}\right)\right\}, \qquad -\frac{3}{\alpha_3}<z<\infty,\ \alpha_3>0,$$
$$=0 \quad \text{elsewhere.} \tag{8.4.3}$$

Note that $g(0)=1/\sqrt{2\pi}$ for all values of α_3, and that the normal pdf (0, 1) is the limiting form of (8.4.3) as $\alpha_3 \rightarrow 0$. The mode of the standardized IG distribution obtained by equating the first derivative of the pdf to zero is

$$M_0(Z)=\frac{\sqrt{36+\alpha_3^4}-(\alpha_3^2+6)}{2\alpha_3}. \tag{8.4.4}$$

B. *Modified Moment Estimators*

As estimating equations we employ $E(X)=\bar{x}$, $V(X)=s^2$, and $E[F(X_1)]=F(x_1)$. Since $E[F(X_1)]=1/(n+1)$, these equations reduce to

$$\begin{aligned}\hat{\gamma}+\hat{\mu}&=\bar{x},\\ \hat{\sigma}^2&=s^2,\\ G(z_1;0,1,\hat{\alpha}_3)&=1/(n+1),\end{aligned} \tag{8.4.5}$$

where

$$G(z;0,1,\alpha_3)=\Phi\left[\frac{z}{\sqrt{1+(\alpha_3/3)z}}\right]+\exp(18/\alpha_3^2)\cdot\Phi\left[\frac{-(z+6/\alpha_3)}{\sqrt{1+(\alpha_3/3)z}}\right] \tag{8.4.6}$$

is the cdf of the standard IG distribution $(0, 1, \alpha_3)$, and $\Phi(\cdot)$ is the cdf of the standard normal distribution (0, 1). We note that $F(x_1)=G(z_1)$ where

$$z_1=(x_1-\bar{x})/s. \tag{8.4.7}$$

The third equation of (8.4.5) can be solved for $\hat{\alpha}_3$, and estimators for σ, μ, and γ follow as

$$\begin{aligned}\hat{\sigma}&=s,\\ \hat{\mu}&=3s/\hat{\alpha}_3,\\ \hat{\gamma}&=\bar{x}-\hat{\mu}.\end{aligned} \tag{8.4.8}$$

Tables of the cdf of the standard IG distribution such as those given as Appendix A.3.3 of Cohen and Whitten (1988) can be employed to advantage in solving the third equation of (8.4.5) for $\hat{\alpha}_3$. However, it is probably more expedient to use tables with entries of α_3 as a function of z_1 and n, given by Cohen and Whitten (1988) as Table 5.2. For a quicker and perhaps slightly less accurate determination of $\hat{\alpha}_3$, the chart reproduced here as Fig. 8.4.1 and given originally by Cohen and Whitten (1985) is recommended.

Maximum likelihood estimates of the IG parameters can, when required, be calculated as described in Chapter 5 or as described by Chan *et al.* (1984) and by Cohen and Whitten (1985, 1988).

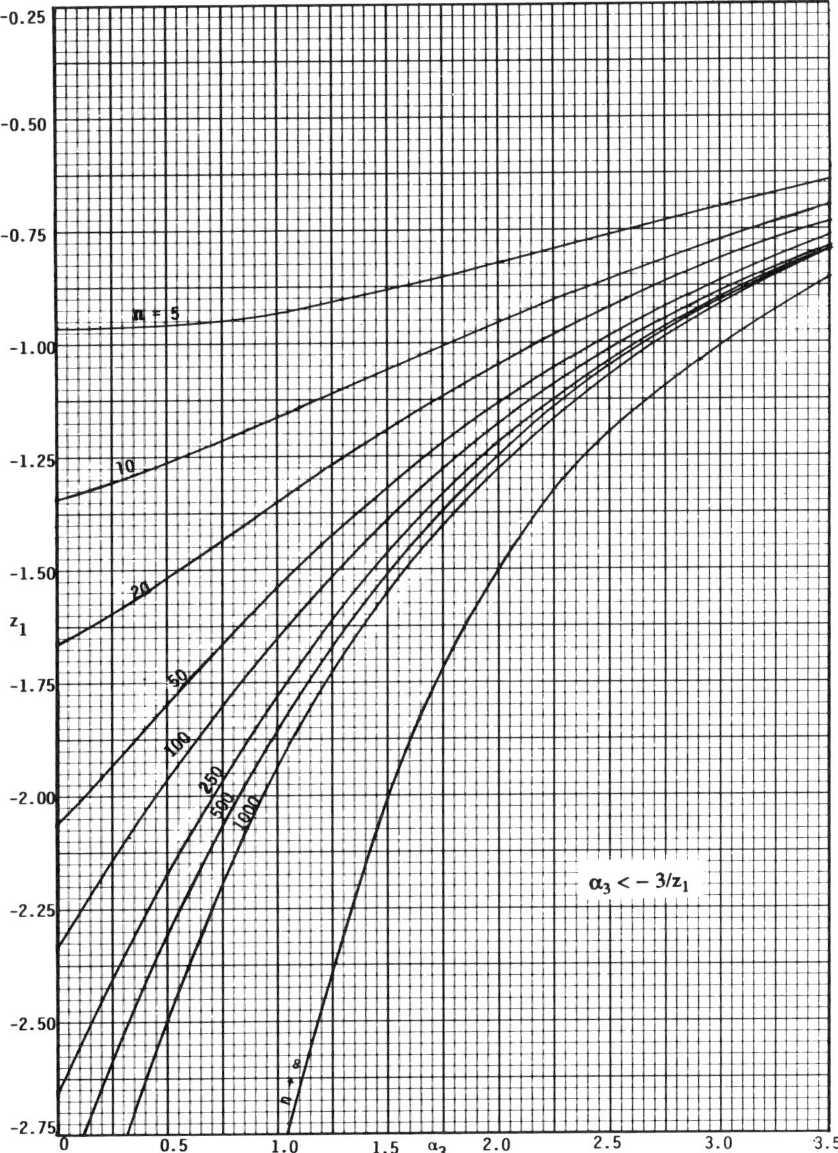

FIGURE 8.4.1. Graphs of α_3 as a function of z_1 and n for the inverse Gaussian distribution. Reproduced from Cohen and Whitten (1985), Figure 1, page 151, with permission of the American Society for Quality Control.

8.5. The Gamma Distribution

The gamma is another positively skewed distribution that is frequently employed as a model for life spans, reaction times, and related phenomena. The three-parameter gamma distribution is well known as Type III in the Pearson system of frequency curves. This distribution has previously been considered by numerous writers. Among them are Johnson and Kotz (1970), who give an excellent account of its properties. One hundred and thirty references are listed at the end of their discussion.

Maximum likelihood estimators of gamma parameters suffer from regularity problems. Moment estimators, in the three-parameter gamma distribution, exhibit large sampling errors that result from use of the third sample moment, a problem that also arises in other skewed distributions. These considerations contribute to the attractiveness of modified moment estimators (MME), which are offered as replacements.

A. *The Gamma Distribution and Its Properties*

The pdf of a random variable that is distributed in accordance with the gamma distribution is

$$f(x;\gamma,\rho,\beta)=\frac{(x-\gamma)^{\rho-1}}{\beta^{\rho}\Gamma(\rho)}\times\exp\left\{-\left(\frac{x-\gamma}{\beta}\right)\right\},\qquad \gamma<x<\infty,\ \rho>0,\ \beta>0, \tag{8.5.1}$$

where $\Gamma(\cdot)$ is the gamma function as defined by Eq. (8.2.5). The expected value, mode, variance, third standard moment, and fourth standard moment are

$$\begin{aligned} E(X)&=\gamma+\rho\beta, & V(X)&=\rho\beta^2,\\ M_0(X)&=\gamma+\beta(\rho-1), & \alpha_3(X)&=2/\sqrt{\rho},\\ \alpha_4(X)&=3+\tfrac{3}{2}\alpha_3^2. \end{aligned} \tag{8.5.2}$$

The cdf is

$$F(x;\gamma,\rho,\beta)=\int_{\gamma}^{x} f(y;\gamma,\rho,\beta)\,dy. \tag{8.5.3}$$

In the notation employed here, γ, β, and ρ are the threshold, scale and shape parameters, respectively. The frequency curve of the gamma distribution is bell-shaped; that is, it has a discernible mode for $\rho>1$, and thus for

$\alpha_3 < 2$. It assumes a reverse J-shape for $\rho \leq 1$. When $\rho = 1$, the exponential distribution is obtained as a special case of the gamma distribution.

When we make the standardizing transformation $Z = [X - E(X)]/\sqrt{V(X)}$, the pdf of the resulting standard gamma distribution $(0, 1, \alpha_3)$, with α_3 as the shape parameter, becomes

$$g(z; 0, 1, \alpha_3) = \frac{1}{\Gamma(4/\alpha_3^2)} \left(\frac{2}{\alpha_3}\right)^{(4/\alpha_3^2)} \left(z + \frac{2}{\alpha_3}\right)^{(4/\alpha_3^2)-1} \exp\left[-\frac{2}{\alpha_3}\left(z + \frac{2}{\alpha_3}\right)\right],$$

$$-\frac{2}{\alpha_3} < z < \infty,$$

$$= 0 \quad \text{elsewhere.} \tag{8.5.4}$$

The cdf of the standardized distribution is

$$G(z; 0, 1, \alpha_3) = \int_{-2/\alpha_3}^{z} g(t; 0, 1, \alpha_3)\, dt. \tag{8.5.5}$$

B. *Modified Moment Estimators*

The MME provide an attractive alternative to the ME and the MLE for estimating parameters of the gamma distribution for the same reasons that made them attractive in connection with the Weibull, lognormal, inverse Gaussian, and other skewed distributions. They are free from regularity problems, and they are valid over the entire parameter space for both bell-shaped and reverse J-shaped distributions. Basic estimating equations are the same as those for the inverse Gaussian distribution; that is, $E(X) = \bar{x}$, $V(X) = s^2$, and $E[F(X_1)] = F(x_1)$. Since $F(x_1) = G(z_1)$, and since $E[F(X_1)] = 1/(n+1)$, we can determine $\hat{\alpha}_3$ from

$$G(z_1; 0, 1, \hat{\alpha}_3) = 1/(n+1). \tag{8.5.6}$$

With $z_1 = (x_1 - \bar{x})/s$ and n available from sample data, $\hat{\alpha}_3$ can be obtained by inverse interpolation in tables of the cdf of the gamma distribution such as those provided by Salvosa (1930), Cohen *et al.* (1969), and Cohen and Whitten (1988). It might be possible to calculate $\hat{\alpha}_3$ with greater accuracy by interpolating in tables of Cohen and Whitten (1988) that give α_3 directly as a function of z_1 and n. A chart consisting of graphs with α_3 as a function of z_1 and n was given by Cohen and Whitten (1985). This chart, which permits $\hat{\alpha}_3$ to be read graphically with slightly less accuracy than calculations based on interpolation in the tables, is reproduced here as Fig. 8.5.1. With $\hat{\alpha}_3$ determined either by interpolation from tables or by reading from the

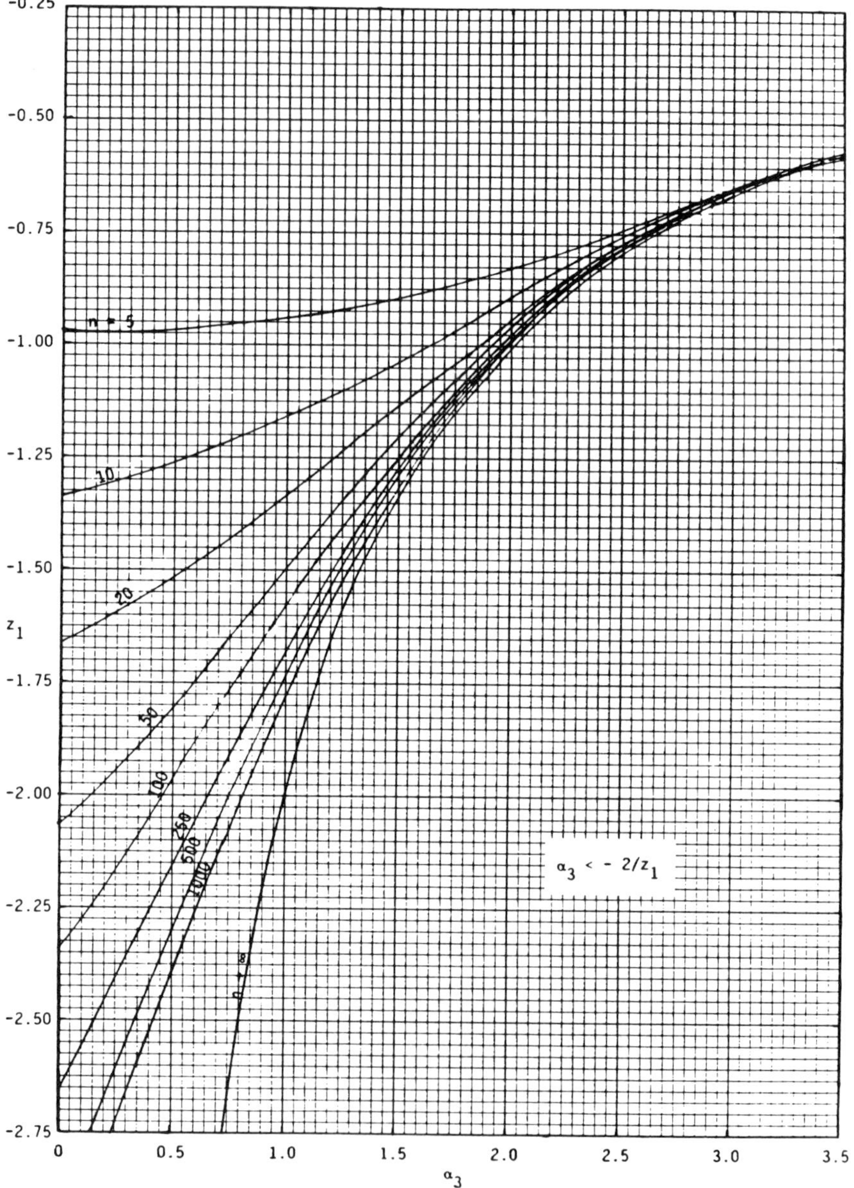

FIGURE 8.5.1. Graphs of α_3 as a function of z_1 and n for the gamma distribution. Reproduced from Cohen and Whitten (1986), Figure 1, page 55, with permission of the American Society for Quality Control.

chart, estimates $\hat{\beta}$, $\hat{\gamma}$, and $\hat{\rho}$ follow from (8.5.2) after a few simple algebraic operations as

$$\begin{aligned}\hat{\beta} &= s\hat{\alpha}_3/2,\\ \hat{\gamma} &= \bar{x} - 2s/\hat{\alpha}_3, \qquad (8.5.7)\\ \hat{\rho} &= 4/\hat{\alpha}_3^2.\end{aligned}$$

8.6. The Exponential Distribution

The exponential distribution can be obtained as a special case from either the Weibull or the gamma distribution. We let $\delta = 1$ in the Weibull distribution and $\rho = 1$ in the gamma distribution. It then follows from (8.2.1) for the Weibull distribution and from (8.5.1) for the gamma distribution that the pdf of the two-parameter exponential distribution becomes

$$\begin{aligned}f(x;\gamma,\beta) &= \frac{1}{\beta}\exp\left\{-\left(\frac{x-\gamma}{\beta}\right)\right\}, \qquad \gamma < x < \infty,\ \beta > 0,\\ &= 0 \quad \text{elsewhere}, \qquad (8.6.1)\end{aligned}$$

and the cdf becomes

$$F(x;\gamma,\beta) = 1 - \exp\{-(x-\gamma)/\beta\}. \qquad (8.6.2)$$

The expected value and the variance become

$$E(X) = \gamma + \beta \quad \text{and} \quad V(X) = \beta^2. \qquad (8.6.3)$$

The pdf of the first order statistic in a sample of size n from an exponential distribution with pdf (8.6.1) follows as

$$\begin{aligned}f_{1:n}(x;\beta) &= \frac{n}{\beta}\exp\left\{-n\left(\frac{x_1-\gamma}{\beta}\right)\right\}, \qquad \gamma < x_1 < \infty,\ \beta > 0,\\ &= 0 \quad \text{elsewhere}. \qquad (8.6.4)\end{aligned}$$

For the expected value and the variance, we have

$$E(X_1) = \gamma + \beta/n \quad \text{and} \quad V(X_1) = (\beta/n)^2. \qquad (8.6.5)$$

The MME in this case are $E(X) = \bar{x}$ and $E(X_1) = x_1$. From (8.6.3) and from (8.6.5), it follows that

$$\hat{\gamma}+\hat{\beta}=\bar{x},$$
$$\hat{\gamma}+\hat{\beta}/n=x_1. \tag{8.6.6}$$

We solve the two equations of (8.6.6) simultaneously and obtain

$$\hat{\gamma}=(nx_1-\bar{x})/(n-1),$$
$$\hat{\beta}=n(\bar{x}-x_1)/(n-1). \tag{8.6.7}$$

These estimators are recognized as the well-known best linear unbiased estimators (BLUE) and as the minimum variance unbiased estimators (MVUE) for parameters of this distribution. The exact variances and covariance of these estimators have been shown to be (cf. Cohen and Helm (1973))

$$V(\hat{\beta})=\beta^2/(n-1),$$
$$V(\hat{\gamma})=\beta^2/n(n-1), \tag{8.6.8}$$
$$\mathrm{Cov}(\hat{\beta},\hat{\gamma})=-\beta^2/n(n-1).$$

8.7. Illustrative Examples

Five examples have been selected to illustrate the practical application of modified moment estimators presented in this chapter. Data for Example 8.7.1 consist of observations of maximum flood levels in millions of cubic feet per second for the Susquehanna River at Harrisburg, Pennsylvania over 20 four-year intervals from 1890 until 1969. These data were made available by Professor Byron Reich of Pennsylvania State University for use by Dumonceaux and Antle (1973), who employed them to illustrate parameter estimation in the Type 1 extreme value for maximums distribution. Example 8.7.2 consists of observations of fatigue life in hours of a certain type of bearing as given by McCool (1974). Example 8.7.3 consists of 100 computer-generated random observations from an inverse Gaussian population for which $\gamma=10$, $\mu=6$, and $\sigma=3$. Accordingly $E(X)=16$, $V(X)=9$, and $\alpha_3(X)=1.5$. Sample data for these examples are listed in Tables 8.7.1, 8.7.2, and 8.7.3. Data for Examples 8.7.4 and 8.7.5 are given on pp. 295 and 296.

Parameter estimates are calculated under the assumptions that each of these samples are in turn from a Weibull, lognormal, inverse Gaussian, and gamma distribution. Although Examples 8.7.1 and 8.7.2 are from real-life

TABLE 8.7.1
Maximum Flood Levels of Susquehanna River at Harrisburg, PA over Four-Year Periods (1890–1969) in Millions of Cubic Feet per Second.[a] Example 8.7.1

0.654	0.613	0.315	0.449	0.297
0.402	0.379	0.423	0.379	0.3225
0.269	0.740	0.418	0.412	0.494
0.416	0.338	0.392	0.484	0.265

[a] Summary: $n=20$, $\bar{x}=0.423125$, $s=0.1252789$, $a_3=1.0673243$, $x_1=0.265$, $s^2/(\bar{x}-x_1)^2=0.6277$, $z_1=(x_1-\bar{x})/s=-1.2622$. Note that $s^2/(\bar{x}-x_1)^2=(1/z_1)^2$.

situations, and Example 8.7.3 is from a known (inverse Gaussian) distribution with known parameter values, they are presented here solely to illustrate computational procedures and to provide comparisons that demonstrate the extent to which estimates of threshold and shape parameters, γ and α_3, vary with the choice of distribution. In the paragraphs that follow, we examine some of the details of parameter estimation.

Example 8.7.1. Data for this example are summarized as: $n=20$, $\bar{x}=0.423125$, $s=0.1252789$, $a_3=1.0673243$, $x_1=0.265$, $s^2/(\bar{x}-x_1)^2=0.6277$, $z_1=(x_1-\bar{x})/s=-1.2622$. We first assume this sample to be from a *Weibull* distribution, and obtain an estimate of δ from the graphs of Fig. 8.2.2. With $n=20$ and $W(20,\hat{\delta})=0.6277$, we read $\hat{\delta}=1.48$. With $\hat{\delta}=1.48$, we enter the graphs of Fig. 8.2.1 and read $C(1.48)=1.62$ and $D(1.48)=1.45$. We then employ Eq. (8.2.11) to calculate $\hat{\beta}=0.201$ and $\hat{\gamma}=0.241$. From the third equation of (8.2.3) with $\hat{\delta}=1.48$, we calculate $\hat{\alpha}_3(X)=1.10$. Subsequent computer calculations made by using a Fortran program given by Cohen

TABLE 8.7.2
Fatigue Life in Hours of a Certain Type of Bearing.[a] Example 8.7.2

152.7	204.7
172.0	216.5
172.5	234.9
173.3	262.6
193.0	422.6

[a] Summary: $n=10$, $\bar{x}=220.48$, $s=78.405638$, $a_3=1.8635835$, $x_1=152.7$, $s^2/(\bar{x}-x_1)^2=1.338$, $z_1=(x_1-\bar{x})/s=-0.8645$.

TABLE 8.7.3
Random Sample from an Inverse Gaussian Distribution: $E(X)=16$, $V(X)=9$, $\alpha_3(X)=1.5$.[a] Example 8.7.3

14.5380	17.4308	18.7610	12.4239
13.2012	14.8387	14.3782	13.4204
12.7371	25.9530	14.8808	13.1681
26.8858	13.9679	15.2530	17.0418
14.5085	13.3312	14.9689	13.3223
14.5715	19.6787	18.8643	12.1711
13.7798	14.5391	11.8642	18.3044
13.2607	20.6495	15.6568	13.6053
12.2354	12.9256	24.2755	22.2760
17.6481	13.2901	19.6653	15.7814
18.2857	18.7098	16.6141	18.1240
17.5578	18.5687	14.5865	16.9805
18.0316	14.0325	15.3300	12.8743
19.6564	14.2242	16.3579	13.5962
12.4371	16.2804	15.2886	19.6399
17.5331	14.0965	15.6275	16.5744
12.5355	13.2092	12.9744	19.9570
13.5962	12.5098	14.0212	12.4156
15.9772	14.2432	17.9305	23.7831
11.3557	17.8338	13.7886	12.1849
14.8192	17.2500	13.8121	15.3410
18.6036	22.9645	13.9231	16.3003
13.9699	15.8471	13.8853	15.3016
14.0373	16.5082	12.1504	13.5269
14.3480	12.6309	14.9820	14.5507

[a] Summary: $n=100$, $\bar{x}=15.7415$, $s=3.1276$, $a_3=1.3230$, $x_1=11.3557$, $s^2/(\bar{x}-x_1)^2=0.50854$, $z_1=(x_1-\bar{x})/s=-1.4023$.

and Whitten (1988) yield estimates; $\hat{\delta}=1.4794$, $\hat{\beta}=0.2014$, $\hat{\gamma}=0.2410$ and $\hat{\alpha}_3=1.966$.

Let us now assume this sample to be from a *lognormal* distribution. With $n=20$ and $J(n, \hat{\delta})=0.6277$, we employ the graphs of Fig. 8.3.1 to read $\hat{\sigma}=0.47$. From (8.3.4), we calculate $\hat{\omega}=1.247$, and from (8.3.5), $\hat{\alpha}_3=1.61$. With $\hat{\omega}=1.247$, we employ the last two equations of (8.3.13) to calculate $\hat{\gamma}=0.171$, and $\hat{\beta}=0.226$. From (8.3.15), we calculate $\hat{\mu}=-1.487$.

At this juncture, we assume our sample to be from an *inverse Gaussian* distribution. With $n=20$ and $z_1=-1.2622$, we employ the graphs of Fig. 8.4.1 to read $\hat{\alpha}_3=1.25$. With $\hat{\alpha}_3$ thus determined, we employ (8.4.8) to calculate $\hat{\sigma}=0.125$, $\hat{\mu}=0.301$, and $\hat{\gamma}=0.122$.

Finally we assume this sample to be from a *gamma* distribution. With $n = 20$ and $z_1 = -1.2622$, we employ the graphs of Fig. 8.5.1 to read $\hat{\alpha}_3 = 1.17$. With $\hat{\alpha}_3$ thus determined, we employ (8.5.7) to calculate $\hat{\beta} = 0.074$, $\hat{\rho} = 2.90$, $\hat{\gamma} = 0.210$. The various estimates from this example are displayed in Table 8.7.4.

Example 8.7.2. Data for this example are summarized as: $n = 10$, $\bar{x} = 220.48$, $s = 78.405638$, $a_3 = 1.8635835$, $x_1 = 152.7$, $s^2/(\bar{x} - x_1)^2 = 1.338$, $z_1 = (x_1 - \bar{x})/s = -0.8645$. Estimates based on this example were calculated as described above for Example 8.7.1. A summary of estimates thus calculated is presented in Table 8.7.5.

Example 8.7.3. Data for this example are summarized as: $n = 100$, $\bar{x} = 15.7415$, $s = 3.1276$, $a_3 = 1.3230$, $x_1 = 11.3557$, $s^2/(\bar{x} - x_1)^2 = 0.50854$, $z_1 = (x_1 - \bar{x})/s = -1.4023$. A summary of estimates for this example, calculated as described for Examples 8.7.1 and 8.7.2, is displayed in Table 8.7.6.

Exponential Distribution (Example 8.7.2). As displayed in Table 8.7.5, we note that $\hat{\delta} = 0.949$ for the Weibull distribution and $\hat{\rho} = 0.865$ for the gamma distribution. Since those estimates are approximately equal to 1, it seems reasonable to assume that perhaps this sample is actually from an *exponential distribution.* Accordingly, we employ Eq. (8.6.7) to calculate $\hat{\beta} = 75.31$, and

TABLE 8.7.4
Summary of Estimates for Example 8.7.1. $\bar{x} = 0.423125$, $s = 0.1252789$, $a_3 = 1.0673243$, $x_1 = 0.265$, $n = 20$, $s^2/(\bar{x} - x_1)^2 = 0.6277$, $z_1 = -1.2622$

Distribution	$\hat{\gamma}$	$\hat{\alpha}_3$	$E(\hat{X})$	$\sqrt{V(\hat{X})}$	Primary Estimates
Weibull	0.241	1.097	0.4231	0.125	$\hat{\delta} = 1.48$ $\hat{\beta} = 0.201$
Lognormal	0.171	1.616	0.4231	0.125	$\hat{\sigma} = 0.47$, $\hat{\beta} = 0.225$ $\hat{\omega} = 1.25$, $\hat{\mu} = -1.49$
Inverse Gaussian	0.122	1.247	0.4231	0.125	$\hat{\mu} = 0.301$ $\hat{\sigma} = 0.125$
Gamma	0.210	1.174	0.4231	0.125	$\hat{\rho} = 2.90$ $\hat{\beta} = 0.074$
Exponential	NA	NA	NA	NA	NA

TABLE 8.7.5
Summary of Estimates for Example 8.7.2. $\bar{x} = 220.48$, $s = 78.405638$, $a_3 = 1.8635835$, $x_1 = 152.7$, $n = 10$, $s^2/(\bar{x} - x_1)^2 = 1.338$, $z_1 = -0.8645$

Distribution	$\hat{\gamma}$	$\hat{\alpha}_3$	$\hat{E(X)}$	$\sqrt{\hat{V(X)}}$	Primary Estimates
Weibull	146.1	2.17	220.5	78.41	$\hat{\delta} = 0.949$ $\hat{\beta} = 72.6$
Lognormal	132.4	3.37	220.5	78.41	$\hat{\sigma} = 0.76$, $\hat{\beta} = 65.8$ $\hat{\omega} = 1.79$, $\hat{\mu} = 4.19$
Inverse Gaussian	124.3	2.45	220.5	78.41	$\hat{\mu} = 96.2$ $\hat{\sigma} = 78.41$
Gamma	147.6	2.15	220.5	78.41	$\hat{\rho} = 0.865$ $\hat{\beta} = 84.3$
Exponential	145.2	2	220.5	75.31	$\hat{\beta} = 75.31$

$\hat{\gamma} = 145.2$. We subsequently employ (8.6.8) to calculate standard errors of these estimators as $\sigma_{\hat{\beta}} = \sqrt{V(\hat{\beta})} = 25.1$ and $\sigma_{\hat{\gamma}} = \sqrt{V(\hat{\gamma})} = 7.94$. For 95% confidence interval on β, we employ (8.6.9) to calculate $43 < \beta < 165$.

Example 8.7.4. This example was employed by Menon (1963) to illustrate estimation in the two-parameter Weibull distribution. Sample data consists

TABLE 8.7.6
Summary of Estimates for Example 8.7.3. $\bar{x} = 15.7415$, $s = 3.1276$, $a_3 = 1.3230$, $x_1 = 11.3557$, $n = 100$, $s^2/(\bar{x} - x_1)^2 = 0.50854$, $z_1 = -1.4023$[a]

Distribution	$\hat{\gamma}$	$\hat{\alpha}_3$	$\hat{E(X)}$	$\sqrt{\hat{V(X)}}$	Primary Estimates
Weibull	11.08	1.10	15.74	3.13	$\hat{\delta} = 1.497$ $\hat{\beta} = 5.07$
Lognormal	9.87	1.75	15.74	3.13	$\hat{\sigma} = 0.50$, $\hat{\beta} = 5.18$ $\hat{\omega} = 1.284$, $\hat{\mu} = 1.64$
Inverse Gaussian	9.36	1.47	15.74	3.13	$\hat{\mu} = 6.383$ $\hat{\sigma} = 3.13$
Gamma	12.91	1.25	15.74	3.13	$\hat{\rho} = 2.56$ $\hat{\beta} = 1.106$
Exponential	NA	NA	NA	NA	NA

[a] Population values: $E(X) = 16$, $\sqrt{v(X)} = 3$, $\alpha_3(X) = 1.5$, $\gamma = 10$.

of $n = 20$ random observations from a reverse J-shaped Weibull population in which $\delta = 0.5$, and $\theta = \sqrt{e} = 1.649$ ($\beta = 2.719$), as tabulated below.

0.806	57.628	1.550	7.057
0.664	1.033	9.098	2.046
0.345	3.532	0.470	0.185
0.001	0.970	0.505	0.435
0.469	0.071	0.030	1.550

In summary, $n = 20$, $\bar{x} = 4.42225$, $s^2 = 154.402$, $x_1 = 0.001$, and $s^2/(\bar{x} - x_1)^2 = 8.3145692$. From the chart of Fig. 8.2.2, we read the Cohen–Whitten estimate $\hat{\delta} = 0.43$. Interpolation in Table 3.2 of Cohen and Whitten (1988) enables us to calculate a more accurate estimate as $\hat{\delta} = 0.432$. From the second equation of (8.2.8) we calculate $\hat{\beta} = 2.5497$. This value might also have been calculated from the first equation of (8.2.11) with $C(\hat{\delta})$ read from the chart of Fig. 8.2.1. It follows that $\hat{\theta} = \hat{\beta}^{\hat{\delta}} = 1.498$. Following is a summary comparison of these estimates with moment and Menon estimates as given by Menon (1963).

Parameters	Population values	Moment estimates	Menon estimates	Cohen–Whitten estimates
δ	0.500	0.43	0.57	0.432
θ	1.649	1.23	1.40	1.498
β	2.719	1.62	1.80	2.550

Note that the C–W estimates of θ and β are in closer agreement with corresponding population parameters than either the moment or the Menon estimates. The C–W estimate of δ is too small and the Menon estimate is too large, whereas the average of the two is almost equal to the population value.

Example 8.7.5. This example was given by Cohen and Whitten (1982b). Sample data, which are tabulated below, consist of $n = 100$ random observations from a bell-shaped Weibull population for which $\gamma = 0$, $\beta = 1$, and $\delta = 1.5639$.

1.5102	1.7300	0.2724	1.0941
0.4386	1.1273	1.3218	0.8744
0.4081	0.1257	0.7082	0.2159
0.7649	1.6378	0.6412	1.0712
0.5281	2.7173	0.4157	1.2513
1.0634	1.7553	1.2008	1.2513
1.4176	0.9209	0.3690	0.7680
2.1820	1.0258	0.3809	1.3782
0.6674	0.1009	1.0501	0.5327
0.6942	0.6404	1.6469	0.5520
0.8721	0.4124	0.3857	0.8587
2.4596	0.7950	0.8949	1.3545
0.9890	1.5888	0.5006	0.2827
0.7589	1.6565	0.9640	0.5513
2.3628	1.1701	1.1651	2.5450
0.4272	0.1854	0.6158	0.2795
0.4869	0.3329	0.8367	1.0735
0.1461	0.5719	0.2908	1.8398
0.4874	0.6130	0.6993	0.6460
1.5160	0.7077	1.0976	0.4527
0.7245	1.1605	0.9569	2.6288
0.8242	1.0593	0.2980	1.2859
0.8230	1.6695	1.0899	1.5082
0.4573	2.2271	0.4852	0.8858
1.7600	0.9194	0.5685	0.3604

In summary, $n = 100$, $\bar{x} = 0.9602$, $s = 0.5961$, $x_1 = 0.1009$, and $s^2/(\bar{x} - x_1)^2 = 0.4812$. From the chart of Fig. 8.2.2, we read the Cohen-Whitten estimate, $\hat{\delta} = 1.55$. Interpolation in Table 3.2 of Cohen and Whitten (1988) yields the more accurate value, $\hat{\delta} = 1.552$. By using the second equation of (8.2.8), we calculate $\hat{\beta} = 1.016$, a value that could also be calculated from the first equation of (8.2.11) with $C(\hat{\delta})$ read from the chart of Fig. 8.2.1. It follows that $\hat{\theta} = \hat{\beta}^{\hat{\delta}} = 1.025$. From the second equation of (8.2.11), we calculate $\hat{\gamma} = 0.55$, where $D(\hat{\delta})$ is read from the chart of Fig. 8.2.1. Following is a summary tabulation of the Cohen–Whitten estimates together with moment and maximum likelihood estimates for this sample as given by Cohen and Whitten (1988).

Parameters	Population values	Moment estimates	Maximum likelihood estimates	Cohen–Whitten estimates
δ	1.5639	1.560	1.501	1.552
θ	1.0000	1.020	0.955	1.025
β	1.0000	1.013	0.970	1.016
γ	0	0.050	0.084	0.055

For this example, the moment estimates are closer to the corresponding population parameters, but the C–W estimates differ only slightly from the moment estimates, and of course, the C–W estimates are easier to calculate.

Concluding Comments: Example 8.7.1. Estimates recorded in Table 8.7.4 indicate that either the Weibull or the gamma distribution is most suitable as a model for this example. In both cases, the estimate of α_3 exceeds the sample value of a_3. This, however, is typical in that in most instances $a_3 < \alpha_3$.

Example 8.7.2. Although both the Weibull and the gamma distribution provide an acceptable model for this example, the exponential distribution is clearly the superior model. Again, we observe that $\hat{\alpha}_3 > a_3$.

Example 8.7.3. It is certainly no surprise that the inverse Gaussian model provides estimates that are in closest agreement with the known population parameters. It is of interest to note that the lognormal model is almost as good as the inverse Gaussian, whereas the Weibull and the gamma distributions are less satisfactory.

Example 8.7.4. The interesting feature of this example is the fact that the parent population is reverse J-shaped Weibull, and the known origin is at zero. It is noted that the Cohen–Whitten estimates represent a slight improvement over both the moment and the Menon estimates for these data. Since $\delta < 1$, maximum likelihood estimates are not applicable.

Example 8.7.5. This sample is from a three-parameter bell-shaped Weibull population. As previously noted, moment estimates are in closer agreement with corresponding population parameters than either the MLE or the C–W estimates. However, the C–W estimates differ only slightly from the moment estimates, and they are somewhat easier to calculate.

Chapter 9 Estimation in Regression Models

9.1. Introduction

In this chapter, we describe some methods of estimation based on order statistics for the parameters in linear regression models. In Section 2, we present the best linear unbiased estimators (BLUEs) based on complete and Type-II censored samples for the parameters of a general linear model (with multiple measurements) as derived by Hamouda (1988). The results of Moussa (1972), Leone and Hamouda (1973), and Hamouda and Leone (1974) on the best linear unbiased estimation of the parameters of the simple linear regression model and the K-sample situation with equal and unequal scale parameters are then deduced as special cases and illustrated with an example. In Section 3, we present the modified maximum likelihood estimators (MMLEs) based on Type-II censored samples for the parameters of a general linear model (with multiple measurements) as derived by Tiku (1973, 1981) and also explained in detail by Tiku *et al.* (1986). The results for the simple linear regression model with multiple measurements are also presented and explained with an illustrative example. Finally, in Section 4 we present the recent work of Balakrishnan and Ambagaspitiya (1990a) on the modified maximum likelihood estimation of the parameters of a simple linear regression model with single measurements, and we demonstrate the robustness of the estimators derived by this method. The MML method of estimation described in this section is also explained with two illustrative examples.

9.2. Best Linear Unbiased Estimation with Multiple Measurements

The best linear unbiased estimators (BLUEs) of the location and scale parameters of a population based on complete as well as Type-II censored samples have been discussed in great detail in Chapter 4. The BLUEs of the parameters in a simple linear regression model and in the K-sample situation with equal and unequal scale parameters have been derived by Moussa (1972), Leone and Hamouda (1973), and Hamouda and Leone (1974). These derivations have been given for complete as well as Type-II censored samples under the assumption of normality and also several other nonnormal distributions for the error component. Moussa (1972) and Leone and Hamouda (1973) have also discussed the robustness features of these estimators. Recently, Hamouda (1988) has derived the BLUEs of the parameters in a general linear model and then deduced the results for the simple linear regression model and the K-sample situations as special cases. We shall present this development here in detail and illustrate with an example.

Let us start with the situation of K independent random samples of sizes n, each drawn from K distributions; that is, let y_{ij} be independent observations from $F(y/\sigma_i)$ for $j=1, 2, \ldots, n$ and $i=1, 2, \ldots, K$. We wish to fit a general linear model of the form

$$y_{ij} = \gamma_1 x_{i1} + \gamma_2 x_{i2} + \cdots + \gamma_p x_{ip} + \varepsilon_{ij}, \tag{9.2.1}$$

where ε_{ij}s and independent variables from $F(\cdot)$ with $E(\varepsilon_{ij})=0$ and $\text{Var}(\varepsilon_{ij}) = \sigma_i^2$, $j=1, 2, \ldots, n$, $i=1, 2, \ldots, K$.

Let us denote the order statistics in the ith sample by

$$y_{i(1)} \le y_{i(2)} \le \cdots \le y_{i(n)},$$

and let

$$z_{i(j)} = \left(y_{i(j)} - \sum_{l=1}^{p} \gamma_l x_{il} \right) \Big/ \sigma_i \tag{9.2.2}$$

denote the jth standardized order statistic from the ith sample. In other words, $z_{i(j)}$ is the jth order statistic from the standardized distribution $F(\cdot)$ with mean zero and variance one. Further, let us denote the mean $E(z_{i(j)})$ by $\alpha_{i(j)}$ and the covariance $\text{Cov}(z_{i(j)}, z_{i(k)})$ by $\beta_{i(j,k)}$ for $1 \le j \le k \le n$ and $i=1, 2, \ldots, K$.

Then, the linear model in (9.2.1) may be expressed as

$$y_{i(j)} = \sum_{l=1}^{p} \gamma_l x_{il} + \sigma_i z_{i(j)}, \tag{9.2.3}$$

or, equivalently,

$$E(y_{i(j)}) = \sum_{l=1}^{p} \gamma_l x_{il} + \sigma_i \alpha_{i(j)},$$

and (9.2.4)

$$\mathrm{Cov}(y_{i(j)}, y_{i(k)}) = \sigma_i^2 \beta_{i(j,k)}$$

for $1 \le j, k \le n$, and $i = 1, 2, \ldots, K$.

If we now denote

$$\mathbf{0}' = (0 \quad 0 \quad \cdots \quad 0)_{1\times n},$$
$$\mathbf{1}' = (1 \quad 1 \quad \cdots \quad 1)_{1\times n},$$
$$\mathbf{x}_{il} = x_{il}\mathbf{1},$$
$$\boldsymbol{\alpha}_i' = (\alpha_{i(1)} \quad \alpha_{i(2)} \quad \cdots \quad \alpha_{i(n)}),$$
$$\mathbf{B}_i = ((\beta_{i(j,k)}))_{n\times n},$$

and

$$\boldsymbol{\theta}' = (\gamma_1 \quad \gamma_2 \quad \cdots \quad \gamma_p \quad \sigma_1 \quad \sigma_2 \quad \cdots \quad \sigma_K)_{(p+K)\times 1},$$

the model in (9.2.4) may be rewritten in matrix notation as

$$E(\mathbf{y}_0) = \mathbf{A}\boldsymbol{\theta} \quad \text{and} \quad \mathrm{Var}(\mathbf{y}_0) = \mathbf{C}, \tag{9.2.5}$$

where the subscript 0 of $\mathbf{y}$ denotes the ordered sample, and

$$\mathbf{A} = \begin{bmatrix} \mathbf{x}_{11} & \mathbf{x}_{12} & \cdots & \mathbf{x}_{1p} & \boldsymbol{\alpha}_1 & \mathbf{0} & \cdots & \mathbf{0} \\ \mathbf{x}_{21} & \mathbf{x}_{22} & \cdots & \mathbf{x}_{2p} & \mathbf{0} & \boldsymbol{\alpha}_2 & \cdots & \mathbf{0} \\ \vdots & \vdots & \cdots & \vdots & \vdots & \vdots & \cdots & \vdots \\ \mathbf{x}_{K1} & \mathbf{x}_{K2} & \cdots & \mathbf{x}_{Kp} & \mathbf{0} & \mathbf{0} & \cdots & \boldsymbol{\alpha}_K \end{bmatrix}_{nK\times(p+K)} \tag{9.2.6}$$

and

$$\mathbf{C} = \begin{bmatrix} \sigma_1^2\mathbf{B}_1 & & & 0 \\ & \sigma_2^2\mathbf{B}_2 & & \\ & & \ddots & \\ 0 & & & \sigma_K^2\mathbf{B}_K \end{bmatrix}_{nK\times nK}. \tag{9.2.7}$$

By using the generalized Gauss–Markov theorem, we derive the BLUE of $\boldsymbol{\theta}$ in the model in (9.2.5) to be

$$\boldsymbol{\theta}_0^* = (\mathbf{A}' \quad \mathbf{C}^{-1} \quad \mathbf{A})^{-1} \quad \mathbf{A}' \quad \mathbf{C}^{-1} \quad \mathbf{y}_0. \tag{9.2.8}$$

The results for the Type-II censored case are obtained from (9.2.8) by replacing the samples by censored samples and the mean vectors $\boldsymbol{\alpha}_i$ and the variance–covariance matrices $\mathbf{B}_i$ by the corresponding quantities based on the censored samples.

A. Results for the Simple Linear Regression Model

Moussa (1972) considered the simple linear regression model of the form

$$y_{ij} = \alpha + \beta(x_i - \bar{x}) + \varepsilon_{ij} \tag{9.2.9}$$

for $j = 1, 2, \ldots, n$ and $i = 1, 2, \ldots, K$, where ε_{ij}s are independent variables having distribution $F(\varepsilon/\sigma)$ which may not be symmetric but depends on an unknown parameter σ.

According to our earlier notations, we have in this case

$$p = 2, \ \sigma_1 = \sigma_2 = \cdots = \sigma_K = \sigma,$$

$$\boldsymbol{\alpha}_i = \boldsymbol{\alpha} \quad \text{and} \quad \mathbf{B}_i = \mathbf{B} \quad \text{for } i = 1, 2, \ldots, K,$$

and, thence, the model in (9.2.5) becomes

$$E(\mathbf{y}_0) = \mathbf{A}\boldsymbol{\theta} \quad \text{and} \quad \operatorname{Var}(\mathbf{y}_0) = \sigma^2\mathbf{D}, \tag{9.2.10}$$

where

$$\mathbf{A} = \begin{bmatrix} \mathbf{1} & (x_1 - \bar{x})\mathbf{1} & \boldsymbol{\alpha} \\ \mathbf{1} & (x_2 - \bar{x})\mathbf{1} & \boldsymbol{\alpha} \\ \vdots & \vdots & \vdots \\ \mathbf{1} & (x_K - \bar{x})\mathbf{1} & \boldsymbol{\alpha} \end{bmatrix}_{nK \times 3}, \tag{9.2.11}$$

$$\mathbf{D} = \begin{bmatrix} \mathbf{B} & & & 0 \\ & \mathbf{B} & & \\ & & \ddots & \\ 0 & & & \mathbf{B} \end{bmatrix}_{nK \times nK}, \tag{9.2.12}$$

and

$$\boldsymbol{\theta}' = (\alpha \quad \beta \quad \sigma).$$

Then, the BLUE of $\boldsymbol{\theta}$ in the model in (9.2.10) is given by

$$\boldsymbol{\theta}_0^* = (\mathbf{A}' \ \mathbf{D}^{-1} \ \mathbf{A})^{-1} \ \mathbf{A}' \ \mathbf{D}^{-1} \ \mathbf{y}_0, \tag{9.2.13}$$

and we have the variance–covariance matrix of this estimator to be

$$\operatorname{Var}(\boldsymbol{\theta}_0^*) = \sigma^2(\mathbf{A}' \ \mathbf{D}^{-1} \ \mathbf{A})^{-1}. \tag{9.2.14}$$

It is possible to write the results in (9.2.13) and (9.2.14) explicitly, and to do so, let us set

$$a = \mathbf{1}' \ \mathbf{B}^{-1} \ \mathbf{1}, \qquad b = \mathbf{1}' \ \mathbf{B}^{-1} \ \boldsymbol{\alpha},$$

$$c = \boldsymbol{\alpha}' \ \mathbf{B}^{-1} \ \boldsymbol{\alpha}, \quad \text{and} \quad d = \frac{1}{K}\sum_{i=1}^{K}(x_i - \bar{x})^2.$$

It may then be easily verified that

$$\mathbf{A}' \quad \mathbf{D}^{-1} \quad \mathbf{A} = K\begin{bmatrix} a & 0 & b \\ 0 & ad & 0 \\ b & 0 & c \end{bmatrix},$$

$$(\mathbf{A}' \quad \mathbf{D}^{-1} \quad \mathbf{A})^{-1} = \frac{1}{Kad(ac-b^2)}\begin{bmatrix} acd & 0 & -abd \\ 0 & ac-b^2 & 0 \\ -abd & 0 & a^2d \end{bmatrix},$$

and

$$\mathbf{A}' \quad \mathbf{D}^{-1} \quad \mathbf{y}_0 = \begin{bmatrix} \mathbf{1}' \quad \mathbf{B}^{-1}\sum_{i=1}^{K}\mathbf{y}_i \\ \mathbf{1}' \quad \mathbf{B}^{-1}\sum_{i=1}^{K}(x_i-\bar{x})\mathbf{y}_i \\ \boldsymbol{\alpha}' \quad \mathbf{B}^{-1}\sum_{i=1}^{K}\mathbf{y}_i \end{bmatrix}.$$

We then obtain the BLUEs of α, β, and σ from (9.2.13) to be

$$\alpha_0^* = \left\{\frac{(c\mathbf{1}'-b\boldsymbol{\alpha}') \quad \mathbf{B}^{-1}}{K(ac-b^2)}\right\}\sum_{i=1}^{K}\mathbf{y}_i, \tag{9.2.15}$$

$$\beta_0^* = \left\{\frac{\mathbf{1}' \quad \mathbf{B}^{-1}}{Kad}\right\}\sum_{i=1}^{K}(x_i-\bar{x})\mathbf{y}_i, \tag{9.2.16}$$

and

$$\sigma_0^* = \left\{\frac{(a\boldsymbol{\alpha}'-b\mathbf{1}') \quad \mathbf{B}^{-1}}{K(ac-b^2)}\right\}\sum_{i=1}^{K}\mathbf{y}_i; \tag{9.2.17}$$

the variances and the covariances of the above estimators are obtained from (9.2.14) to be

$$\text{Var}(\alpha_0^*) = \sigma^2\left\{\frac{c}{K(ac-b^2)}\right\}, \tag{9.2.18}$$

$$\text{Var}(\beta_0^*) = \sigma^2/(Kad), \tag{9.2.19}$$

$$\text{Var}(\sigma_0^*) = \sigma^2\left\{\frac{a}{K(ac-b^2)}\right\}, \tag{9.2.20}$$

$$\text{Cov}(\alpha_0^*, \beta_0^*) = \text{Cov}(\beta_0^*, \sigma_0^*) = 0, \tag{9.2.21}$$

and

$$\text{Cov}(\alpha_0^*, \sigma_0^*) = -\sigma^2\left\{\frac{b}{K(ab-c^2)}\right\}. \tag{9.2.22}$$

If the population distribution F is symmetric, then it is easy to show that $b = \mathbf{1}' \mathbf{B}^{-1} \boldsymbol{\alpha} = 0$, in which case $(\mathbf{A}' \mathbf{D}^{-1} \mathbf{A})^{-1}$ becomes a diagonal matrix

$$(\mathbf{A}' \mathbf{D}^{-1} \mathbf{A})^{-1} = \text{Diag}\left(\frac{1}{Ka} \quad \frac{1}{Kad} \quad \frac{1}{Kc}\right).$$

In this case, the estimators in (9.2.15) through (9.2.17) become

$$\alpha_0^* = \left\{\frac{\mathbf{1}' \mathbf{B}^{-1}}{Ka}\right\} \sum_{i=1}^{K} \mathbf{y}_i = \mathbf{f}'_\alpha \sum_{i=1}^{K} \mathbf{y}_i, \tag{9.2.23}$$

$$\beta_0^* = \left\{\frac{\mathbf{1}' \mathbf{B}^{-1}}{Kad}\right\} \sum_{i=1}^{K} (x_i - \bar{x})\mathbf{y}_i = \mathbf{f}'_\beta \sum_{i=1}^{K} (x_i - \bar{x})\mathbf{y}_i, \tag{9.2.24}$$

and

$$\sigma_0^* = \left\{\frac{\boldsymbol{\alpha}' \mathbf{B}^{-1}}{Kc}\right\} \sum_{i=1}^{K} \mathbf{y}_i = \mathbf{f}'_\sigma \sum_{i=1}^{K} \mathbf{y}_i, \tag{9.2.25}$$

where

$$\mathbf{f}'_\alpha = \frac{\mathbf{1}' \mathbf{B}^{-1}}{Ka}, \qquad \mathbf{f}'_\beta = \frac{\mathbf{1}' \mathbf{B}^{-1}}{Kad}, \quad \text{and} \quad \mathbf{f}'_\sigma = \frac{\boldsymbol{\alpha}' \mathbf{B}^{-1}}{Kc}, \tag{9.2.26}$$

are the coefficients of the BLUEs of α, β, and σ, respectively. Tables of these coefficients have been constructed by Moussa (1972) for n up to 20 and K up to 10 for the normal, the standard double exponential, the scale contaminated normal with $\sigma = 3$ and mixing proportions 0.01, 0.05 and 0.10, and the Cauchy distributions (with at least two observations symmetrically censored for the Cauchy). Moreover, the variances and the covariances in (9.2.18) through (9.2.22) become in this case

$$\text{Var}(\alpha_0^*) = \frac{\sigma^2}{Ka} = \nu_\alpha \sigma^2, \tag{9.2.27}$$

$$\text{Var}(\beta_0^*) = \frac{\sigma^2}{Kad} = \nu_\beta \sigma^2, \tag{9.2.28}$$

and

$$\text{Var}(\sigma_0^*) = \frac{\sigma^2}{Kc} = \nu_\sigma \sigma^2. \tag{9.2.29}$$

The coefficients ν_α, ν_β, and ν_σ have also been tabulated by Moussa (1972).

We shall illustrate the method of estimation with the following example given by Hamouda (1988).

Example 9.2.1. It is believed that the sales of new employees increase with experience. A company collected samples of sales (y) of five employees for the first five months on the job (x), and the data are given below in Table 9.2.1.

The regression model considered here is

$$y_{ij} = \alpha + \beta(x_i - \bar{x}) + \varepsilon_{ij}, \qquad 1 \le i \le 5,\ 1 \le j \le 5.$$

Then, the BLUEs of α, β and σ and their variances were computed from Eqs. (9.2.23) through (9.2.29) for the normal, the scale contaminated normal with $\sigma = 3$ and mixing proportions $\pi = 0.01$, 0.05 and 0.10, and the double exponential distributions, by using the tables prepared by Moussa (1972) for the case $K = 5$ and $n = 5$. These values are presented in Table 9.2.2. From this table, we observe that the values of the estimates of α, β, and σ and their variances are very close between the distributions, and particularly so for the normal and the contaminated normal with small mixing proportion.

B. *Results for the K-Sample Situation with Equal Variance*

Moussa (1972) examined the one-way layout for both equal and unequal variance cases, and these are derived once again as special cases here. Let us consider the model

$$y_{ij} = \gamma_i + \varepsilon_{ij}, \qquad j = 1, 2, \ldots, n_i, \qquad i = 1, 2, \ldots, K, \qquad (9.2.30)$$

where ε_{ij}s are independent variables from distribution $F(\varepsilon/\sigma)$, which may or may not be symmetric but depends on the scale parameter σ.

TABLE 9.2.1
Sales (y) by Months on the Job (x)

$x =$	1		2		3		4		5	
	y_{ij}	$y_{i(j)}$	y_{ij}	$y_{i(j)}$	y_{ij}	$y_{i(j)}$	y_{ij}	$y_{i(j)}$	y_{ij}	$y_{i(j)}$
	0.60	0.40	1.30	1.30	1.60	1.60	1.85	1.85	2.64	2.21
	0.40	0.52	2.40	1.70	2.70	1.90	2.63	2.34	2.21	2.64
	0.70	0.60	1.70	2.10	1.90	2.70	2.34	2.63	4.30	3.75
	0.75	0.70	2.80	2.40	3.20	3.20	3.67	3.67	3.75	4.30
	0.52	0.75	2.10	2.80	3.80	3.80	4.21	4.21	5.62	5.62

Reprinted from E. Moussa-Hamouda, *Commun. Statist.-Theor. Meth.* (1988) **17(7)**, pp. 2343–2367, by courtesy of Marcel Dekker, Inc.

TABLE 9.2.2
BLUEs of α, β, and σ and Their Variances for Different Distributions

	Normal	Scale contaminated normal $\sigma=3$			Double exponential
		$\pi=0.01$	$\pi=0.05$	$\pi=0.10$	
α_0^*	2.3876	2.3753	2.3553	2.3479	2.3509
β_0^*	0.6940	0.6800	0.6580	0.6520	0.6611
σ_0^*	0.8769	0.8662	0.8301	0.7724	0.6924
$\mathrm{Var}(\alpha_0^*)$	0.0308	0.0321	0.0344	0.0344	0.0304
$\mathrm{Var}(\beta_0^*)$	0.0154	0.0160	0.0172	0.0172	0.0152
$\mathrm{Var}(\sigma_0^*)$	0.0205	0.0231	0.0278	0.0277	0.0220

Reprinted from E. Moussa-Hamouda, *Commun. Statist.-Theor. Meth.* (1988) **17(7)**, pp. 2343–2367, by courtesy of Marcel Dekker, Inc.

The model in (9.2.30) may be written in matrix notation as

$$E(\mathbf{y}_0)=\mathbf{A}\boldsymbol{\theta} \quad \text{and} \quad \mathrm{Var}(\mathbf{y}_0)=\sigma^2\mathbf{D}, \tag{9.2.31}$$

where

$$\mathbf{A}=\begin{bmatrix} \mathbf{1}_1 & \mathbf{0} & \cdots & \mathbf{0} & \boldsymbol{\alpha}_1 \\ \mathbf{0} & \mathbf{1}_2 & \cdots & \mathbf{0} & \boldsymbol{\alpha}_2 \\ \vdots & \vdots & \cdots & \vdots & \vdots \\ \mathbf{0} & \mathbf{0} & \cdots & \mathbf{1}_K & \boldsymbol{\alpha}_K \end{bmatrix}_{\sum_{i=1}^K n_i\times(K+1)} \tag{9.2.32}$$

and

$$D=\begin{bmatrix} \mathbf{B}_1 & & & 0 \\ & \mathbf{B}_2 & & \\ & & \ddots & \\ 0 & & & \mathbf{B}_K \end{bmatrix}_{\sum_{i=1}^K n_i\times\sum_{i=1}^K n_i}, \tag{9.2.33}$$

with $\mathbf{1}_i'=(1\ 1\ \cdots\ 1)_{1\times n_i}$, $\mathbf{B}_i=\mathrm{Var}(\mathbf{z}_i)$ and $\boldsymbol{\theta}'=(\gamma_1\ \gamma_2\ \cdots\ \gamma_K\ \sigma)$. Now let us set for $i=1,2,\ldots,K$,

$$a_i=\mathbf{1}_i'\ \mathbf{B}_i^{-1}\ \mathbf{1}_i, \qquad b_i=\mathbf{1}_i'\ \mathbf{B}_i^{-1}\ \boldsymbol{\alpha}_i, \quad \text{and} \quad c=\sum_{i=1}^K \boldsymbol{\alpha}_i'\ \mathbf{B}_i^{-1}\ \boldsymbol{\alpha}_i.$$

Then the matrix $\mathbf{A}' \ \mathbf{D}^{-1} \ \mathbf{A}$ may be written as

$$\mathbf{A}' \ \mathbf{D}^{-1} \ \mathbf{A} = \begin{bmatrix} a_1 & 0 & \cdots & 0 & b_1 \\ 0 & a_2 & \cdots & 0 & b_2 \\ \vdots & \vdots & \cdots & \vdots & \vdots \\ 0 & 0 & \cdots & a_K & b_K \\ b_1 & b_2 & \cdots & b_K & c \end{bmatrix},$$

whose inverse is given by

$$(\mathbf{A}' \ \mathbf{D}^{-1} \ \mathbf{A})^{-1} = \begin{bmatrix} (a_1 + Rb_1^2)/a_1^2 & b_1 b_2 R/a_1 a_2 & \cdots & b_1 b_K R/a_1 a_K & -Rb_1/a_1 \\ b_1 b_2 R/a_1 a_2 & (a_2 + Rb_2^2)/a_2^2 & \cdots & b_2 b_K R/a_2 a_K & -Rb_2/a_2 \\ \vdots & \vdots & \cdots & \vdots & \vdots \\ b_1 b_K R/a_1 a_K & b_2 b_K R/a_2 a_K & \cdots & (a_K + Rb_K^2)/a_K^2 & -Rb_K/a_K \\ -Rb_1/a_1 & -Rb_2/a_2 & \cdots & -Rb_K/a_K & R \end{bmatrix},$$

where $R = 1/\{c - \sum_{i=1}^{K} b_i^2/a_i\}$. The BLUE of $\boldsymbol{\theta}$ in the model in (9.2.31) is obtained in this case to be

$$\boldsymbol{\theta}_0^* = (\mathbf{A}' \ \mathbf{D}^{-1} \ \mathbf{A})^{-1} \ \mathbf{A}' \ \mathbf{D}^{-1} \ \mathbf{y}_0$$

$$= (\mathbf{A}' \ \mathbf{D}^{-1} \ \mathbf{A})^{-1} \begin{bmatrix} \mathbf{1}_1' \ \mathbf{B}_1^{-1} \ \mathbf{y}_1 \\ \mathbf{1}_2' \ \mathbf{B}_2^{-1} \ \mathbf{y}_2 \\ \vdots \\ \mathbf{1}_K' \ \mathbf{B}_K' \ \mathbf{y}_K \\ \sum_{i=1}^{K} \boldsymbol{\alpha}_i' \ \mathbf{B}_i^{-1} \ \mathbf{y}_i \end{bmatrix},$$

which yields, for $i = 1, 2, \ldots, K$,

$$\gamma_{0,i}^* = \left(\frac{a_i + Rb_i^2}{a_i^2}\right)\left(\mathbf{1}_i' \ \mathbf{B}_i^{-1} \ \mathbf{y}_i\right) + \left(\frac{Rb_i}{a_i}\right)\left\{\sum_{j=1, j\neq i}^{K} \frac{b_j}{a_j}(\mathbf{1}_j' \ \mathbf{B}_j^{-1} \ \mathbf{y}_j) - \sum_{i=1}^{K} \boldsymbol{\alpha}_i' \ \mathbf{B}_i^{-1} \ \mathbf{y}_i\right\}, \tag{9.2.34}$$

and

$$\sigma_0^* = R\left\{\sum_{i=1}^{K} \boldsymbol{\alpha}_i' \ \mathbf{B}_i^{-1} \ \mathbf{y}_i - \sum_{j=1}^{K} \frac{b_j}{a_j}(\mathbf{1}_j' \ \mathbf{B}_j^{-1} \ \mathbf{y}_j)\right\}. \tag{9.2.35}$$

The variances and covariances of these estimators are obtained to be

$$\mathrm{Var}(\gamma_{0,i}^*) = \sigma^2\left(\frac{a_i + Rb_i^2}{a_i^2}\right), \qquad i = 1, 2, \ldots, K, \tag{9.2.36}$$

$$\mathrm{Var}(\sigma_0^*) = \sigma^2 R, \tag{9.2.37}$$

$$\text{Cov}(\gamma_{0,i}^*, \gamma_{0,j}^*) = \sigma^2\left(\frac{b_i}{a_i}\frac{b_j}{a_j}\right), \qquad i \neq j = 1, 2, \ldots, K, \tag{9.2.38}$$

and

$$\text{Cov}(\gamma_{0,i}^*, \sigma_0^*) = -\sigma^2\left(\frac{Rb_i}{a_i}\right), \qquad i = 1, 2, \ldots, K. \tag{9.2.39}$$

In the case when the population distribution is symmetric, we have as before $b_i = \mathbf{1}_i' \ \mathbf{B}_i^{-1} \ \boldsymbol{\alpha}_i = 0$ for $i = 1, 2, \ldots, K$, and we obtain the BLUEs from (9.2.34) and (9.2.35) to be

$$\gamma_{0,i}^* = \left(\frac{\mathbf{1}_i' \ \mathbf{B}_i^{-1}}{\mathbf{1}_i' \ \mathbf{B}_i^{-1} \ \mathbf{1}_i}\right)\mathbf{y}_i, \qquad i = 1, 2, \ldots, K, \tag{9.2.40}$$

and

$$\sigma_0^* = R \sum_{i=1}^{K} \boldsymbol{\alpha}_i' \ \mathbf{B}_i^{-1} \ \mathbf{Y}_i = \frac{\sum_{i=1}^{K} \boldsymbol{\alpha}_i' \ \mathbf{B}_i^{-1} \ \mathbf{y}_i}{\sum_{i=1}^{K} \boldsymbol{\alpha}_i' \ \mathbf{B}_i^{-1} \ \boldsymbol{\alpha}_i}. \tag{9.2.41}$$

Further, the variances of the above estimators are obtained from (9.2.36) and (9.2.37) to be

$$\text{Var}(\gamma_{0,i}^*) = \frac{\sigma^2}{\mathbf{1}_i' \ \mathbf{B}_i^{-1} \ \mathbf{1}_i}, \qquad i = 1, 2, \ldots, K, \tag{9.2.42}$$

and

$$\text{Var}(\sigma_0^*) = \frac{\sigma^2}{\sum_{i=1}^{K} \boldsymbol{\alpha}_i' \ \mathbf{B}_i^{-1} \ \boldsymbol{\alpha}_i}. \tag{9.2.43}$$

The estimators are all uncorrelated in this case. Furthermore, we observe that the estimators of γ_i are exactly the same as in the single-sample situation (see Chapter 4).

C. Results for the K-Sample Situation with Unequal Variances

In this case, we may write the model to be

$$E(\mathbf{y}_0) = \mathbf{A}\boldsymbol{\theta} \quad \text{and} \quad \text{Var}(\mathbf{y}_0) = \mathbf{D}, \tag{9.2.44}$$

where

$$\mathbf{A} = \begin{bmatrix} \mathbf{1}_1 & \mathbf{0} & \cdots & \mathbf{0} & \boldsymbol{\alpha}_1 & \mathbf{0} & \cdots & \mathbf{0} \\ \mathbf{0} & \mathbf{1}_2 & \cdots & \mathbf{0} & \mathbf{0} & \boldsymbol{\alpha}_2 & \cdots & \mathbf{0} \\ \vdots & \vdots & \cdots & \vdots & \vdots & \vdots & \cdots & \vdots \\ \mathbf{0} & \mathbf{0} & \cdots & \mathbf{1}_K & \mathbf{0} & \mathbf{0} & \cdots & \boldsymbol{\alpha}_K \end{bmatrix},$$

$$D = \begin{bmatrix} \sigma_1^2\mathbf{B}_1 & & & 0 \\ & \sigma_2^2\mathbf{B}_2 & & \\ & & \ddots & \\ 0 & & & \sigma_K^2\mathbf{B}_K \end{bmatrix},$$

and

$$\boldsymbol{\theta}' = (\gamma_1 \quad \gamma_2 \quad \cdots \quad \gamma_K \quad \sigma_1 \quad \sigma_2 \quad \cdots \quad \sigma_K).$$

The BLUEs may be derived in this case to be

$$\gamma_{0,i}^* = \frac{1}{a_i c_i - b_i^2}(c_i \quad \mathbf{1}_i' \quad \mathbf{B}_i^{-1} \quad \mathbf{y}_i - b_i \quad \boldsymbol{\alpha}_i' \quad \mathbf{B}_i^{-1} \quad \mathbf{y}_i), \qquad i = 1, 2, \ldots, K, \tag{9.2.45}$$

and

$$\sigma_{0,i}^* = \frac{1}{a_i c_i - b_i^2}(a_i \quad \boldsymbol{\alpha}_i' \quad \mathbf{B}_i^{-1} \quad \mathbf{y}_i - b_i \quad \mathbf{1}_i' \quad \mathbf{B}_i^{-1} \quad \mathbf{y}_i), \qquad i = 1, 2, \ldots, K, \tag{9.2.46}$$

and the variances of the above estimators are given by

$$\mathrm{Var}(\gamma_{0,i}^*) = \sigma_i^2 \left\{ \frac{c_i}{a_i c_i - b_i^2} \right\}, \qquad i = 1, 2, \ldots, K, \tag{9.2.47}$$

and

$$\mathrm{Var}(\sigma_{0,i}^*) = \sigma_i^2 \left\{ \frac{a_i}{a_i c_i - b_i^2} \right\}, \qquad i = 1, 2, \ldots, K. \tag{9.2.48}$$

In the case in which the population distribution is symmetric, since $b_i = \mathbf{1}_i' \ \mathbf{B}_i^{-1} \ \boldsymbol{\alpha}_i = 0$ for $i = 1, 2, \ldots, K$, the BLUEs in (9.2.45) and (9.2.46) reduce to

$$\gamma_{0,i}^* = \left(\frac{\mathbf{1}_i' \quad \mathbf{B}_i^{-1}}{\mathbf{1}_i' \quad \mathbf{B}_i^{-1} \quad \mathbf{1}_i} \right) \mathbf{y}_i, \qquad i = 1, 2, \ldots, K, \tag{9.2.49}$$

and

$$\sigma_{0,i}^* = \left(\frac{\boldsymbol{\alpha}_i' \quad \mathbf{B}_i^{-1}}{\boldsymbol{\alpha}_i' \quad \mathbf{B}_i^{-1} \quad \boldsymbol{\alpha}_i} \right) \mathbf{y}_i, \qquad i = 1, 2, \ldots, K, \tag{9.2.50}$$

and their variances in (9.2.47) and (9.2.48) reduce to

$$\mathrm{Var}(\gamma_{0,i}^*) = \frac{\sigma_i^2}{\mathbf{1}_i' \;\; \mathbf{B}_i^{-1} \;\; \mathbf{1}_i}, \qquad i = 1, 2, \ldots, K, \tag{9.2.51}$$

and

$$\mathrm{Var}(\sigma_{0,i}^*) = \frac{\sigma_i^2}{\boldsymbol{\alpha}_i' \;\; \mathbf{B}_i^{-1} \;\; \boldsymbol{\alpha}_i}, \qquad i = 1, 2, \ldots, K. \tag{9.2.52}$$

These results may be noted to be exactly the same as in the single-sample situation handled in Chapter 4.

Some asymptotic properties of the BLUEs and inference procedures based on the BLUEs have been discussed by Hamouda (1988).

9.3. Modified Maximum Likelihood Estimation with Multiple Measurements

For a doubly Type-II censored sample from a normal $N(\mu, \sigma^2)$ population, Tiku (1967, 1970) derived the modified maximum likelihood estimators (MMLEs) of μ and σ. These estimators are the same as the AMLEs of μ and σ developed in Section 6.3, except that the coefficients α, β, γ, and δ involved in approximating the likelihood function are determined differently. For the symmetrically censored sample with r observations censored on either side, Tiku (1967, 1980) has given the formulas

$$\delta = \beta = \frac{h(k_2) - h(k_1)}{k_2 - k_1} \quad \text{and} \quad \gamma = \alpha = h(k_1) - k_1\delta, \tag{9.3.1}$$

where

$$k_1 = F^{-1}\left\{1 - q - \sqrt{\frac{q(1-q)}{n}}\right\},$$

$$k_2 = F^{-1}\left\{1 - q + \sqrt{\frac{q(1-q)}{n}}\right\},$$

$$q = \frac{r}{n}, \quad \text{and} \quad h(z) = \frac{f(z)}{1 - F(z)},$$

with $f(\cdot)$ and $F(\cdot)$ denoting the density and the cumulative distribution functions of a standard normal variable, respectively. By using these estimators, Tiku (1973, 1981, 1982) handled the Type-II censoring situations in regression and other experimental design problems. Interested readers may refer to the book by Tiku *et al.* (1986) for more details on this subject.

In this section, we describe the MML method of estimation of parameters in a general linear model with multiple measurements. The details given here are the case when the samples are symmetrically Type-II censored; however, the method of estimation is available in general and may be secured from the original source.

Let us start with the simple linear regression model

$$y_{ij} = \alpha + \beta x_i + \varepsilon_{ij}, \qquad j = 1, 2, \ldots, n_i, \qquad i = 1, 2, \ldots, K, \qquad (9.3.2)$$

where x_is are the explanatory variables, and ε_{ij}s are independent and identically distributed error variables with mean 0 and common unknown variance σ^2. Further, let us assume that we have available the following symmetrically censored samples:

$$y_{i(r_i+1)} \le y_{i(r_i+2)} \le \cdots \le y_{i(n_i-r_i)}, \qquad i = 1, 2, \ldots, K, \qquad (9.3.3)$$

where the smallest and the largest r_i observations in the ith sample have been censored. Let us denote, for $i = 1, 2, \ldots, K$,

$$A_i = n_i - 2r_i,$$

$$m_i = A_i + 2r_i\delta_i,$$

$$D_i = \frac{1}{m_i}\left\{\sum_{j=r_i+1}^{n_i-r_i} y_{i(j)} + r_i\delta_i(y_{i(r_i+1)} + y_{i(n_i-r_i)})\right\},$$

$$B_i = r_i\gamma_i(y_{i(n_i-r_i)} - y_{i(r_i+1)}),$$

and

$$C_i = \sum_{j=r_i+1}^{n_i-r_i} y_{i(j)}^2 + r_i\delta_i(y_{i(r_i+1)}^2 + y_{i(n_i-r_i)}^2) - m_i K_i^2.$$

In addition, let us write

$$A = \sum_{i=1}^{K} A_i, \qquad M = \sum_{i=1}^{K} m_i, \qquad B = \sum_{i=1}^{K} B_i,$$

$$\bar{D} = \frac{1}{M}\sum_{i=1}^{K} m_i D_i, \qquad \bar{x} = \frac{1}{M}\sum_{i=1}^{K} m_i x_i,$$

$$S_{x,x} = \sum_{i=1}^{K} m_i(x_i - \bar{x})x_i = \sum_{i=1}^{K} m_i(x_i - \bar{x})^2,$$

and

$$S_{x,y} = \sum_{i=1}^{K} m_i(x_i - \bar{x})D_i = \sum_{i=1}^{K} m_i(x_i - \bar{x})(D_i - \bar{D}).$$

With these notations, Tiku (1981) and Tiku *et al.* (1986) have derived the

MMLEs of α, β and σ to be

$$\hat{\alpha} = \bar{D} - \hat{\beta}\bar{x}, \tag{9.3.4}$$

$$\hat{\beta} = S_{x,y}/S_{x,x}, \tag{9.3.5}$$

and

$$\hat{\sigma} = \{B + (B^2 + 4AC)^{1/2}\}/2\{A(A-2)\}^{1/2}; \tag{9.3.6}$$

here,

$$C = \sum_{i=1}^{K} C_i + \sum_{i=1}^{K} m_i\{(D_i - \bar{D}) - \hat{\beta}(x_i - \bar{x})\}^2. \tag{9.3.7}$$

Furthermore, under the assumption of normality for the error component, Tiku (1981) and Tiku *et al.* (1986) have shown that the estimators $\hat{\alpha}$ and $\hat{\beta}$ are exactly unbiased for α and β, respectively, and $\hat{\sigma}$ is asymptotically unbiased for σ, and that

$$\text{Var}(\hat{\alpha}) \simeq \sigma^2\left\{\frac{1}{M} + \frac{\bar{x}^2}{S_{x,x}}\right\}, \tag{9.3.8}$$

$$\text{Var}(\hat{\beta}) \simeq \frac{\sigma^2}{S_{x,x}}, \tag{9.3.9}$$

$$\text{Var}(\hat{\sigma}) \simeq \frac{\sigma^2}{2(A-2)}, \tag{9.3.10}$$

and

$$\text{Cov}(\hat{\alpha}, \hat{\beta}) \simeq -\sigma^2\left(\frac{\bar{x}}{S_{x,x}}\right). \tag{9.3.11}$$

Some asymptotic distributional properties of these estimators and inference procedures based on them have been discussed in great detail by Tiku *et al.* (1986).

We now present an example, taken from Tiku *et al.* (1986), in order to illustrate the MML method of estimation described above.

Example 9.3.1. In a time-mortality experiment, five doses of a drug were administered to five different groups of sample specimens. However, a few specimens died before the observation began, and the experiment was terminated when some specimens still remained alive. The log survival times (y) of the specimens were recorded during the full periods of observations, and the explanatory variable x_i here is log(10 dose). These values are presented in Table 9.3.1.

TABLE 9.3.1
The Log Survival Times of the Specimens at Five Doses

Group	x_i					$y_{i(j)}$					
1	0.310	—	1.159	1.173	2.142	2.168	2.220	2.479	2.932	3.449	—
2	0.477	—	—	1.745	2.483	2.553	2.724	3.482	3.542	—	—
3	0.602	2.624	2.672	3.213	3.523	3.609					
4	1.000	3.466	3.533	3.549	3.689						
5	1.398	—	—	3.081	3.320	3.362	3.594	3.642	3.859	—	—

For the tabulated data, we computed various quantities required for the evaluation of the MMLEs, and these are summarized in Table 9.3.2. We then find:

$$\sum_{i=1}^{5} m_i(x_i - \bar{x})^2 = 6.839,$$

$$\sum_{i=1}^{5} m_i(D_i - \bar{D})^2 = 9.660,$$

$$\sum_{i=1}^{5} m_i(x_i - \bar{x})(D_i - \bar{D}) = 7.295,$$

$$\hat{\beta} = \frac{7.295}{6.839} = 1.067,$$

$$\hat{\alpha} = \bar{D} - \hat{\beta}\bar{x} = 2.92285 - (1.067 \times 0.73197) = 2.1418,$$

TABLE 9.3.2
Computational Tableau for the Evaluation of the MMLEs

	Groups				
i	1	2	3	4	5
n_i	10	10	5	4	10
r_i	1	2	0	0	2
γ_i	0.6737	0.7493	0	0	0.7493
δ_i	0.8611	0.7914	1	1	0.7914
m_i	9.722	9.166	5	4	9.166
D_i	2.231	2.716	3.128	3.559	3.474
x_i	0.301	0.477	0.602	1.000	1.398

and

$$\hat{\sigma} = \{B + (B^2 + 4AC)^{1/2}\}/2\{A(A-2)\}^{1/2}$$
$$= \frac{5.402 + (29.178 + 1748.108)^{1/2}}{2\sqrt{29 \times 27}} = 0.8498.$$

The above-described MML estimation method may be readily extended to a general linear model of the form

$$y_{ij} = \alpha + \beta_1 x_{1i} + \beta_2 x_{2i} + \cdots + \beta_p x_{pi} + \varepsilon_{ij},$$
$$j = 1, 2, \ldots, n_i, \qquad i = 1, 2, \ldots, K, \tag{9.3.12}$$

where $x_{1i}, x_{2i}, \ldots, x_{pi}$ $(i = 1, 2, \ldots, K)$ are the explanatory variables and ε_{ij}s are the independent and identically distributed error variables with mean zero and common unknown variance σ^2. Once again, we shall present here only the results for the case in which the samples are symmetrically Type-II censored, even though this method of estimation is applicable in general. To this end, let us assume that we have available the following symmetrically censored samples:

$$y_{i(r_i+1)} \leq y_{i(r_i+2)} \leq \cdots \leq y_{i(n_i-r_i)}, \qquad i = 1, 2, \ldots, K, \tag{9.3.13}$$

where the smallest and the largest r_i observations in the ith sample have been censored. Let us now denote, for $i = 1, 2, \ldots, K$,

$$A_i = n_i - 2r_i,$$
$$m_i = A_i + 2r_i\delta_i,$$
$$D_i = \frac{1}{m_i}\left\{\sum_{j=r_i+1}^{n_i-r_i} y_{i(j)} + r_i\delta_i(y_{i(r_i+1)} + y_{i(n_i-r_i)})\right\},$$
$$B_i = r_i\gamma_i(y_{i(n_i-r_i)} - y_{i(r_i+1)}),$$
$$A = \sum_{i=1}^{K} A_i, \qquad M = \sum_{i=1}^{K} m_i, \qquad B = \sum_{i=1}^{K} B_i,$$

and

$$\bar{D} = \frac{1}{M}\sum_{i=1}^{K} m_i D_i;$$

further, for $j, k = 1, 2, \ldots, p$, let us set

$$\bar{x}_j = \frac{1}{M}\sum_{i=1}^{K} m_i x_{ji},$$
$$S_{j,k} = \sum_{i=1}^{K} m_i(x_{ji} - \bar{x}_j)x_{ki} = \sum_{i=1}^{K} m_i(x_{ji} - \bar{x}_j)(x_{ki} - \bar{x}_k),$$

and

$$T_j = \sum_{i=1}^{K} m_i(x_{ji} - \bar{x}_j)D_i = \sum_{i=1}^{K} m_i(x_{ji} - \bar{x}_j)(D_i - \bar{D}).$$

With these notations, the MMLEs of the parameters in the general linear model in (9.3.12) may be shown to be as follows (see Tiku *et al.*, 1986):

$$\hat{\alpha} = \bar{K} - \sum_{j=1}^{p} \hat{\beta}_j \bar{x}_j, \tag{9.3.14}$$

$(\hat{\beta}_1, \hat{\beta}_2, \ldots, \hat{\beta}_p)$ are the solutions of the system of equations

$$\sum_{j=1}^{p} S_{i,j}\hat{\beta}_j = T_i, \qquad i = 1, 2, \ldots, p, \tag{9.3.15}$$

and

$$\hat{\sigma} = \{B + (B^2 + 4AC)^{1/2}\}/2\{A(A - p - 1)\}^{1/2}, \tag{9.3.16}$$

where

$$\begin{aligned} C = \sum_{i=1}^{K} \Bigg[\sum_{j=r_i+1}^{n_i-r_i} \left(y_{i(j)} - \hat{\alpha} - \sum_{k=1}^{p} \hat{\beta}_k x_{ki} \right)^2 & \\ + r_i\delta_i \Bigg\{ \left(y_{i(r_i+1)} - \hat{\alpha} - \sum_{k=1}^{p} \hat{\beta}_k x_{ki} \right)^2 & \\ + \left(y_{i(n_i-r_i)} - \hat{\alpha} - \sum_{k=1}^{p} \hat{\beta}_k x_{ki} \right)^2 \Bigg\} \Bigg]. & \end{aligned} \tag{9.3.17}$$

Tiku *et al.* (1986) have discussed some asymptotic distributional properties and robustness features of the above-given MML estimators, and they also have developed inference procedures based on these estimators.

9.4. Modified Maximum Likelihood Estimation with Single Measurements

In this section, we consider a simple linear regression model with single measurements, and we describe the modified maximum likelihood estimation of the parameters based on a Type-II symmetrically censored sample of normalized residuals as given by Balakrishnan and Ambagaspitiya (1990a). We present the simulated values of the bias and the variance of these MMLEs under both normal and mixture-normal models for the error component and compare them with the corresponding values of the least-squares estimators. Through these simulated values, we display that the

MMLEs are highly efficient and robust to departures from normality of the error component, while the corresponding least-squares estimators are extremely sensitive to such departures. We also consider two examples given by Daniel and Wood (1971) and Mendenhall (1983) and illustrate the method of estimation presented in this section.

It should be mentioned here that Yale and Forsythe (1976) first introduced a Winsorized estimation procedure for the parameters of a linear regression model. By considering a simple linear regression model and then assuming a mixture-normal distribution for the error component, they demonstrated that the Winsorized estimators are more efficient than the ordinary least-squares estimators. Tan and Tabatabai (1988) extended the work of Yale and Forsythe (1976) by introducing the r-Winsorized estimators. Tan and Tabatabai (1988) also presented some modified Winsorized estimators for the parameters of a linear regression model and subsequently showed that these estimators are close to the MMLEs. In addition, they incorporated the PRESS method of Allen (1971) in their approaches and made some comparisons of these estimation methods by considering the same models as those given by Yale and Forsythe (1976). A more detailed comparison of a nonparametric method based on ranks proposed by Adichie (1967a, b), an M-estimation method proposed by Andrews (1973, 1974), and all the methods mentioned above has been made recently by Balakrishnan and Ambagaspitiya (1990b).

Let us consider a simple linear regression model

$$y_i = \alpha + \beta x_i + \varepsilon_i, \qquad i = 1, 2, \ldots, n, \tag{9.4.1}$$

where α and β are the usual regression parameters, $(x_1, x_2, \ldots, x_n)$ are the values of the explanatory variable, and ε_i is the random error involved in measuring y_i. We shall assume that the ε_is are independently normally distributed with mean zero and unknown variance σ^2.

Let us denote

$$Z_i = (y_i - \alpha - \beta x_i)/\sigma, \qquad i = 1, 2, \ldots, n, \tag{9.4.2}$$

and their order statistics by

$$Z_{1:n} \le Z_{2:n} \le \cdots \le Z_{n:n}. \tag{9.4.3}$$

Further, let us denote the (x, y) pair corresponding to the order statistic $Z_{i:n}$ by $(x_{[i]}, y_{[i]})$ for $i = 1, 2, \ldots, n$. Let us now consider the Type-II symmetrically censored sample

$$Z_{r+1:n} \le Z_{r+2:n} \le \cdots \le Z_{n-r:n}, \tag{9.4.4}$$

where r is chosen to be $[0.5 + 0.1n]$ ($[g]$ denotes the integer part of g) to achieve robustness to most departures from normality; see, for example, Tiku *et al.* (1986). Then, under the assumption of normality for the error variable, we have the likelihood function based on the symmetrically censored sample in (9.4.4) to be

$$L = \frac{n!}{(r!)^2} \sigma^{-A} \{F(Z_{r+1:n})\}^r \{1 - F(Z_{n-r:n})\}^r \prod_{i=r+1}^{n-r} f(Z_{i:n}), \qquad (9.4.5)$$

where $A = n - 2r$, and f and F are the density function and the cumulative distribution function of a standard normal variable, respectively. From (9.4.5) we have

$$\begin{aligned} \ln L = \text{Const} - A \ln \sigma + r \ln F(Z_{r+1:n}) \\ + r \ln\{1 - F(Z_{n-r:n})\} - \frac{1}{2} \sum_{i=r+1}^{n-r} Z_{i:n}^2. \end{aligned} \qquad (9.4.6)$$

Upon differentiating (9.4.6) with respect to α, β, and σ, we obtain the following likelihood equations:

$$\frac{\partial \ln L}{\partial \alpha} = \frac{1}{\sigma} \left[-r \frac{f(Z_{r+1:n})}{F(Z_{r+1:n})} + r \frac{f(Z_{n-r:n})}{1 - F(Z_{n-r:n})} + \sum_{i=r+1}^{n-r} Z_{i:n} \right]$$

$$= 0, \qquad (9.4.7)$$

$$\frac{\partial \ln L}{\partial \beta} = \frac{1}{\sigma} \left[-r x_{[r+1]} \frac{f(Z_{r+1:n})}{F(Z_{r+1:n})} + r x_{[n-r]} \frac{f(Z_{n-r:n})}{1 - F(Z_{n-r:n})} + \sum_{i=r+1}^{n-r} x_{[i]} Z_{i:n} \right]$$

$$= 0, \qquad (9.4.8)$$

and

$$\frac{\partial \ln L}{\partial \sigma} = \frac{1}{\sigma} \left[-A - r Z_{r+1:n} \frac{f(Z_{r+1:n})}{F(Z_{r+1:n})} + r Z_{n-r:n} \frac{f(Z_{n-r:n})}{1 - F(Z_{n-r:n})} + \sum_{i=r+1}^{n-r} Z_{i:n}^2 \right]$$

$$= 0. \qquad (9.4.9)$$

Eqs. (9.4.7) through (9.4.9) do not provide explicit estimators for α, β, and σ. However, by adopting Tiku's approximations as explained in the preceding section, we obtain the following modified likelihood equations:

$$\frac{\partial \ln L}{\partial \alpha} \simeq \frac{\partial \ln L^*}{\partial \alpha} = \frac{1}{\sigma} \left[-r\{\gamma - \delta Z_{r+1:n}\} + r\{\gamma + \delta Z_{n-r:n}\} + \sum_{i=r+1}^{n-r} Z_{i:n} \right]$$

$$= 0, \qquad (9.4.10)$$

$$\frac{\partial \ln L}{\partial \beta} \simeq \frac{\partial \ln L^*}{\partial \beta} = \frac{1}{\sigma}\Bigg[-rx_{[r+1]}\{\gamma - \delta Z_{r+1:n}\} + rx_{[n-r]}\{\gamma + \delta Z_{n-r:n}\} + \sum_{i=r+1}^{n-r} x_{[i]} Z_{i:n}\Bigg] = 0, \tag{9.4.11}$$

and

$$\frac{\partial \ln L}{\partial \sigma} \simeq \frac{\partial \ln L^*}{\partial \sigma} = \frac{1}{\sigma}\Bigg[-A - rZ_{r+1:n}\{\gamma - \delta Z_{r+1:n}\} + rZ_{n-r:n}\{\gamma + \delta Z_{n-r:n}\} + \sum_{i=r+1}^{n-r} Z_{i:n}^2\Bigg] = 0. \tag{9.4.12}$$

From (9.4.10) we derive the MML estimator of α as

$$\hat{\alpha} = \tilde{y} - \hat{\beta}\tilde{x}, \tag{9.4.13}$$

where

$$m = n - 2r + 2r\delta, \tag{9.4.14}$$

$$\tilde{x} = \frac{1}{m}\left\{\sum_{i=r+1}^{n-r} x_{[i]} + r\delta(x_{[r+1]} + x_{[n-r]})\right\}, \tag{9.4.15}$$

and

$$\tilde{y} = \frac{1}{m}\left\{\sum_{i=r+1}^{n-r} y_{[i]} + r\delta(y_{[r+1]} + y_{[n-r]})\right\}. \tag{9.4.16}$$

Next, from (9.4.11) we derive the MML estimator of β as

$$\hat{\beta} = \frac{S^*_{x,y}}{S^*_{x,x}} + \hat{\sigma}\left\{\frac{r\gamma(x_{[n-r]} - x_{[r+1]})}{S^*_{x,x}}\right\}, \tag{9.4.17}$$

where

$$S^*_{x,x} = \sum_{i=r+1}^{n-r} x_{[i]}^2 + r\delta(x_{[r+1]}^2 + x_{[n-r]}^2) - m\tilde{x}^2 \tag{9.4.18}$$

and

$$S^*_{x,y} = \sum_{i=r+1}^{n-r} x_{[i]}y_{[i]} + r\delta(x_{[r+1]}y_{[r+1]} + x_{[n-r]}y_{[n-r]}) - m\tilde{x}\tilde{y}. \tag{9.4.19}$$

Finally, upon substituting the expressions of $\hat{\alpha}$ and $\hat{\beta}$ in (9.4.13) and (9.4.17), respectively, in (9.4.12) and simplifying the resulting equation, we obtain

$$A\sigma^2 - B\sigma - C = 0, \tag{9.4.20}$$

where

$$B = r\gamma(y_{[n-r]} - y_{[r+1]}) - \frac{S^*_{x,y}}{S^*_{x,x}}\{r\gamma(x_{[n-r]} - x_{[r+1]})\} \tag{9.4.21}$$

and

$$C = S^*_{y,y} - \frac{S^{*2}_{x,y}}{S^*_{x,x}} = S^*_{y,y}\left\{1 - \frac{S^{*2}_{x,y}}{S^*_{x,x}S^*_{y,y}}\right\}; \tag{9.4.22}$$

here,

$$S^*_{y,y} = \sum_{i=r+1}^{n-r} y^2_{[i]} + r\delta(y^2_{[r+1]} + y^2_{[n-r]}) - m\tilde{y}^2. \tag{9.4.23}$$

Now, upon using the fact that $\delta \geq 0$ and the result that

$$S^{*2}_{x,y} \leq S^*_{x,x}S^*_{y,y},$$

noted by a simple application of the Cauchy–Schwarz inequality, it follows from (9.4.22) that $C \geq 0$. Hence, the quadratic equation in (9.4.20) admits only one positive root, thus yielding the MML estimator of σ to be

$$\hat{\sigma} = \{B + (B^2 + 4AC)^{1/2}\}/(2A). \tag{9.4.24}$$

Since $r\gamma(x_{[n-r]} - x_{[r+1]})/S^*_{x,x}$ will usually be small, (9.4.17) essentially gives the simplified estimator

$$\hat{\beta} \simeq S^*_{x,y}/S^*_{x,x}. \tag{9.4.25}$$

It is of interest to note here that the simplified estimator $\hat{\beta}$ is of exactly the same form as the classical least-squares estimator of β. Furthermore, following the suggestion of Tiku and Stewart (1977) and Tiku *et al.* (1986), the estimator $\hat{\sigma}$ in (9.4.24) may be corrected for bias by defining

$$\hat{\sigma} = \{B + (B^2 + 4AC)^{1/2}\}/2\{A(A-2)\}^{1/2}. \tag{9.4.26}$$

We may observe here that when $r = 0$ (no observations censored), the MML estimators $\hat{\alpha}$, $\hat{\beta}$, and $\hat{\sigma}$ in (9.4.13), (9.4.25), and (9.4.26), respectively, simply reduce to the corresponding classical estimators.

A. *Determination of $x_{[i]}$ and $y_{[i]}$*

The MMLEs of α, β, and σ derived above are all based on the order statistics of Z_i defined in (9.4.2). We realize that the variables Z_i themselves involve the unknown parameters α, β, and σ. Therefore, one needs to use some initial estimates of these parameters in order to start the ordering of these Z_i. For this purpose, Yale and Forsythe (1976) and Tan and Tabatabai (1988) have used two methods; one is based on the least-squares estimators, while the other is based on a procedure similar to the PRESS method of Allen (1971). Balakrishnan and Ambagaspitiya (1990a, b) proceeded as follows to determine $x_{[i]}$ and $y_{[i]}$.

By denoting the ordinary least-squares estimators of α and β by $\tilde{\alpha}$ and $\tilde{\beta}$, we have the estimated residuals to be

$$\tilde{\varepsilon}_i = y_i - \tilde{\alpha} - \tilde{\beta}x_i, \qquad i = 1, 2, \ldots, n. \tag{9.4.27}$$

Under the assumption that the error variables ε_i in (9.4.1) are distributed as $N(0, \sigma^2)$, we have the estimated residuals $\tilde{\varepsilon}_i$ in (9.4.27) to be normally distributed with mean zero and variance given by

$$\operatorname{Var}(\tilde{\varepsilon}_i) = \sigma^2 \left\{ \frac{n-1}{n} - \frac{(x_i - \bar{x})^2}{\sum_{j=1}^{n} (x_j - \bar{x})^2} \right\}. \tag{9.4.28}$$

We observe from (9.4.28) that the estimated residuals $\tilde{\varepsilon}_i$ have a larger variation if the chosen x_i are closer to their mean $\bar{x}$. Hence, we pointed out by Behnken and Draper (1972) and Barnett and Lewis (1978), we need to take this "ballooning effect" of the estimated residuals into account while examining the size of the estimated residuals $\tilde{\varepsilon}_i$ for possible identification of outliers among them.

By taking (9.4.28) into account, we may therefore use the "normalized estimated residuals" defined as

$$\frac{\tilde{\varepsilon}_i}{s_i} = \tilde{\varepsilon}_i \Bigg/ \left[s \left\{ \frac{n-1}{n} - \frac{(x_i - \bar{x})^2}{\sum^{n} (x_j - \bar{x})^2} \right\}^{1/2} \right], \tag{9.4.29}$$

where $s^2 = 1/(n-2) \sum_{i=1}^{n} \tilde{\varepsilon}_i^2$ is an unbiased estimator of σ^2, and s_i^2 is an unbiased estimator of $\operatorname{Var}(\tilde{\varepsilon}_i)$ in (9.4.28).

Following the notations of Theil (1971), Hoaglin and Welsch (1978), and Belsley *et al.* (1980), we may note here, as pointed out by Balakrishnan and Ambagaspitiya (1990a), that the "normalized estimated residuals" defined in (9.4.29) are the same as their "standardized residuals" given by

$$\frac{\tilde{\varepsilon}_i}{s(1-h_i)^{1/2}}, \tag{9.4.30}$$

where h are simply the diagonal elements of their hat matrix.

In order to determine the $x_{[i]}$ and $y_{[i]}$ that are necessary for the computation of the MMLEs of α, β, and σ, Balakrishnan and Ambagaspitiya (1990a) suggested the following three steps:

(1) From the given data, first compute the least-squares estimates of α and β;

(2) Next, from (9.4.29) compute the normalized estimated residuals for all the n pairs of observations and then arrange them in increasing order of magnitude; and

(3) $(x_{[i]}, y_{[i]})$ is taken to be the (x, y) pair corresponding to the ith largest normalized estimated residual for $i = 1, 2, \ldots, n$.

B. *Asymptotic Variance-Covariance Matrix and Distributional Result*

From Eqs. (9.4.10) through (9.4.12), we obtain

$$E\left(-\frac{\partial^2 \ln L}{\partial\alpha^2}\right) \simeq E\left(-\frac{\partial^2 \ln L^*}{\partial\alpha^2}\right) = \frac{m}{\sigma^2}, \tag{9.4.31}$$

$$E\left(-\frac{\partial^2 \ln L}{\partial\alpha\,\partial\beta}\right) \simeq E\left(-\frac{\partial^2 \ln L^*}{\partial\alpha\,\partial\beta}\right) = \frac{m\tilde{x}}{\sigma^2}, \tag{9.4.32}$$

$$E\left(-\frac{\partial^2 \ln L}{\partial\alpha\,\partial\sigma}\right) \simeq E\left(-\frac{\partial^2 \ln L^*}{\partial\alpha\,\partial\sigma}\right) = 0, \tag{9.4.33}$$

$$E\left(-\frac{\partial^2 \ln L}{\partial\beta^2}\right) \simeq E\left(-\frac{\partial^2 \ln L^*}{\partial\beta^2}\right) = \frac{mD_1}{\sigma^2}, \tag{9.4.34}$$

$$E\left(-\frac{\partial^2 \ln L}{\partial\beta\,\partial\sigma}\right) \simeq E\left(-\frac{\partial^2 \ln L^*}{\partial\beta\,\partial\sigma}\right) = \frac{mD_2}{\sigma^2}, \tag{9.4.35}$$

and

$$E\left(-\frac{\partial^2 \ln L}{\partial\sigma^2}\right) \simeq E\left(-\frac{\partial^2 \ln L^*}{\partial\sigma^2}\right) = \frac{mD_3}{\sigma^2}, \tag{9.4.36}$$

where

$$D_1 = \frac{1}{m}\left\{\sum_{i=r+1}^{n-r} x_{[i]}^2 + r\delta(x_{[r+1]}^2 + x_{[n-r]}^2)\right\}, \tag{9.4.37}$$

$$D_2 = \frac{2}{m}\left\{\sum_{i=r+1}^{n-r} x_{[i]}\alpha_{i:n} + r\delta(x_{[r+1]}\alpha_{r+1:n} + x_{[n-r]}\alpha_{n-r:n})\right\} + \frac{r\gamma}{m}(x_{[n-r]} - x_{[r+1]}), \tag{9.4.38}$$

and

$$D_3 = \frac{3}{m}\left[\sum_{i=r+1}^{n-r} \alpha_{i:n}^{(2)} + r\delta\{\alpha_{r+1:n}^{(2)} + \alpha_{n-r:n}^{(2)}\}\right] + \frac{2r\gamma}{m}(\alpha_{n-r:n} - \alpha_{r+1:n}) - \frac{A}{m}. \tag{9.4.39}$$

In the above formulas, $\alpha_{i:n}$ and $\alpha_{i:n}^{(2)}$ $(1 \le i \le n)$ denote the first and the second moment, respectively, of the ith order statistic in a sample of size n from the standard normal distribution. Then, by using standard results of maximum likelihood estimation (Kendall and Stuart, 1973; Rao, 1975), we have the following theorem.

Theorem 9.4.1. *Asymptotically, that is, for large $A = n - 2r$ (hence, large n), $(\hat{\alpha}, \hat{\beta}, \hat{\sigma})$ has a trivariate normal distribution with mean vector $(E(\hat{\alpha}) = \alpha$, $E(\hat{\beta}) = \beta$, $E(\hat{\sigma}) = \sigma)$ and variance-covariance matrix*

$$V = \frac{\sigma^2}{m(D_1D_3 - D_2^2 - \bar{x}^2 D_3)}\begin{bmatrix} D_1D_3 - D_2^2 & -\bar{x}D_3 & \bar{x}D_2 \\ & D_3 & -D_2 \\ & & D_1 - \bar{x}^2 \end{bmatrix}. \tag{9.4.40}$$

C. *Robustness of the MMLEs*

In order to assess the robustness features of the classical estimators and also the MMLEs of (α, β, σ), Balakrishnan and Ambagaspitiya (1990a) followed the lines of Yale and Forsythe (1976) and Tan and Tabatabai (1988) and let the explanatory variables x_i in the simple linear regression model be taken from a standard normal population, and the error variables ε_i be taken independently from a normal population with mean 0 and variance σ^2, but $\Pr(\sigma = 1) = p$ and $\Pr(\sigma = \sigma') = 1 - p$. The sample sizes $n = 10$ and $n = 20$ and the choices of parameters $\alpha = 0$, $\beta = 1$, $\sigma' = 4$, 6 and $p = 1.0$, 0.9, 0.8, 0.7, and 0.6, were considered by Balakrishnan and Ambagaspitiya (1990a) in their simulation study. A somewhat different mixture model has been applied by Elashoff (1972) in this context.

The values of

(1) Bias of $\hat{\alpha}$,
(2) Bias of $\hat{\beta}$,
(3) Bias of $\hat{\sigma}$,
(4) Variance of $\hat{\alpha}$,
(5) Variance of $\hat{\beta}$, and
(6) Variance of $\hat{\sigma}$ for the classical estimators and also for the MML estimators

were simulated (based on 2,000 Monte Carlo runs) for $n = 10$ and $n = 20$ by Balakrishnan and Ambagaspitiya (1990a). These values, for some selected choices of p and σ', are presented in Tables 9.4.1 and 9.4.2, respectively. It is of interest to mention here that it may be easily shown that, under normality of the error component, the bias in all these estimators is zero for large n. It is clear from Tables 9.4.1 and 9.4.2 that the MMLEs of α and β have a bias of almost zero under both normal and mixture-normal models. Under normality, the MMLE of σ has a little bias for small n. We also see that under normality, the MMLEs of α, β, and σ have larger variances than the corresponding MLEs, as one would expect. However, the MLEs of α and β continue to have a bias of almost zero even under the mixture-normal models, although their variances increase substantially as the nonnormality becomes pronounced. Furthermore, we observe from Tables 9.4.1 and 9.4.2 that the effect of nonnormality on the MLE of σ is quite drastic, and even a small departure from normality is seen to increase its bias as well as variance very drastically. On the other hand, the MMLE of σ, viz., $\hat{\sigma}$ in (9.4.26), displays its robustness feature through much smaller bias and much smaller variance. It should be mentioned here that some other robust and nonparametric methods of estimation are available, and

TABLE 9.4.1
Values of Bias and Variance of the MLEs and the MMLEs of α, β, and σ; $n = 10$

p	σ'	Estimator	(1)	(2)	(3)	(4)	(5)	(6)
			Bias			Variance		
1.0	1.0	MLE	−0.0083	0.0023	−0.0338	0.1204	0.1292	0.0596
		MMLE	−0.0076	0.0057	−0.1167	0.1423	0.2577	0.0792
0.9	4.0	MLE	−0.0150	0.0012	0.1903	0.1985	0.2185	0.2862
		MMLE	−0.0073	0.0171	−0.0424	0.1628	0.2867	0.1176
0.8	4.0	MLE	−0.0232	−0.0045	0.4166	0.2819	0.3368	0.4724
		MMLE	−0.0107	0.0142	0.0310	0.1999	0.3193	0.1656
0.8	6.0	MLE	−0.0332	−0.0090	0.7874	0.4957	0.6129	1.2650
		MMLE	−0.0124	0.0188	0.1001	0.2415	0.3852	0.3034
0.7	6.0	MLE	−0.0324	−0.0004	1.1624	0.7053	0.8364	1.5579
		MMLE	−0.0225	0.0086	0.2849	0.3666	0.5190	0.5718
0.6	4.0	MLE	−0.0130	−0.0034	0.8188	0.4639	0.5374	0.6805
		MMLE	−0.0110	0.0172	0.2701	0.3289	0.5131	0.3559
0.6	6.0	MLE	−0.0161	−0.0072	1.4949	0.9221	1.0832	1.7561
		MMLE	−0.0099	0.0128	0.5002	0.5376	0.8129	0.8304

TABLE 9.4.2
Values of Bias and Variance of the MLEs and the MMLEs of α, β, and σ; $n=20$

p	σ'	Estimator	(1)	(2)	(3)	(4)	(5)	(6)
			Bias			Variance		
1.0	1.0	MLE	−0.0012	0.0003	−0.0141	0.0556	0.0594	0.0282
		MMLE	−0.0015	0.0030	−0.0608	0.0613	0.1173	0.0403
0.9	4.0	MLE	−0.0036	0.0000	0.2405	0.0935	0.0995	0.1829
		MMLE	−0.0012	0.0027	0.0032	0.0705	0.1272	0.0520
0.8	4.0	MLE	−0.0015	−0.0126	0.4870	0.1377	0.1430	0.2778
		MMLE	0.0007	0.0085	0.0842	0.0866	0.1415	0.0731
0.8	6.0	MLE	−0.0017	−0.0212	0.9244	0.2437	0.2548	0.7750
		MMLE	−0.0012	0.0062	0.1375	0.1020	0.1465	0.1165
0.7	6.0	MLE	0.0014	−0.0222	1.2882	0.3257	0.3665	0.9052
		MMLE	0.0014	0.0017	0.2876	0.1497	0.1988	0.2274
0.6	4.0	MLE	0.0011	−0.0121	0.8875	0.2139	0.2340	0.3723
		MMLE	0.0023	0.0048	0.2876	0.1414	0.1900	0.1510
0.6	6.0	MLE	0.0026	−0.0204	1.6251	0.4179	0.4660	0.9651
		MMLE	0.0005	−0.0012	0.4696	0.2190	0.2672	0.3656

interested readers may refer to Adichie (1967a, b), Andrews (1973, 1974), Hinich and Talwar (1975), Schweder (1976), Yale and Forsythe (1976), Tan and Tabatabai (1988), and Singh and Gupta (1989). Recently, Balakrishnan and Ambagaspitiya (1990b) have carried out an extensive study comparing the MMLE method presented in this section with most of the estimation methods mentioned above. By considering various sample sizes and a wide range of nonnormal distributions, they have examined the efficiency and robustness features of the estimators of the parameters α, β, and σ derived by all these methods.

Example 9.4.1. Let us consider here the example given by Daniel and Wood (1971), and utilized by Yale and Forsythe (1976) and Tan and Tabatabai (1988), in order to illustrate the method of MML estimation given in this section. In this example, the explanatory variable (x) is the

TABLE 9.4.3
Normalized Residuals and Their Ranks for Daniel and Wood's Data

i	y_i	x_i	Normalized residual	Rank
1	76	123	0.82426	16
2	70	109	−0.42855	8
3	55	62	−0.33815	9
4[a]	71	104	1.74711	19
5	55	57	1.03315	17
6	48	37	0.56270	15
7	50	44	0.33906	12
8[a]	66	100	−1.35069	1
9	41	16	0.36098	13
10	43	28	−1.30185	3
11[a]	82	138	1.82567	20
12	68	105	−1.02365	4
13	88	159	1.21433	18
14[a]	58	75	−1.32829	2
15	64	88	0.20085	11
16	88	164	−0.17566	10
17	89	169	−0.70965	7
18	88	167	−1.01908	5
19	84	149	0.53110	14
20	88	167	−1.01908	6

[a] These observations are censored for the calculation of the MMLEs of α, β, and σ.

acid number of a chemical determined by titration, and the dependent variable (y) is its organic acid content determined by extraction and weighing. The original data, as given by Daniel and Wood (1971), is presented in Table 9.4.3.

For the data presented in Table 9.4.3, Daniel and Wood (1971) showed that the simple linear regression model provides an excellent fit. The least-squares estimates of α and β are 35.458 and 0.322, respectively, and the estimate of σ is 1.230.

In Table 9.4.3, we have also presented the values of the normalized estimated residuals calculated from (9.4.29) and their ranks. By censoring the pairs corresponding to the two smallest and the two largest normalized estimated residuals, we obtain the following:

$$n = 20, \qquad r = 2, \qquad A = 16,$$

$$\gamma = 0.6901867, \qquad \delta = 0.83086, \qquad m = 19.3235,$$

$$\tilde{x} = 101.159, \qquad \tilde{y} = 67.9323,$$

$$S^*_{x,x} = 61506.2, \qquad S^*_{x,y} = 19965.1, \qquad S^*_{y,y} = 6499.93,$$

$$B = 3.4188, \qquad C = 19.2208,$$

$$\hat{\alpha} = 35.096, \qquad \hat{\beta} = 0.325, \quad \text{and} \quad \hat{\sigma} = 1.294.$$

We note that the MMLEs are quite close to the least-squares estimates. However, we shall see next that this situation changes dramatically when one or two outliers are introduced in the data.

In order to illustrate the Winsorization method, Yale and Forsythe (1976) changed the first y observation from 76 to 100, and also the third y observation from 55 to 75, to cause a contamination in the data given by Daniel and Wood (1971). Tan and Tabatabai (1988) recently used the same contaminated data for illustrating their modified Winsorized method.

In Table 9.4.4, we have presented the values of the normalized estimated residuals calculated from (9.4.29) and their ranks for the above-explained contaminated data. It is important to point out here that the first and third observations, which originally had ranks 16 and 9, respectively, now take on ranks 20 and 19 after the contamination is introduced. Hence, we note that when the observations corresponding to the two smallest and the two largest normalized estimated residuals are censored, the two outliers deliberately introduced get eliminated, thus demonstrating the robustness of the MMLEs. The least-squares estimates of α and β are 38.377 and 0.315, respectively, and the estimate of σ is 7.209. We note here that the introduction of two outliers in the data has had a devastating effect on the classical estimate of σ. On the other hand, for the MMLE method we have the following:

$$\tilde{x} = 112.221, \qquad \tilde{y} = 71.723,$$

$$S^*_{x,x} = 40949.9, \qquad S^*_{x,y} = 13082.4, \qquad S^*_{y,y} = 4212.21,$$

$$B = 5.32797, \qquad C = 32.7401,$$

$$\hat{\alpha} = 35.871, \qquad \hat{\beta} = 0.319, \quad \text{and} \quad \hat{\sigma} = 1.719.$$

We observe here that the presence of the two outliers has had a very minimal effect on the MMLEs of α, β, and σ.

TABLE 9.4.4
Normalized Residuals and Their Ranks for Daniel and Wood's Data with Contamination

i	y_i	x_i	Normalized residual	Rank
1[a]	100	123	3.27661	20
2	70	109	−0.38045	8
3[a]	75	62	2.48129	19
4	71	104	−0.01408	17
5	55	57	−0.19104	14
6	48	37	−0.30180	11
7	50	44	−0.32867	10
8	66	100	−0.54664	3
9	41	16	−0.37509	9
10[a]	43	28	−0.63560	1
11	82	138	0.02938	18
12	68	105	−0.48586	4
13	88	159	−0.05946	16
14[a]	58	75	−0.57054	2
15	64	88	−0.29460	12
16	88	164	−0.29327	13
17	89	169	−0.38113	7
18	88	167	−0.43522	5
19	84	149	−0.18313	15
20	88	167	−0.43522	6

[a] These observations are censored for the calculation of the MMLEs of α, β, and σ.

Example 9.4.2. Let us now consider the EPA 1980 mileage rating data given by Mendenhall (1983, p. 410). In this example, the explanatory variable (x) is the cylinder volume of the car, and the dependent variable (y) is the combined mileage per gallon of the subcompact cars. The data, consisting of nine pairs of observations, is presented in Table 9.4.5. This example has been utilized by Tan and Tabatabai (1988) to study the performance of their modified Winsorized estimators for data with such a small sample size.

For the data presented in Table 9.4.5, Mendenhall (1983) has shown that the simple linear regression model provides an excellent fit. The least-squares estimates of α and β are 34.009 and −0.084, respectively, and the estimate of σ is 1.259. By censoring the pairs corresponding to the smallest and the

TABLE 9.4.5
Mendenhall's EPA 1980 Mileage Rating Data

Car	VW Rabbit	Datsun 210	Chevette	Dodge Omni	Mazda 626
y	24	29	26	24	24
x	97	85	98	105	120

Car	Oldsmobile Starfire	Mercury Capri	Toyota Celica	Datsun 810
y	22	23	23	21
x	151	140	134	146

largest normalized estimated residuals, we obtain the MML estimates of α, β, and σ as 31.398, −0.064, and 0.907, respectively.

In order to illustrate the robustness features of the modified Winsorization method, Tan and Tabatabai (1988) contaminated the data in Table 9.4.5 by changing the third y-observation from 26 to 48. In this case, the classical estimates of α, β, and σ are obtained as 48.531, −0.185, and 7.578, respectively. We observe once again that the introduction of just one outlier in the data has had a devastating effect on the classical estimators. On the other hand, by censoring the pairs corresponding to the smallest and the largest normalized estimated residuals, we find the MML estimates of α, β, and σ to be 34.601, −0.087, and 1.453, respectively. As in the last example, we observe that the presence of an outlier in the data has had a very minimal effect on the MML estimates of α, β, and σ.

Chapter 10 A Sample Completion Technique for Censored Samples

10.1. Introductory Remarks

This chapter is concerned with a general iterative procedure developed by Whitten *et al.* (1988) that employs order statistics to transform (i.e., complete) singly right-censored samples from various distributions into pseudo-complete samples. Parameter estimates can then be calculated from the "completed sample" by using appropriate complete sample estimators, which have been described in preceding chapters.

Samples under consideration consist of n full-term (complete) ordered observations $x_{1:N} \leq x_{2:N} \leq \cdots \leq x_{n:N} \leq T$, plus c partial-term (censored) observations about which it is known only that they exceed the point of censoring, T. The total sample size is $N = n + c$. In Type I censoring, T is a predetermined constant and c is the observed value of a random variable; in Type II censoring, $T = x_{n:N}$ is the nth order statistic in a sample of size N and c is a predetermined constant. In order to complete a censored sample of the types described above, we employ an iterative procedure to calculate estimates $\hat{x}_{(n+i):N}$, $i = 1, 2, \ldots, c$, which are then combined with the n full-term observations to create a "completed" sample. Each of the censored observations, of course, exceeds T.

Let $x^{(j)}_{(n+i):N}$, $i = 1, 2, \ldots, c$, designate the jth iterant to $\hat{x}_{(n+i):N}$. For the first iteration, it is convenient to let $x^{(1)}_{(n+1):N} = x^{(1)}_{(n+2):N} = \cdots = x^{(1)}_{(n+c):N} = T$.

Of course, T is a lower bound on the $x_{(n+i):N}$, and we might save a few iterations by selecting approximations that slightly exceed this value. However, in most applications, T is a convenient starting point, and unless c is large, it usually results in reasonably rapid convergence to the final estimates.

Accordingly, first approximations to the mean and standard deviation of a "completed" sample of size N are

$$\bar{x}_N^{(1)} = \left[\sum_{i=1}^{n} x_{i:N} + cT\right] \Big/ N,$$
$$s_N^{(1)} = \sqrt{\left[\sum_{i=1}^{n} (x_{i:N} - \bar{x}_N^{(1)})^2 + c(T - \bar{x}_N^{(1)})^2\right] \Big/ (N-1)}. \qquad (10.1.1)$$

For subsequent iterations, i.e., for $j = 2, \ldots, k$,

$$\bar{x}_N^{(j)} = \left[\sum_{i=1}^{n} x_{i:N} + \sum_{i=1}^{c} x_{(n+i):N}^{(j)}\right] \Big/ N,$$
$$s_N^{(j)} = \sqrt{\left[\sum_{i=1}^{n} (x_{i:N} - \bar{x}_N^{(j)})^2 + \sum_{i=1}^{c} (x_{(n+i):N}^{(j)} - \bar{x}_N^{(j)})^2\right] \Big/ (N-1)}. \qquad (10.1.2)$$

We must, of course, calculate $x_{(n+i):N}^{(j)}$, $i = 1, 2, \ldots, c$ before $\bar{x}_N^{(j)}$ and $s_N^{(j)}$ can be evaluated, and these calculations are dependent upon the type of distribution that is involved.

10.2. Censored Samples from the Normal Distribution

As an estimator for $x_{(n+i):N}$, we employ

$$\hat{x}_{(n+i):N} = T + \hat{\sigma}[\xi_{(n+i):N} - \xi_{n:N}], \qquad i = 1, 2, \ldots, c, \qquad (10.2.1)$$

where

$$\xi_{(n+i):N} = \Phi^{-1}\left(\frac{n+i}{N+1}\right) \quad \text{and} \quad \hat{\sigma} = s_N, \qquad (10.2.2)$$

and where $\Phi(\ )$ is the cdf of the standard normal distribution (0, 1). Readers are reminded that in Type II censoring, $T = x_{n:N}$, but in Type I censoring, $T \geq x_{n:N}$.

With T as a first approximation to the censored observations, we employ (10.1.1) to calculate first approximations $\bar{x}_N^{(1)}$ and $s_N^{(1)}$. Since $x_{(n+i):N} > T$, we now calculate second approximations as

$$x_{(n+i):N}^{(2)} = T + s_N^{(1)}[\xi_{(n+i):N} - \xi_{n:N}], \qquad i = 1, 2, \ldots, c. \qquad (10.2.3)$$

We substitute these values into (10.1.2) with $j=2$ and calculate second approximations $\bar{x}_N^{(2)}$ and $s_N^{(2)}$. The cycle of iterations is continued as necessary until for some final kth cycle, the absolute values $|\bar{x}_n^{(k)}-\bar{x}_n^{(k-1)}|$ and $|s_N^{(k)}-s_N^{(k-1)}|$ are both less than prescribed maximum errors. For the jth cycle of iterations,

$$x_{(n+i):N}^{(j)}=T+s_N^{(j-1)}[\xi_{(n+i):N}-\xi_{n:N}], \qquad i=1,2,\ldots,c. \qquad (10.2.4)$$

In most practical applications, convergence of the iterative process is reasonably rapid for small values of the ratio c/N. Even when this ratio exceeds 0.3, the number of iterations required to attain final estimates would not be considered excessive.

For parameters μ and σ of the normal distribution, moment and maximum likelihood estimators are identical, and the iterative estimates in this case converge to

$$\hat{\mu}=\bar{x}_N^{(k)} \quad \text{and} \quad \hat{\sigma}=s_N^{(k)}, \qquad (10.2.5)$$

where (k) designates the final stage of iteration.

Note that in the normal distribution, $\xi_{n:N}$ is a function of n and N only. Accordingly, it remains invariant from one iteration to the next. In skewed distributions, however, this quantity is also a function of the shape parameter, and consequently it varies between successive iterations. These changes are recognized by affixing the iteration symbol (j) to read $\xi_{n:N}^{(j)}$ when we consider estimation in skewed distributions.

10.3. Censored Samples from Skewed Distributions

The three-parameter lognormal, Weibull, gamma, and inverse Gaussian distributions are positively skewed with a threshold parameter γ that must be estimated in addition to the mean and variance. Each of these distributions has been standardized (*cf.* Chapters 5 and 8) with zero mean, unit variance, and a shape parameter. In the interest of uniformity, we employ α_3 as a common shape parameter. The threshold parameter for each of the distributions named here is a function of the shape parameter.

As a consequence of the added threshold parameter, the standardized order statistic $\xi_{(n+i):N}$ must be approximated anew for each iteration. At the end of each cycle of the iterative process, it is necessary that we estimate, not only the mean and variance, but the third standard moment, α_3, as well. For the estimation of distribution parameters at the conclusion of each cycle of iteration, we are at liberty to choose any applicable complete sample estimator. Possible choices include moment, maximum likelihood, modified estimators, and various others. In most applications, the modified estimators of Cohen *et al.* (1984, 1985) and of Cohen and Whitten (1985, 1986), described in Chapter 8, might be preferred. These estimators employ the first two moments plus the first order statistic. They are easy to calculate, and for complete samples, they are unbiased with respect to the mean and variance. They are applicable over the entire parameter space and are devoid of regularity problems. Although their sampling behavior might not always be optimal, it is at least near optimal.

Except for differences in the estimation of parameters at the conclusion of each cycle of iteration, procedures are essentially identical for each of the skewed distributions considered here.

A. *The Inverse Gaussian Distribution*

The pdf of this distribution in the parameterization of Chan *et al.* (1983) is given in Chapter 8, Eq. (8.4.1). The expected value, variance, and third standard moment are

$$E(X)=\gamma+\mu,\ V(X)=\sigma^2 \quad \text{and} \quad \alpha_3(X)=3\sigma/\mu. \tag{10.3.1}$$

The pdf and the cdf of the standardized IG distribution $(0, 1, \alpha_3)$ are included in Chapter 8 as equations (8.4.3) and (8.4.6). Modified moment estimating equations for complete samples of size N are

$$E(X)=\bar{x}_N,\ V(X)=s_N^2 \quad \text{and} \quad G(z_{1:N}; 0, 1, \alpha_3)=1/(N+1), \tag{10.3.2}$$

where $z_{1:N}=(x_{1:N}-\bar{x}_N)/s_N$.

Equations (10.3.2) are subsequently reduced to

$$\hat{\sigma}=s_N,\ \hat{\mu}=3s_N/\hat{\alpha}_3 \quad \text{and} \quad \hat{\gamma}=\bar{x}_N-\hat{\mu}. \tag{10.3.3}$$

We obtain $\hat{\alpha}_3$ as a solution of the third equation of (10.3.2). To facilitate calculations involved in this solution, Cohen and Whitten (1985) presented tables and a chart of α_3 as a function of N and $z_{1:N}$. The chart is reproduced in Chapter 8 with permission of the publisher as Fig. 8.4.1.

First approximations $\bar{x}_N^{(1)}$ and $s_N^{(1)}$ are calculated from Eqs. (10.1.1) as in the case of the normal distribution. A first approximation to $z_{1:n}$ is calculated as

$$z_{1:N}^{(1)}=[x_{1:N}-\bar{x}_N^{(1)}]/s_N^{(1)}. \tag{10.3.4}$$

With N given and $z_{1:N}^{(1)}$ calculated from (10.3.4), we obtain a first approximation to α_3 from

$$\int_{-3/\alpha_3^{(1)}}^{z_{1:N}^{(1)}} g(z; 0, 1, \alpha_3^{(1)})\, dz=\frac{1}{N+1}. \tag{10.3.5}$$

With $z_{1:N}^{(1)}$ and N as arguments, this can be accomplished with a degree of accuracy that is sufficient for most practical purposes from the chart of Fig. 8.4.1. Accuracy might be improved by interpolation in tables provided by Cohen and Whitten (1985, 1988). With $\alpha_3^{(1)}$ thus determined, it follows from (10.3.3) that $\mu^{(1)} = 3s_N^{(1)}/\alpha_3^{(1)}$, and $\gamma^{(1)} = \bar{x}_N^{(1)} - \mu^{(1)}$.

First approximations, $\xi_{(n+i):N}^{(1)}$, $i = 0, 1, 2, \ldots, c$, to order statistics of the standardized IG distribution $(0, 1, \alpha_3)$ are obtained as solutions to

$$G(\xi_{(n+i):N}^{(1)}; 0, 1, \alpha_3^{(1)}) = \frac{n+i}{N+1}, \qquad i = 0, 1, \ldots, c, \tag{10.3.6}$$

where the standardized cdf, as given by Eq. (8.4.6), is the sum of two standard normal components. Second approximations $x_{(n+i):N}^{(2)}$, $i = 1, 2, \ldots, c$, then follow as

$$x_{(n+i):N}^{(2)} = T + s_N^{(1)}[\xi_{(n+i):N}^{(1)} - \xi_{n:N}^{(1)}]. \tag{10.3.7}$$

Note that (10.3.7) differs from (10.2.1) only in that $\xi_{(n+i):N}$ and $\xi_{n:N}$ have been replaced by $\xi_{(n+i):N}^{(1)}$ and $\xi_{n:N}^{(1)}$. Iterations are continued as necessary. At the conclusion of the jth cycle, $\bar{x}_N^{(j)}$ and $s_N^{(j)}$ are given by Eqs. (10.1.2). It follows that

$$z_{1:N}^{(j)} = (x_{1:N} - \bar{x}_N^{(j)})/s_N^{(j)}. \tag{10.3.8}$$

Iterations are continued until after k cycles, $|\bar{x}_n^{(k)} - \bar{x}_N^{(k-1)}|$ and $|s_N^{(k)} - s_N^{(k-1)}|$ are less than prescribed errors. Final estimates of the distribution parameters are

$$\hat{\alpha}_3 = \alpha_3^{(k)}, \; \hat{\sigma} = s_N^{(k)}, \; \hat{\mu} = 3\hat{\sigma}/\hat{\alpha}_3, \; \hat{\gamma} = \bar{x}_N^{(k)} - \hat{\mu}. \tag{10.3.9}$$

B. *The Gamma Distribution*

The procedure for estimating parameters of the gamma distribution is essentially the same as that described for the IG distribution. First approximations $\bar{x}_N^{(1)}$, $s_N^{(1)}$, and $z_{1:N}^{(1)}$ are calculated precisely as for the IG distribution. A first approximation to α_3 of the gamma distribution is calculated from

$$\int_{-2/\alpha_3^{(1)}}^{z_{1:N}^{(1)}} g(z; 0, 1, \alpha_3^{(1)})\, dz = \frac{1}{N+1}, \tag{10.3.10}$$

where the pdf, $g(z; 0, 1, \alpha_3)$, is given in Chapter 8 as Eq. (8.5.4). With $z_{1:N}^{(1)}$ and N as arguments, (10.3.10) can be solved for $\alpha_3^{(1)}$ as described for complete samples in Chapter 8. In most practical applications, $\alpha_3^{(1)}$ can be read with sufficient accuracy from the graphs of Fig. 8.5.1. Improved accuracy, when needed, can be attained by interpolation in tables of Cohen and Whitten (1986, 1988).

With $\alpha_3^{(1)}$ thus calculated, approximations $\xi_{(n+i):N}^{(1)}$, $i=0, 1, 2, \ldots, c$, can be obtained as solutions to

$$G(\xi_{(n+i):N}^{(1)}; 0, 1, \alpha_3^{(1)}) = \frac{n+i}{N+1}, \qquad i=0, 1, 2, \ldots, c, \qquad (10.3.11)$$

where $G(\ )$ is the cdf of the standardized gamma distribution as given in Chapter 8 by Eq. (8.5.5).

Second approximations to the censored observations follow from (10.3.7) as in the case of samples from the IG distribution. The iterative procedure is continued through additional cycles until the prescribed accuracy is achieved. The required estimates then become

$$\hat{\mu} = \bar{x}_N^{(k)}, \ \hat{\sigma} = s_N^{(k)} \quad \text{and} \quad \hat{\gamma} = \hat{\mu} - \hat{\sigma}(2/\hat{\alpha}_3), \qquad (10.3.12)$$

where k designates the final cycle of iteration. If estimates $\hat{\beta}$ and $\hat{\rho}$ are required, they may be calculated as

$$\hat{\rho} = 4/\hat{\alpha}_3^2 \quad \text{and} \quad \hat{\beta} = \hat{\sigma}/\sqrt{\hat{\rho}}. \qquad (10.3.13)$$

C. *The Weibull Distribution*

The procedure for calculating estimates from Weibull distribution samples is similar to that for the IG and the gamma distribution. At the end of the jth cycle of iterations, we calculate $\bar{x}_N^{(j)}$, $s_N^{(j)}$, and $W(N, \delta^{(j)}) = [s_N^{(j)}/(\bar{x}_N^{(j)} - x_{1:N})]^2$. As described in Chapter 8, we can either read $\delta^{(j)}$ from the chart of Fig. 8.2.2 or obtain it by interpolation in tables provided by Cohen *et al.* (1985) or by Cohen and Whitten (1988). With $\delta^{(j)}$ thus determined, approximations $\beta^{(j)}$ and $\gamma^{(j)}$ follow as

$$\beta^{(j)} = s_N^{(j)}/\sqrt{\Gamma_2^{(j)} - (\Gamma_1^{(j)})^2} \quad \text{and} \quad \gamma^{(j)} = \bar{x}_N^{(j)} - \beta^{(j)}\Gamma_1^{(j)}. \qquad (10.3.14)$$

Approximations $\xi_{(n+i):N}^{(j)}$ can be obtained by interpolation in tables of the cdf of the standardized Weibull distribution $(0, 1, \alpha_3)$ provided by Cohen and Whitten (1988), where

$$G(\xi_{(n+i):N}^{(j)}; 0, 1, \alpha_3^{(j)}) = \frac{n+i}{N+1}, \qquad i=0, 1, 2, \ldots, c, \qquad (10.3.15)$$

and where $\alpha_3^{(j)}$, expressed as a function of $\delta^{(j)}$, is

$$\alpha_3^{(j)} = \frac{\Gamma_3^{(j)} - 3\Gamma_2^{(j)}\Gamma_1^{(j)} + 2(\Gamma_1^{(j)})^3}{[\Gamma_2^{(j)} - (\Gamma_1^{(j)})^2]^{3/2}}. \qquad (10.3.16)$$

With $\xi^{(j)}_{(n+i):N}$ thus determined, corresponding estimates of the censored observations can be calculated by substitution in (10.3.7).

Iteration is continued until the desired accuracy is achieved. Let k denote the final cycle of iteration, and with $\hat{\delta} = \delta^{(k)}$, we calculate

$$\begin{aligned} \hat{\beta} &= \beta^{(k)} = \frac{s_N^{(k)}}{\sqrt{\Gamma_2^{(k)} - (\Gamma_1^{(k)})^2}}, \\ \hat{\gamma} &= \gamma^{(k)} = \bar{x}_N^{(k)} - \beta^{(k)}\Gamma_1^{(k)}. \end{aligned} \tag{10.3.17}$$

D. *The Lognormal Distribution*

For the jth cycle of iteration, we calculate $\bar{x}_N^{(j)}$, $s_N^{(j)}$, and $J(N, \sigma^{(j)}) = [s_N^{(j)}/(\bar{x}_N^{(j)} - x_{1:N})]^2$ as previously described. With the aid of the chart of Fig. 8.3.1 or tables provided by Cohen *et al.* (1985) or by Cohen and Whitten (1988), we calculate $\sigma^{(j)}$. We then use this value to calculate corresponding approximations

$$\omega^{(j)} = \exp(\sigma^{(j)})^2 \quad \text{and} \quad \alpha_3^{(j)} = (\omega^{(j)} + 2)\sqrt{\omega^{(j)} - 1}. \tag{10.3.18}$$

Approximations $\xi^{(j)}_{(n+i):N}$ can be calculated with the aid of tables of the cdf of the standardized lognormal distribution $(0, 1, \alpha_3)$, such as those provided by Cohen and Whitten (1988), from the relation

$$G(\xi^{(j)}_{(n+i):N}; 0, 1, \alpha_3^{(j)}) = \frac{n+i}{N+1}, \qquad i = 0, 1, 2, \ldots, c. \tag{10.3.19}$$

With $\xi^{(j)}_{(n+i):N}$ thus determined, corresponding approximations to estimates of the censored observations follow by substitution in (10.3.7). Iteration is continued until the desired accuracy is attained. Again, we let k designate the final cycle of iteration, and our estimates become

$$\begin{aligned} &\hat{\sigma} = \sigma^{(k)}, && \hat{\omega} = \exp(\sigma^{(k)})^2, \\ &\hat{\beta} = s_N^{(k)}/\sqrt{\hat{\omega}(\hat{\omega} - 1)}, && \hat{\gamma} = \bar{x}_N^{(k)} - \hat{\beta}\sqrt{\hat{\omega}}. \end{aligned} \tag{10.3.20}$$

10.4. Illustrative Examples

In order to illustrate the practical application of the pseudo-complete sample technique, two examples from Whitten *et al.* (1988) are reproduced here with permission of the publisher, Marcel Dekker, Inc.

Example 10.4.1. A random sample of size $N = 100$ was generated from a normal population with mean $\mu = 60$, and standard deviation $\sigma = 10$. Data for the complete sample are listed in Table 10.4.1 in order of magnitude.

The complete sample mean and standard deviation are $\bar{x} = 60.0646$ and $s = 10.1434$. We create Type II singly right-censored samples by censoring at $x_{70} = 65.31073$ with $c = 30$ ($h = 0.30$); at $x_{75} = 66.51777$ with $c = 25$ ($h = 0.25$); at $x_{80} = 67.94096$ with $c = 20$ ($h = 0.20$); at $x_{85} = 70.63933$ with $c = 15$ ($h = 0.15$); at $x_{90} = 72.73769$ with $c = 10$ ($h = 0.10$); and at $x_{95} = 78.51110$ with $c = 5$ ($h = 0.05$). Pseudo-complete sample estimates were calculated from each of these samples as previously described for the normal distribution. For comparison, maximum likelihood estimates were calculated as described in Chapter 5. The resulting estimates are entered in Table 10.4.2.

It is interesting to note that as shown in Table 10.4.2, agreement between the PC estimates and the MLE is quite close for each of the censored samples. Furthermore, agreement between complete and censored sample estimates is reasonably close even for $h = 0.30$.

TABLE 10.4.1
Random Sample from a Normal Distribution; $N = 100$, $\mu = 60$, and $\sigma = 10^a$

39.45954	41.49674	44.69999	44.78551	45.05776
45.31829	45.43818	46.00102	46.38647	46.54045
46.81214	47.03041	47.36126	47.61510	48.59502
48.74420	49.11825	49.37493	49.68881	50.71081
50.81868	51.04614	51.19043	51.54267	51.71816
51.78690	52.54264	52.64338	53.03228	53.42319
53.43807	53.55375	54.29137	54.96132	55.09467
55.38929	56.03945	56.16351	56.94655	57.04456
57.22293	57.48425	58.20325	58.20744	58.30598
58.55287	58.91081	59.45376	59.65105	59.77864
60.66803	60.77887	60.86139	61.04561	61.07243
61.71520	61.81030	61.92270	61.99394	62.64923
62.80568	63.00968	63.41527	63.45314	63.46503
63.71842	64.41524	64.80918	65.24510	65.31073
65.54905	65.61684	65.72280	65.89674	66.51777
67.17174	67.18133	67.54990	67.81920	67.94096
68.19722	69.19740	70.32907	70.55229	70.63933
71.32172	71.96223	71.97013	72.10827	72.73769
73.04883	74.89332	75.04623	78.06891	78.51110
79.76774	80.35937	80.80418	82.27799	84.85814

[a] $\bar{x} = 60.0646$, $s = 10.1434$, $x_{1:100} = 39.45954$, $N = 100$.
Reproduced from Whitten *et al.* (1988), Table II, p. 2250, by courtesy of Marcel Dekker, Inc.

TABLE 10.4.2
Censored Sample Estimates of Normal Distribution Parameters

Proportion censored h	$\bar{x}_n$	$T = x_{n:100}$	s_n^2	$\hat{\mu}$ PC[a]	$\hat{\mu}$ MLE[b]	$\hat{\sigma}$ PC[a]	$\hat{\sigma}$ MLE[b]
0.05	58.9304	78.5111	81.3193	60.1027	60.112	10.231	10.221
0.10	57.9869	72.7377	68.6851	59.898	59.926	9.831	9.864
0.15	57.1614	70.6393	60.4481	59.915	59.939	9.863	9.894
0.20	56.3726	67.9410	53.5911	59.718	59.758	9.567	9.631
0.25	55.6286	66.5178	48.3005	59.745	59.780	9.610	9.670
0.30	54.8977	65.3107	43.7294	59.830	59.854	9.716	9.764
Complete sample estimates				60.0646		10.1434	
Population values				60		10	

[a] PC: estimates from pseudo-complete samples; $N = 100$.
[b] MLE: maximum likelihood estimates from censored samples. Cohen estimates, calculated as described in Chapter 5.

Reproduced from Whitten *et al.* (1988), Table III, p. 2250, by courtesy of Marcel Dekker, Inc.

TABLE 10.4.3
Successive Iterations of Pseudo-Complete Sample Estimates of Normal Distribution Parameters; $N = 100$, $n = 95$, $c = 5$, $h = 0.05$

Item	Iteration number (j) 1	2	3	4	5	Complete sample summary
$\bar{x}_N^{(j)}$	59.9094	60.0940	60.1013	60.1017	60.1017	60.065
$s_N^{(j)}$	9.8198	10.2126	10.2299	10.2307	10.2307	10.143
$x_{96}^{(j)}$	78.5111	79.3938	79.4291	79.4307	79.4308	79.768
$x_{97}^{(j)}$	78.5111	80.4311	80.5079	80.5113	80.5114	80.359
$x_{98}^{(j)}$	78.5111	81.7064	81.8343	81.8399	81.8401	80.804
$x_{99}^{(j)}$	78.5111	83.4021	83.5978	83.6064	83.6068	82.278
$x_{100}^{(j)}$	78.5111	86.0753	86.3779	86.3912	86.3918	84.858

$\xi_{95} = 1.5598$, $\xi_{96} = 1.6497$, $\xi_{97} = 1.7553$, $\xi_{98} = 1.8852$, $\xi_{99} = 2.0579$, $\xi_{100} = 2.3301$.

Reproduced from Whitten *et al.* (1988), Table IV, p. 2251, by courtesy of Marcel Dekker, Inc.

TABLE 10.4.4
Random Sample from an Inverse Gaussian Distribution; $\gamma = 10$, $\mu = 6$, $\sigma = 5$ ($\alpha_3 = 2.5$)[a]

10.8884	11.1417	11.1562	11.3311	11.3493
11.4726	11.5578	11.5839	11.6037	11.6113
11.6640	11.7501	11.8600	12.0766	12.1246
12.1868	12.2967	12.3286	12.3296	12.3327
12.3504	12.4423	12.5053	12.6288	12.6423
12.6492	12.6268	12.7462	12.8156	12.8532
13.1523	13.1642	13.1807	13.2550	13.3709
13.4310	13.5195	13.5336	13.5786	13.5987
13.6522	13.6539	13.6596	13.6825	13.7392
13.7657	13.7764	13.8136	13.8846	14.0068
14.0490	14.0568	14.3218	14.5941	14.6496
14.6811	14.7629	15.0891	15.2358	15.3520
15.4604	15.5547	15.6059	15.7032	15.7977
15.8906	15.8931	15.9440	16.1099	16.1922
16.2001	16.4352	16.4605	17.2224	17.2458
17.3007	17.4365	17.8224	18.1382	18.3316
18.4063	18.6581	18.8229	19.0166	19.0801
19.2521	19.3555	21.0255	22.1120	23.0403
24.3057	24.9244	25.2042	25.2134	25.8634
26.3125	27.5814	29.3053	32.5742	42.4409

[a] $\bar{x} = 15.9236$, $s = 5.1460$, $x_{1:100} = 10.8884$, $N = 100$.
Reproduced from Whitten *et al.* (1988), Table V, p. 2252, by courtesy of Marcel Dekker, Inc.

TABLE 10.4.5
Summary of Estimates of Inverse Gaussian Distribution Parameters

Estimator	PARAMETER				
	γ	μ	E(X)	σ	α_3
Population	10	6	16	5	2.5
MME (Complete Sample)	10.009	5.915	15.924	5.146	2.610
PC (h=.05)	9.905	5.954	15.859	4.828	2.433
PC (h=.10)	9.911	5.940	15.851	4.829	2.439
PC (h=.15)	9.517	5.872	15.389	3.702	1.892
PC (h=.20)	9.576	5.878	15.454	3.837	1.958
PC (h=.25)	9.461	5.851	15.312	4.565	1.828
PC (h=.30)	9.207	5.835	15.042	3.094	1.591

Reproduced from Whitten *et al.* (1988), Table VI, p. 2253, by courtesy of Marcel Dekker, Inc.

TABLE 10.4.6
Successive Iterations of Pseudo-Complete Sample Estimates of Inverse Gaussian Parameters; $\gamma = 10$, $\mu = 6$, $\sigma = 5$, $E(X) = 16$, $\alpha_3 = 2.5$, $N = 100$, $h = 0.05$.

Item	Iteration Number (j)									Complete Sample Summary
	1	2	3	4	5	6	7	8	9	
$\bar{x}_N^{(j)}$	15.6350	15.8134	15.8484	15.8567	15.8587	15.8592	15.8593	15.8593	15.8593	15.924
$\mu^{(j)}$	5.9760	5.9590	5.9553	5.9545	5.9544	5.9543	5.9543	5.9543	5.9543	5.915
$s_N^{(j)}$	5.1485	4.7899	4.7899	4.8186	4.8257	4.8274	4.8278	4.8279	4.8280	5.146
$\gamma^{(j)}$	9.6590	9.8931	9.8931	9.9021	9.9043	9.9049	9.9050	9.9050	9.9050	10.009
$\alpha_3^{(j)}$	2.0826	2.3515	2.4129	2.4277	2.4314	2.4322	2.4324	2.4325	2.4325	2.610
$z_1^{(j)}$	-1.1442	-1.0544	-1.0355	-1.0311	-1.0300	-1.0297	-1.0296	-1.0296	-1.0296	-0.9785
$\xi_{95}^{(j)}$	1.7737	1.7648	1.7619	1.7612	1.7610	1.7610	1.7609	1.7609	1.7609	
$\xi_{96}^{(j)}$	1.9576	1.9580	1.9572	1.9570	1.9569	1.9569	1.9569	1.9569	1.9569	
$\xi_{97}^{(j)}$	2.1847	2.1979	2.1998	2.2003	2.2004	2.2004	2.2004	2.2004	2.2004	
$\xi_{98}^{(j)}$	2.4809	2.5123	2.5183	2.5197	2.5200	2.5201	2.5201	2.5201	2.5102	
$\xi_{99}^{(j)}$	2.9043	2.9648	2.9771	2.9801	2.9808	2.9809	2.9810	2.9810	2.9810	
$\xi_{100}^{(j)}$	3.6431	3.7609	3.7860	3.7920	3.7935	3.7938	3.7939	3.7940	3.7940	
$x_{96}^{(j)}$	25.8634	26.6263	26.7661	26.7988	26.8067	26.8087	26.8092	26.8093	26.8093	26.3125
$x_{97}^{(j)}$	25.8634	27.5687	27.8865	27.9610	27.9791	27.9836	27.9847	27.9849	27.9850	27.5814
$x_{98}^{(j)}$	25.8634	28.7973	29.3552	29.4863	29.5182	29.5260	29.5280	29.5284	29.5285	29.3053
$x_{99}^{(j)}$	25.8634	30.5537	31.4684	31.6841	31.7367	31.7495	31.7527	31.7535	31.7537	32.5742
$x_{100}^{(j)}$	25.8634	33.6188	31.1869	35.5587	35.6493	35.6715	35.6770	35.6783	35.6787	42.4409

Reproduced from Whitten *et al.* (1988), Table VII, p. 2253, by courtesy of Marcel Dekker, Inc.

As an illustration of the iterative procedure, results of the five cycles of iteration required to reach final estimates for $h = 0.05$ are entered in Table 10.4.3, along with corresponding complete sample estimates.

Attention is invited to the close agreement between corresponding results from censored and complete samples as displayed in the two right-hand columns of Table 10.4.3.

Example 10.4.2. A random sample of size $N = 100$ is generated from an IG distribution with $\gamma = 10$, $\mu = 6$, $\sigma = 5$, and thus with $\alpha_3 = 2.5$. The complete sample data are listed in order of magnitude in Table 10.4.4.

For the complete sample, $N = 100$, $\bar{x} = 15.9236$, $s = 5.1460$, $a_3 = 2.3142$, and $x_1 = 10.8884$. Again for illustrative purposes, we create Type II singly right-censored samples by censoring at $x_{70} = 16.1922$ with $c = 30$ $(h = 0.30)$; at $x_{75} = 17.2458$ with $c = 25$ $(h = 0.25)$; at $x_{80} = 18.3316$ with $c = 20$ $(h = 0.20)$; at $x_{85} = 19.0801$ with $c = 15$ $(h = 0.15)$; at $x_{90} = 23.0403$ with $c = 10$ $(h = 0.10)$; and at $x_{95} = 25.8634$ with $c = 5$ $(h = 0.05)$.

For each of these censored samples, PC estimates were calculated as described for the IG distribution. These results are displayed in Table 10.4.5, along with corresponding complete sample modified moment estimates.

As an illustration of the iterative process, results of the nine cycles of iteration required to reach final estimates for $h = 0.05$ are entered in Table 10.4.6, along with a summary of corresponding complete sample results.

Comments. The PC technique converges quite rapidly for samples from a normal distribution even when the proportion of censored observations is somewhat large. As shown in Table 10.4.2, the agreement between PC and ML estimates is close for values of h as large as 0.30. Estimates in skewed distributions, however, seem to be more sensitive to increases in the proportions of censored observations. As shown in Table 10.4.5, estimates might be considered generally satisfactory for $h \leq 0.15$, but σ and α_3 appear to be underestimated for larger values of h. In some applications, convergence problems have been encountered in calculating estimates of skewed distribution parameters when α_3, though positive, is near zero. Consequently, PC estimation in skewed distributions might not be appropriate if there is reason to expect that $\alpha_3 < 0.3$ (approximately). This restriction, however, is unlikely to be a hindrance in practical applications, since the normal distribution, which is free from convergence problems, is a more appropriate model when α_3 is small. It is expected that the PC technique will be particularly useful in the analysis of lifespan, reaction time, and survival data.

Bibliography

ABDEL-ATY, S. H. (1954). Ordered variables in discontinuous distributions, *Statist. Neerlandica* **8,** 61–82.

ABE, S. (1971). Simplified linear unbiased estimators for location and scale parameters in doubly censored samples, *Rep. Statist. Appl. Res.*, JUSE, **18,** 43–55 and 83–96.

ABRAMOWITZ, M., and STEGUN, I. A. (eds.) (1965). Handbook of Mathematical Functions with Formulas, Graphs, and Mathematical Tables, Dover Publications, New York.

ADATIA, A., and CHAN, L. K. (1981). Some relations between stratified, grouped and selected order statistics samples, *Scand. Actuar. J.* 1981, 193–202.

ADICHIE, J. N. (1967a). Asymptotic efficiency of a class of non-parametric tests for regression parameters, *Ann. Math. Statist.* **38,** 884–893.

ADICHIE, J. N. (1967b). Estimates of regression parameters based on rank tests, *Ann. Math. Statist.* **38,** 894–903.

AITCHISON, J., and BROWN, J. A. C. (1957). The Lognormal Distribution, Cambridge University Press, Cambridge, England.

ALI, M. M., and CHAN, L. K. (1964). On Gupta's estimates of the parameters of the normal distribution, *Biometrika* **51,** 498–501.

ALI, M. M., UMBACH, D., and HASSANEIN, K. M. (1981a). Estimation of quantiles of exponential and double exponential distributions based on two order statistics, *Commun. Statist.—Theor. Meth.*, **10,** 1921–1932.

ALI, M. M., UMBACH, D., and HASSANEIN, K. M. (1981b). Small sample quantile estimation of Pareto populations using two order statistics, *Aligarh J. Statist.* **1,** 139–164.

ALI, M. M., UMBACH, D., and SALEH, A. K. MD. E. (1982). Small sample quantile estimation of the exponential distribution using optimal spacings, *Sakhyā B* **44,** 135–142.

ALI, M. M., UMBACH, D., SALEH, A. K. MD. E., and HASSANEIN, K. M. (1983). Estimating quantiles using optimally selected order statistics, *Commun. Statist.—Theor Meth.*, **12**(19), 2261-2271.

ALI, M. M., UMBACH, D., and SALEH, A. K. MD. E. (1985). Tests of significance for the exponential distribution based on selected quantiles, *Sankhyā B* **47**, 310-318.

ALLEN, D. M. (1971). Mean square error of prediction as a criterion for selecting variables, *Technometrics* **13**, 469-476.

ANDREWS, D. F. (1973). Robust estimation for multiple linear regression models, *Bull. Int. Statist. Inst.* **45**, 105-111.

ANDREWS, D. F. (1974). A robust method for multiple linear regression, *Technometrics* **16**, 523-531.

ARNOLD, B. C. (1977). Recurrence relations between expectations of functions of order statistics, *Scand. Actuar. J.* 1977, 169-174.

ARNOLD, B. C. (1983). Pareto Distributions, International Co-operative Publishing House, Fairland, Maryland.

ARNOLD, B. C., and BALAKRISHNAN, N. (1989). Relations, Bounds and Approximations For Order Statistics, Lecture Notes in Statistics No. 53, Springer-Verlag, New York.

ARNOLD, B. C., and MEEDEN, G. (1975). Characterization of distributions by sets of moments of order statistics, *Ann. Statist.* **3**, 754-758.

BAIN, L. J. (1972). Inferences based on censored sampling from the Weibull or extreme-value distribution, *Technometrics* **14**, 693-702.

BAIN, L. J., and ANTLE, C. E. (1967). Estimation of parameters in the Weibull distribution, *Technometrics* **9**, 621-627.

BAIN, L. J., BALAKRISHNAN, N., EASTMAN, J., ENGELHARDT, M., and ANTLE, C. (1990). Reliability estimation based on MLE's for complete and censored samples, *in* The Logistic Distribution (N. Balakrishnan, ed.), Marcel Dekker, New York (in press).

BALAKRISHNAN, N. (1985). Order statistics from the half logistic distribution, *J. Statist. Comput. Simul.* **20**, 287-309.

BALAKRISHNAN, N. (1986). Order statistics from discrete distributions, *Commun. Statist.—Theor. Meth.*, **15**(3), 657-675.

BALAKRISHNAN, N. (ed.) (1988a). Order Statistics and Applications, Special Issue of *Commun. Statist.—Theor. Meth.*, **17**(7).

BALAKRISHNAN, N. (1988b). Recurrence relations for order statistics from n independent and non-identically distributed random variables, *Ann. Inst. Statist. Math.* **40**, 273-277.

BALAKRISHNAN, N. (1989a). Approximate MLE of the scale parameter of the Rayleigh distribution with censoring, *IEEE Trans. on Reliab.* **38**, 355-357.

BALAKRISHNAN, N. (1989b). Approximate maximum likelihood estimation of the mean and standard deviation of the normal distribution based on Type II censored samples, *J. Statist. Comput. Simul.* **32**, 137-148.

BALAKRISHNAN, N. (1989c). A relation for the covariances of order statistics from n independent and non-identically distributed random variables, *Statist. Hefte* **30,** 141–146.

BALAKRISHNAN, N. (1990a). Linear estimation with polynomial coefficients, *in* The Logistic Distribution (N. Balakrishnan, ed.), Marcel Dekker, New York (in press).

BALAKRISHNAN, N. (1990b). Maximum likelihood estimation based on complete and Type-II censored samples, *in* The Logistic Distribution (N. Balakrishnan, ed.), Marcel Dekker, New York (in press).

BALAKRISHNAN, N. (1990c). Approximate maximum likelihood estimation for a generalized logistic distribution, *J. Statist. Plann. Inf.*, **26,** 221–236.

BALAKRISHNAN, N. (1990d). On the maximum likelihood estimation of the location and scale parameters of exponential distribution based on multiply Type II censored samples, *J. Appl. Statist.*, **17,** 55–61.

BALAKRISHNAN, N. (1990e). Best linear unbiased estimates of the mean and standard deviation of normal distribution for complete and censored samples of sizes 21(1)30(5)40, Submitted for publication.

BALAKRISHNAN, N. (1990f). Best linear unbiased estimates of the location and scale parameters of logistic distribution for complete and censored samples of sizes 2(1)25(5)40, Submitted for publication.

BALAKRISHNAN, N., and AMBAGASPITIYA, R. S. (1990a). A robust method of estimation based on the MML estimators for a simple linear regression model, *J. Statist. Plann. Inf.*, in press.

BALAKRISHNAN, N., and AMBAGASPITIYA, R. S. (1990b). An empirical comparison of some robust methods of estimation of parameters in a simple linear regression model, I: Departures from normality, Submitted for publication.

BALAKRISHNAN, N., and JOSHI, P. C. (1983). Single and product moments of order statistics from symmetrically truncated logistic distribution, *Demonstratio Mathematica* **16,** 833–841.

BALAKRISHNAN, N., and JOSHI, P. C. (1984). Product moments of order statistics from doubly truncated exponential distribution, *Naval Res. Logist. Quart.* **31,** 27–31.

BALAKRISHNAN, N., and KOCHERLAKOTA, S. (1985). On the double Weibull distribution: Order statistics and estimation, *Sankhyā B* **47,** 161–178.

BALAKRISHNAN, N., and KOCHERLAKOTA, S. (1986). On the moments of order statistics from doubly truncated logistic distribution, *J. Statist. Plann. Inf.* **13,** 117–129.

BALAKRISHNAN, N., and LEUNG, M. Y. (1988a). Order statistics from the Type I generalized logistic distribution, *Commun. Statist.—Simul. Comput.*, **17**(1), 25–50.

BALAKRISHNAN, N., and LEUNG, M. Y. (1988b). Means, variances and covariances of order statistics, BLUE's for the Type I generalized logistic distribution, and some applications, *Commun. Statist.—Simul. Comput.*, **17**(1), 51–84.

BALAKRISHNAN, N., and MALIK, H. J. (1986). A note on moments of order statistics, *Amer. Statist.* **40,** 147–148.

BALAKRISHNAN, N., and MALIK, H. J. (1990). Means, variances and covariances of logistic order statistics for sample sizes up to fifty, *Selected Tables in Mathematical Statistics*, in press.

BALAKRISHNAN, N., and PUTHENPURA, S. (1986). Best linear unbiased estimators of location and scale parameters of the half logistic distribution, *J. Statist. Comput. Simul.* **25**, 193-204.

BALAKRISHNAN, N., and VARADAN, J. (1990). Approximate maximum likelihood estimation of the location and scale parameters of the extreme value distribution based on complete and censored samples, *IEEE Trans. on Reliab.*, in press.

BALAKRISHNAN, N., and WONG, K. H. T. (1990a). Approximate maximum likelihood estimation of the location and scale parameters of half logistic distribution based on Type-II right-censored samples, *IEEE Trans. on Reliab.*, in press.

BALAKRISHNAN, N., and WONG, K. H. T. (1990b). Approximate maximum likelihood estimation of the location and scale parameters of half logistic distribution based on doubly Type-II censored samples, Submitted for publication.

BALAKRISHNAN, N., and WONG, K. H. T. (1990c). Best linear unbiased estimation of location and scale parameters of the half logistic distribution based on Type II censored samples, *Amer. J. Math. Mgt. Sci.*, in press.

BALAKRISHNAN, N., MALIK, H. J., and PUTHENPURA, S. (1987). Best linear unbiased estimation of location and scale parameters of the log-logistic distribution, *Commun. Statist.—Theor. Meth.*, **16**(12), 3477-3495.

BALAKRISHNAN, N., MALIK, H. J., and AHMED, S. E. (1988). Recurrence relations and identities for moments of order statistics, II: Specific continuous distributions, *Commun. Statist.—Theor. Meth.* (*Statist. Reviews*), **17**(8), 2657-2694.

BALAKRISHNAN, N., AMBAGASPITIYA, R. S., and VARADAN, J. (1990a). ML estimation for the exponential distribution based on doubly Type-II censored samples, *IEEE Trans. on Reliab.*, in press.

BALAKRISHNAN, N., CHAN, P. S., and VARADAN, J. (1990b). Means, variances and covariances of order statistics and best linear unbiased estimates of the location and scale parameters of extreme value distribution for complete and censored samples of size 30 and less, Submitted for publication.

BALMER, D. W., BOULTON, M., and SACK, R. A. (1974). Optimal solutions in parameter estimation problems for the Cauchy distribution, *J. Amer. Statist. Assoc.* **69**, 238-242.

BARNETT, V. D. (1966). Order statistics estimators of the location of the Cauchy distribution, *J. Amer. Statist. Assoc.* **61**, 1205-1218. Correction **63**, 383-385.

BARNETT, V., and LEWIS, T. (1978). Outliers in Statistical Data, John Wiley & Sons, New York.

BASU, D. (1955). On statistics independent of a complete sufficient statistic, *Sankhyā* **15**, 377-380.

BEHNKEN, D. W., and DRAPER, N. R. (1972). Residuals and their variance patterns, *Technometrics* **14**, 101-111.

BELSLEY, D. A., KUH, E., and WELSCH, R. E. (1980). Regression Diagnostics: Identifying Influential Data and Sources of Collinearity, John Wiley & Sons, New York.

BENNETT, C. A. (1952). Asymptotic properties of ideal linear estimators, Ph.D. Thesis, University of Michigan.

BENSON, F. (1949). A note on the estimation of mean and standard deviation from quantiles, *J. Roy. Statist. Soc., Ser. B*, **11,** 21–100.

BERNARDO, J. M. (1976). Psi (digamma) function, Algorithm AS103, *Appl. Statist.* **25,** 315–317.

BEYER, J. N., MOORE, A. H., and HARTER, H. L. (1976). Estimation of the standard deviation of the logistic distribution by the use of selected order statistics, *Technometrics* **18,** 313–332.

BILLMAN, B. R., ANTLE, C. E., and BAIN, L. J. (1972). Statistical inference from censored Weibull samples, *Technometrics* **14,** 831–840.

BIRNBAUM, A., and DUDMAN, J. (1963). Logistic order statistics, *Ann. Math. Statist.* **34,** 658–663.

BIRNBAUM, A., and LASKA, E. M. (1967). Optimally robust linear estimators of location (Abstract), *Ann. Math. Statist.* **38,** 1932.

BIRNBAUM, A., and MIKÉ, V. (1970). Asymptotically robust estimators of location, *J. Amer. Statist. Assoc.* **65,** 1265–1282.

BIRNBAUM, A., LASKA, E. M., and MEISNER, M. (1971). Optimally robust linear estimators of location, *J. Amer. Statist. Assoc.* **66,** 302–310.

BLOCH, D. (1966). A note on the estimation of the location parameter of the Cauchy distribution, *J. Amer. Statist. Assoc.* **61,** 852–855.

BLOM, G. (1958). Statistical Estimates and Transformed Beta-Variables, Almqvist and Wiksell, Uppsala, Sweden.

BLOM, G. (1962). Nearly best linear estimates of location and scale parameters, *in* Contributions to Order Statistics (A. E. Sarhan and B. G. Greenberg, eds.), pp. 34–46, John Wiley & Sons, New York.

BOFINGER, E. (1975). Optimal condensation of distributions and optimal spacing of order statistics, *J. Amer. Statist. Assoc.* **70,** 151–154.

BONDESSON, L. (1976). When is the sample mean BLUE? *Scand. J. Statist.* **3,** 116–120.

BORENIUS, G. (1966). On the limit distribution of an extreme value in a sample from a normal distribution, *Skand. Aktuarietidskr.* 1965, 1–15.

BOSE, R. C., and GUPTA, S. S. (1959). Moments of order statistics from a normal population, *Biometrika* **46,** 433–440.

BREAKWELL, J. V. (1953). On estimating both mean and standard deviation of a normal population from the lowest r out of n observations (Abstract), *Ann. Math. Statist.* **24,** 683.

BREIMAN, L., STONE, C. J., and GINS, J. D. (1979). New methods for estimating tail probabilities and extreme value distributions, TSC-PD-A226-1, Technology Service Corporation, Santa Monica, California.

BREITER, M. C., and KRISHNAIAH, P. R. (1968). Tables for the moments of gamma order statistics, *Sankhyā B* **30,** 59–72.

BURR, I. W. (1955). Calculation of exact sampling distribution of ranges from a discrete population, *Ann. Math. Statist.* **26,** 530–532. Correction **38,** 280.

BURROWS, P. M. (1972). Expected selection differentials for directional selection, *Biometrics* **28,** 1091-1100.

BURROWS, P. M. (1975). Variance of selection differentials in normal samples, *Biometrics* **31,** 125-133.

CALITZ, F. (1973). Maximum likelihood estimation of the parameters of the three-parameter lognormal distribution—a reconsideration, *Austral. J. Statist.* **3,** 185-190.

CANE, G. J. (1974). Linear estimation of parameters of the Cauchy distribution based on sample quantiles, *J. Amer. Statist. Assoc.* **69,** 243-245.

CHAN, L. K. (1969). Linear quantile estimates of the location and scale parameters of the logistic distribution, *Statist. Hefte* **10,** 277-282.

CHAN, L. K. (1970). Linear estimation of the location and scale parameters of the Cauchy distribution based on sample quantiles, *J. Amer. Statist. Assoc.* **65,** 851-859.

CHAN, L. K. (1971). Some asymptotic properties of the linearized maximum likelihood estimate and best linear unbiased estimate, *Ann. Inst. Statist. Math.* **23,** 225-232.

CHAN, L. K. (1974). On the asymptotic normality of some unbiased estimates, *Scand. Actuar. J.* 1974, 151-156.

CHAN, L. K., and CHAN, N. N. (1973). On the optimum best linear unbiased estimates of the parameters of the normal distribution based on selected order statistics, *Skand. Aktuarietidskr.* 1973, 120-128.

CHAN, L. K., and CHENG, S. W. H. (1971). On the Student's test based on sample percentiles from the normal, logistic, and Cauchy distributions, *Technometrics* **13,** 127-137.

CHAN, L. K., and CHENG, S. W. (1972). Optimum spacing for the asymptotically best linear estimate of the location parameter of the logistic distribution when samples are complete or censored, *Statist. Hefte* **13,** 41-57.

CHAN, L. K., and CHENG, S. W. (1973). On the optimum spacing for the asymptotically best linear estimate of the scale parameter of the Pareto distribution, *Tamkang J. Math.* **4,** 1-21.

CHAN, L. K., and CHENG, S. W. (1974). An algorithm for determining the asymptotically best linear estimate of the mean from multiply censored logistic data, *J. Amer. Statist. Assoc.* **69,** 1027-1030.

CHAN, L. K., and CHENG, S. W. (1982). The best linear unbiased estimates of parameters using order statistics, *Soochow J. Math.* **8,** 1-13.

CHAN, L. K., and CHENG, S. W. (1988). Linear estimation of the location and scale parameters based on selected order statistics, *Commun. Statist.—Theor. Meth.* **17**(7), 2259-2278.

CHAN, L. K., and KABIR, A. B. M. L. (1969). Optimum quantiles for the linear estimation of the parameters of the extreme-value distribution in complete and censored samples, *Naval Res. Logist. Quart.* **16,** 381-404.

CHAN, L. K., and MEAD, E. R. (1971a). Linear estimation of the parameters of the extreme-value distribution based on suitably chosen order statistics, *IEEE Trans. on Reliab.* **R-20,** 74-83.

CHAN, L. K., and MEAD, E. R. (1971b). Tables to facilitate calculation of an asymptotically optimal t-test for equality of location parameters of a certain extreme-value distribution, *IEEE Trans. on Reliab.* **R-20,** 235-243.

CHAN, L. K., CHAN, N. N., and MEAD, E. R. (1971). Best linear unbiased estimates of the parameters of the logistic distribution based on selected order statistics, *J. Amer. Statist. Assoc.* **66,** 889-892.

CHAN, L. K., CHENG, S. W. H., and MEAD, E. R. (1972). An optimum t-test for the scale parameter of an extreme-value distribution, *Naval Res. Logist. Quart.* **19,** 715-723.

CHAN, L. K., CHAN, N. N., and MEAD, E. R. (1973a). Linear estimation of the parameters of the Cauchy distribution using selected order statistics, *Utilitas Math.* **3,** 311-318.

CHAN, L. K., CHENG, S. W. H., MEAD, E. R., and PANJER, H. H. (1973b). On a t-test for the scale parameter based on sample percentiles, *IEEE Trans. on Reliab.* **R-22,** 82-87.

CHAN, L. K., CHENG, S. W. H., and MEAD, E. R. (1974). Simultaneous estimation of location and scale parameters of the Weibull distribution, *IEEE Trans. on Reliab.* **R-23,** 335-341.

CHAN, M., COHEN, A. C., and WHITTEN, B. J. (1983). The standardized inverse Gaussian distribution: Tables of the cumulative probability function, *Commun. Statist.—Simul. Comput.* **12,** 423-442.

CHAN, M., COHEN, A. C., and WHITTEN, B. J. (1984). Modified maximum likelihood and modified moment estimators for the three-parameter inverse Gaussian distribution, *Commun. Statist.—Simul. Comput.* **13,** 47-68.

CHAN, P. S. (1989). Half logistic distribution: Type II censoring and estimation, M.Sc. Thesis, McMaster University, Hamilton, Ontario, Canada.

CHENG, S. W. (1975). A unified approach to choosing optimum quantiles for the ABLE's, *J. Amer. Statist. Asssoc.* **70,** 155-159.

CHENG, S. W. (1983). On the most powerful quantile test of the scale parameter, *Ann. Inst. Statist. Math.* **35,** 407-414.

CHENG, R. C. H., and AMIN, N. A. K. (1981). Maximum likelihood estimation of parameters in the inverse Gaussian distribution with unknown origin, *Technometrics* **23,** 257-263.

CHERNOFF, H. (1971). A note on optimal spacings for systematic statistics, Mimeographed Report, Department of Statistics, Stanford University.

CHERNOFF, H., and LIEBERMAN, G. J. (1954). Use of normal probability paper, *J. Amer. Statist. Assoc.* **49,** 778-785.

CHERNOFF, H., GASTWIRTH, J. L., and JOHNS, M. V., Jr. (1967). Asymptotic distribution of linear combinations of functions of order statistics with applications to estimation, *Ann. Math. Statist.* **38,** 52-72.

CHEW, V. (1968). Some useful alternatives to the normal distribution, *Amer. Statist.* **22,** 22-24.

CHHIKARA, R. S., and FOLKS, J. L. (1974). Estimation of the inverse Gaussian distribution function, *J. Amer. Statist. Assoc.* **69,** 250-254.

CHHIKARA, R. S., and FOLKS, J. L. (1988). The Inverse Gaussian Distribution: Theory, Methodology, and Applications, Marcel Dekker, New York.

CHU, J. T., and YA'COUB, K. (1968). Linear order estimates using subsamples, *SIAM J. Appl. Math.* **16,** 162–166.

CLARK, C. E., and WILLIAMS, G. T. (1958). Distributions of the members of an ordered sample, *Ann. Math. Statist.* **29,** 862–870.

COHEN, A., LO, S.-H., and SINGH, K. (1985). Estimating a quantile of a symmetric distribution, *Ann. Statist.* **13,** 1114–1128.

COHEN, A. C. (1951). Estimating parameters of logarithmic-normal distributions by maximum likelihood, *J. Amer. Statist. Assoc.* **46,** 206–212.

COHEN, A. C. (1955). Maximum likelihood estimation of the dispersion parameter of a chi-distributed radial error from truncated and censored samples with applications to target analysis, *J. Amer. Statist. Assoc.* **50,** 884–893.

COHEN, A. C. (1959). Simplified estimators for the normal distribution when samples are singly censored or truncated, *Technometrics* **1,** 217–237.

COHEN, A. C. (1961). Tables for maximum likelihood estimates: singly truncated and singly censored samples, *Technometrics* **3,** 535–541.

COHEN, A. C. (1965). Maximum likelihood estimation in the Weibull distribution based on complete and on censored samples, *Technometrics* **7,** 579–588.

COHEN, A. C. (1975). Multi-censored sampling in the three-parameter Weibull distribution, *Technometrics* **17,** 347–351.

COHEN, A. C. (1976). Progressively censored sampling in the log-normal distribution, *Technometrics* **18,** 99–104.

COHEN, A. C. (1988). Estimation in lognormal distribution, *in* Lognormal distributions—Theory and Applications (E. L. Crow and K. Shimizu, eds.), Chapters 4 and 5, Marcel Dekker, New York.

COHEN, A. C., and HELM, R. (1973). Estimation in the exponential distribution, *Technometrics* **14,** 841–846.

COHEN, A. C., and NORGAARD, N. J. (1977). Progressively censored sampling in the three-parameter gamma distribution, *Technometrics* **19,** 333–340.

COHEN, A. C., and WHITTEN, B. J. (1980). Estimation in the three-parameter lognormal distribution, *J. Amer. Statist. Assoc.* **75,** 399–404.

COHEN, A. C., and WHITTEN, B. J. (1981). Estimation of lognormal distributions, *Amer. J. Math. Mgt. Sci.* **1,** 139–153.

COHEN, A. C., and WHITTEN, B. J. (1982a). Modified moment and maximum likelihood estimators for parameters of the three-parameter gamma distribution, *Commun. Statist.—Simul. Comput.* **11,** 197–216.

COHEN, A. C., and WHITTEN, B. J. (1982b). Modified maximum likelihood and modified moment estimators for the three-parameter Weibull distribution, *Commun. Statist.—Theor. Meth.* **11,** 2631–2656.

COHEN, A. C., and WHITTEN, B. J. (1985). Modified moment estimation for the three-parameter inverse Gaussian distribution, *J. Qual. Tech.* **17,** 147–154.

COHEN, A. C., and WHITTEN, B. J. (1986). Modified moment estimation for the three-parameter gamma distribution, *J. Qual. Tech.* **18,** 53-62.

COHEN, A. C., and WHITTEN, B. J. (1988). Parameter Estimation in Reliability and Life Span Models, Marcel Dekker, New York.

COHEN, A. C., HELM, F. R., and SUGG, M. (1969). Tables of areas of the standardized Pearson Type III density function, NASA Contractor Report CR-61266 Contract NAS8-11175.

COHEN, A. C., WHITTEN, B. J., and DING, Y. (1984). Modified moment estimation for the three-parameter Weibull distribution, *J. Qual. Tech.* **16,** 159-167.

COHEN, A. C., WHITTEN, B. J., and DING, Y. (1985). Modified moment estimation for the three-parameter lognormal distribution, *J. Qual. Tech.* **17,** 92-99.

COLE, R. H. (1951). Relations between moments of order statistics, *Ann. Math. Statist.* **22,** 308-310.

CRAIG, A. T. (1943). A note on the best linear estimate, *Ann. Math. Statist.* **14,** 88-90.

CROW, E. L., and SHIMIZU, K. (eds.) (1988). Lognormal Distributions, Marcel Dekker, New York.

CSÖRGO, M., and REVESZ, P. (1981). Strong Approximations in Probability and Statistics, Academic Press, New York.

D'AGOSTINO, R. B. (1971). Linear estimation of the Weibull parameters, *Technometrics* **13,** 171-182.

D'AGOSTINO, R. B., and LEE, A. F. S. (1975). Asymptotically best linear unbiased estimation of the Rayleigh parameter for complete and tail-censored samples, *IEEE Trans. on Reliab.* **24,** 156-157.

D'AGOSTINO, R. B., and LEE, A. F. S. (1976). Linear estimation of the logistic parameters for complete or tail-censored samples, *J. Amer. Statist. Assoc.* **71,** 462-464.

D'AGOSTINO, R. B., and STEPHENS, M. A. (eds.) (1986). Goodness-of-fit Techniques, Marcel Dekker, New York.

DALENIUS, T. (1950). The problem of optimum stratification, *Skand. Aktuarietidskr.* **33,** 203-213.

DALY, J. F. (1946). On the use of the sample range in an analogue of Student's *t*-test, *Ann. Math. Statist.* **17,** 71-74.

DANIEL, C., and WOOD, F. S. (1971). Fitting Equations to Data, second edition, John Wiley & Sons, New York.

DANZIGER, L. (1970). Planning censored life tests for estimation of the hazard rate of a Weibull distribution with prescribed precision, *Technometrics* **12,** 408-412.

DAVID, F. N., and JOHNSON, N. L. (1954). Statistical treatment of censored data. I. Fundamental formulae, *Biometrika* **41,** 228-240.

DAVID, H. A. (1970). Order Statistics, first edition, John Wiley & Sons, New York.

DAVID, H. A. (1981). Order Statistics, second edition, John Wiley & Sons, New York.

DAVID, H. A., and MISHRIKY, R. S. (1968). Order statistics for discrete populations and for grouped samples, *J. Amer. Statist. Assoc.* **63,** 1390-1398.

DAVID, H. A., and SHU, V. S. (1978). Robustness of location estimators in the presence of an outlier, *in* Contributions to Survey Sampling and Applied Statistics:

Papers in Honor of H. O. Hartley (H. A. David, ed.), pp. 235-250, Academic Press, New York.

DAVID, H. T. (1963). The sample mean among the extreme normal order statistics, *Ann. Math. Statist.* **34,** 33-55.

DAVIS, C. S., and STEPHENS, M. A. (1977). The covariance matrix of normal order statistics, *Commun. Statist. B* **6,** 75-81.

DAVIS, C. S., and STEPHENS, M. A. (1978). Approximating the covariance matrix of normal order statistics, Algorithm AS128, *Appl. Statist.* **27,** 206-212.

DAVIS, D. J. (1952). An analysis of some failure data, *J. Amer. Statist. Assoc.* **47,** 113-150.

DAVIS, H. T. (1935). Tables of the Higher Mathematical Functions, Vols. 1 and 2, Principia Press, Bloomington.

DIXON, W. J. (1957). Estimates of the mean and standard deviation of a normal population, *Ann. Math. Statist.* **28,** 806-809.

DIXON, W. J. (1960). Simplified estimation from censored normal samples, *Ann. Math. Statist.* **31,** 385-391.

DIXON, W. J., and TUKEY, J. W. (1968). Approximate behavior of the distribution of Winsorized t (trimming/Winsorization 2), *Technometrics* **10,** 83-98.

DOWNTON, F. (1953). A note on ordered least squares estimation, *Biometrika* **40,** 457-458.

DOWNTON, F. (1966a). Linear estimates with polynomial coefficients, *Biometrika* **53,** 129-141.

DOWNTON, F. (1966b). Linear estimates of parameters in the extreme value distribution, *Technometrics* **8,** 3-17.

DUBEY, S. D. (1967). Some percentile estimators for Weibull parameters, *Technometrics* **9,** 119-129.

DUMONCEAUX, R., and ANTLE, C. E. (1973). Discrimination between the lognormal and the Weibull distributions, *Technometrics* **15,** 923-926.

DYER, D. D. (1973). Estimation of the scale parameter of the chi distribution based on sample quantiles, *Technometrics* **15,** 489-496.

DYER, D. D., and WHISENAND, C. W. (1973a). Best linear unbiased estimator of the parameter of the Rayleigh distribution—Part I: Small sample theory for censored order statistics, *IEEE Trans. on Reliab.* **R-22,** 27-34.

DYER, D. D., and WHISENAND, C. W. (1973b). Best linear unbiased estimator of the parameter of the Rayleigh distribution—Part II: Optimum theory for selected order statistics, *IEEE Trans. on Reliab.* **R-22,** 229-231.

EISENBERGER, I. (1968). Testing the mean and standard deviation of a normal distribution using quantiles, *Technometrics* **10,** 781-792.

EISENBERGER, I., and POSNER, E. C. (1965). Systematic statistics used for data compression in space telemetry, *J. Amer. Statist. Assoc.* **60,** 97-133.

ELASHOFF, J. D. (1972). A model for quadratic outliers in linear regression, *J. Amer. Statist. Assoc.* **67,** 478-485.

ENGELHARDT, M. (1975). On simple estimation of the parameters of the Weibull or extreme-value distribution, *Technometrics* **17,** 369-374.

ENGELHARDT, M., and BAIN, L. J. (1973). Some complete and censored sampling results for the Weibull or extreme-value distribution, *Technometrics* **15,** 541-549.

ENGELHARDT, M., and BAIN, L. J. (1974). Some results on point estimation for the two parameter Weibull or extreme-value distribution, *Technometrics* **16,** 49-56.

ENGELHARDT, M., and BAIN, L. J. (1977). Simplified statistical procedures for the Weibull or extreme value distribution, *Technometrics* **19,** 323-331.

EPSTEIN, B. (1956). Simple estimators of the parameters of exponential distributions when samples are censored, *Ann. Inst. Statist. Math.* **8,** 15-26.

EPSTEIN, B. (1962). Simple estimates of the parameters of exponential distributions, *in* Contributions to Order Statistics (A. E. Sarhan and B. G. Greenberg, eds.), pp. 361-371, John Wiley & Sons, New York.

EPSTEIN, B., and SOBEL, M. (1953). Life testing, *J. Amer. Statist. Assoc.* **48,** 486-502.

EPSTEIN, B., and SOBEL, M. (1954). Some theorems relevant to life testing from an exponential distribution, *Ann. Math. Statist.* **25,** 373-381.

ERTO, P., and GUIDA, M. (1985). Tables for exact lower confidence limits for reliability and quantiles, based on least-squares estimators of Weibull parameters, *IEEE Trans. on Reliab.* **R-34,** 219-223.

EUBANK, R. L. (1981a). A density-quantile function approach to optimal spacing selection, *Ann. Statist.* **9,** 494-500.

EUBANK, R. L. (1981b). Estimation of the parameters and quantiles of the logistic distribution by linear functions of sample quantiles, *Scand. Actuar. J.* 1981, 229-236.

EUBANK, R. L. (1986). Quantiles, *in* Encyclopedia in Statistical Science (N. L. Johnson and S. Kotz, eds.), Vol. 7, pp. 424-432, John Wiley & Sons, New York.

FAMA, E. F., and ROLL, R. (1971). Parameter estimates for symmetric stable distributions, *J. Amer. Statist. Assoc.* **66,** 331-338.

FISHER, R. A. (1966). The Design of Experiments, eighth edition, Oliver & Boyd, Edinburgh.

FOLKS, J. L., and CHHIKARA, R. S. (1978). The inverse Gaussian distribution and its statistical application—a review, *J. Roy. Statist. Soc., Ser. B* **40,** 263-289.

FRASER, D. A. S. (1968). The Structure of Inference, John Wiley & Sons, New York.

GAJJAR, A. V., and KHATRI, C. G. (1969). Progressively censored samples from log-normal and logistic distributions, *Technometrics* **11,** 793-803.

GALAMBOS, J. (1978). The Asymptotic Theory of Extreme Order Statistics, John Wiley & Sons, New York.

GALAMBOS, J., and KOTZ, S. (1978). Characterizations of Probability Distributions, Lecture Notes in Mathematics No. 675, Springer-Verlag, New York.

GIBBONS, D. E., and MCDONALD, G. C. (1975). Small-sample estimation for the log-normal distribution with unit shape parameter, *IEEE Trans. on Reliab.* **24,** 290-295.

GODWIN, H. J. (1949a). Some low moments of order statistics, *Ann. Math. Statist.* **20,** 279-285.

GODWIN, H. J. (1949b). On the estimation of dispersion by linear systematic statistics, *Biometrika* **36,** 92-100.

GOVINDARAJULU, Z. (1963). On moments of order statistics and quasi-ranges from normal populations, *Ann. Math. Statist.* **34,** 633–651.

GOVINDARAJULU, Z. (1966). Best linear estimates under symmetric censoring of the parameters of a double exponential population, *J. Amer. Statist. Assoc.* **61,** 248–258.

GOVINDARAJULU, Z. (1968). Certain general properties of unbiased estimates of location and scale parameters based on ordered observations, *SIAM J. Appl. Math.* **16,** 533–551.

GOVINDARAJULU, Z., and EISENSTAT, S. (1965). Best estimates of location and scale parameters of a chi (1 d.f.) distribution, using ordered observations, *Rep. Stat. Appl. Res., JUSE* **12,** 149–164.

GOVINDARAJULU, Z., and JOSHI, M. (1968). Best linear unbiased estimation of location and scale parameters of Weibull distribution using ordered observations, *Rep. Stat. Appl. Res., JUSE* **15,** 57–70.

GRAVEL, R., and VAN EEDEN, C. (1981). Best linear unbiased estimators of the location of a double quadratic distribution based on order statistics, *J. Statist. Comput. Simul.* **13,** 225–243.

GREENBERG, B. G., and SARHAN, A. E. (1958). Applications of order statistics to health data, *Amer. J. Publ. Health.* **48,** 1388–1394.

GRUNDY, P. M. (1952). The fitting of grouped truncated and grouped censored normal distributions, *Biometrika* **39,** 252–259.

GUMBEL, E. J. (1958). Statistics of Extremes, Columbia University Press, New York.

GUPTA, A. K. (1952). Estimation of the mean and standard deviation of a normal population from a censored sample, *Biometrika* **39,** 260–273.

GUPTA, S. S. (1960). Order statistics from the gamma distribution, *Technometrics* **2,** 243–262.

GUPTA, S. S. (1962). Gamma distribution, *in* Contributions to Order Statistics (A. E. Sarhan and B. G. Greenberg, eds), pp. 431–450, John Wiley & Sons, New York.

GUPTA, S. S., and GNANADESIKAN, M. (1966). Estimation of the parameters of the logistic distribution, *Biometrika* **53,** 565–570.

GUPTA, S. S., and PANCHAPAKESAN, S. (1979). Multiple Decision Procedures: Theory and Methodology of Selecting and Ranking Populations, John Wiley & Sons, New York.

GUPTA, S. S., and SHAH, B. K. (1965). Exact moments and percentage points of the order statistics and the distribution of the range from the logistic distribution, *Ann. Math. Statist.* **36,** 907–920.

GUPTA, S. S., QUREISHI, A. S., and SHAH, B. K. (1967). Best linear unbiased estimators of the parameters of the logistic distribution using order statistics, *Technometrics* **9,** 43–56.

HALL, I. J. (1975). One-sided tolerance limits for a logistic distribution based on censored samples, *Biometrics* **31,** 873–880.

HALPERIN, M. (1952). Maximum-likelihood estimation in truncated samples, *Ann. Math. Statist.* **23,** 226–238.

HAMMERSLEY, J. M., and MORTON, K. W. (1954). The estimation of location and scale parameters from grouped data, *Biometrika* **41,** 296–301.

HAMOUDA, E. M. (1988). Inference in regression problems based on order statistics, *Commun. Statist.—Theor. Meth.* **17**(7), 2343–2367.

HAMOUDA, E. M., and LEONE, F. C. (1974). The O-BLUE estimators for complete and censored samples in linear regression, *Technometrics* **16,** 441–446.

HARLEY, B. I., and PEARSON, E. S. (1957). The distribution of range in normal samples with $n = 200$, *Biometrika* **44,** 257–260.

HARRELL, F. E., and DAVIS, C. E. (1982). A new distribution-free quantile estimator, *Biometrika* **69,** 635–640.

HARTER, H. L. (1959). The use of sample quasi-ranges in estimating population standard deviation, *Ann. Math. Statist.* **30,** 980–999. Correction **31,** 228.

HARTER, H. L. (1960). Tables of range and studentized range, *Ann. Math. Statist.* **31,** 1122–1147.

HARTER, H. L. (1961a). Expected values of normal order statistics, *Biometrika* **48,** 151–165. Correction **48,** 476.

HARTER, H. L. (1961b). Estimating the parameters of negative exponential populations from one or two order statistics, *Ann. Math. Statist.* **32,** 1078–1090.

HARTER, H. L. (1970a). Order Statistics and Their Uses in Testing and Estimation, Vol. 1, U.S. Government Printing Office, Washington, D.C.

HARTER, H. L. (1970b). Order Statistics and Their Uses in Testing and Estimation, Vol. 2, U.S. Government Printing Office, Washington, D.C.

HARTER, H. L. (1971). Some optimization problems in parameter estimation, *in* Optimizing Methods in Statistics (J. S. Rustagi, ed.), pp. 33–62, Academic Press, New York.

HARTER, H. L. (1978). A Chronological Annotated Bibliography of Order Statistics, Vol. 1: Pre-1950, U.S. Government Printing Office, Washington, D.C.

HARTER, H. L. (1983). A Chronological Annotated Bibliography of Order Statistics, Vol. 2: 1950–1959, American Sciences Press, Columbus, Ohio.

HARTER, H. L., and MOORE, A. H. (1965a). Point and interval estimators, based on m order statistics, for the scale parameter of a Weibull population with known shape parameter, *Technometrics* **7,** 405–422.

HARTER, H. L., and MOORE, A. H. (1965b). Maximum-likelihood estimation of the parameters of gamma and Weibull populations from complete and from censored samples, *Technometrics* **7,** 639–643.

HARTER, H. L., and MOORE, A. H. (1966a). Local-maximum-likelihood estimation of the parameters of three-parameter lognormal populations from complete and censored samples, *J. Amer. Statist. Assoc.* **61,** 842–851.

HARTER, H. L., and MOORE, A. H. (1966b). Iterative maximum-likelihood estimation of the parameters of normal populations from singly and doubly censored samples, *Biometrika* **53,** 205–213.

HARTER, H. L., and MOORE, A. H. (1967). Maximum-likelihood estimation, from censored samples, of the parameters of a logistic distribution, *J. Amer. Statist. Assoc.* **62,** 675–684.

HARTER, H. L., and MOORE, A. H. (1968). Maximum-likelihood estimation, from doubly censored samples, of the parameters of the first asymptotic distribution of extreme values, *J. Amer. Statist. Assoc.* **63,** 889-901.

HASSANEIN, K. M. (1968). Analysis of extreme-value data by sample quantiles for very large samples, *J. Amer. Statist. Assoc.* **63,** 877-888.

HASSANEIN, K. M. (1969a). Estimation of the parameters of the logistic distribution of sample quantiles, *Biometrika* **56,** 684-687.

HASSANEIN, K. M. (1969b). Estimation of the parameters of the extreme value distribution by use of two or three order statistics, *Biometrika* **56,** 429-436.

HASSANEIN, K. M. (1971). Percentile estimators for the parameters of the Weibull distribution, *Biometrika* **58,** 673-676.

HASSANEIN, K. M. (1972). Simultaneous estimation of the parameters of the extreme value distribution by sample quantiles, *Technometrics* **14,** 63-70.

HASSANEIN, K. M. (1977). Simultaneous estimation of the location and scale parameters of the gamma distribution by linear functions of order statistics, *Scand. Actuarial J.* 1977, 88-93.

HASSANEIN, K. M., SALEH, A. K. MD. E., and BROWN, E. F. (1984). Quantile estimates in complete and censored samples from extreme-value and Weibull distributions, *IEEE Trans. on Reliab.* **R-33,** 370-373.

HASSANEIN, K. M., SALEH, A. K. MD. E., and BROWN, E. F. (1985). Best linear unbiased quantile estimators of the logistic distribution using order statistics, *J. Statist. Comput. Simul.* **23,** 123-131.

HASSANEIN, K. M., SALEH, A. K. MD. E., and BROWN, E. F. (1986a). Estimation and testing of quantiles of the extreme-value distribution, *J. Statist. Plann. Inf.* **14,** 389-400.

HASSANEIN, K. M., SALEH, A. K. MD. E., and BROWN, E. F. (1986b). Best linear unbiased estimators for normal distribution quantiles for sample sizes up to 20, *IEEE Trans. on Reliab.* **R-35,** 327-329.

HAWKINS, D. M. (1979). Fractiles of an extended multiple outlier test, *J. Statist. Comput. Simul.* **8,** 227-236.

HERD, R. G. (1956). Estimation of the parameters of a population from a multi-censored sample, Ph.D. Thesis, Iowa State College, Ames, Iowa.

HEYDE, C. C. (1963). On a property of the lognormal distribution, *J. Roy. Statist. Soc., Ser. B,* **25,** 392-393.

HIGUCHI, I. (1956). On the solutions of certain simultaneous equations, *Ann. Inst. Statist. Math.* **5,** 77-90.

HILL, B. M. (1963). The three-parameter lognormal distribution and Bayesian analysis of a point-source epidemic, *J. Amer. Statist. Assoc.* **58,** 72-84.

HINICH, M. J., and TALWAR, P. P. (1975). A simple method for robust estimation, *J. Amer. Statist. Assoc.* **70,** 113-119.

HOAGLIN, D. C., and WELSCH, R. E. (1978). The Hat matrix in regression and ANOVA, *Amer. Statist.* **32,** 17-22, *Corrigenda* **32,** 146.

HOEFFDING, W. (1953). On the distribution of the expected values of the order statistics, *Ann. Math. Statist.* **24,** 93-100.

HOSKING, J. R. M., and WALLIS, J. R. (1987). Parameter and quantile estimation for the generalized Pareto distribution, *Technometrics* **29,** 339-349.

ISIDA, M., and TAGAMI, S. (1959). The bias and precision in the maximum likelihood estimation of the parameters of normal populations from singly truncated sample, *Reports of Statistical Application Research, JUSE* **6,** 105-110.

JOHNS, M. V., JR., and LIEBERMAN, G. J. (1966). An exact asymptotically efficient confidence bound for reliability in the case of the Weibull distribution, *Technometrics* **8,** 135-175.

JOHNSON, N. L., and KOTZ, S. (1970). Continuous Univariate Distributions—1, Houghton Mifflin, Boston, Massachusetts.

JONES, H. L. (1948). Exact lower moments of order statistics in small samples from a normal distribution, *Ann. Math. Statist.* **19,** 270-273.

JOSHI, P. C. (1971). Recurrence relations for the mixed moments of order statistics, *Ann. Math. Statist.* **42,** 1096-1098.

JOSHI, P. C. (1978). Recurrence relations between moments of order statistics from exponential and truncated exponential distributions, *Sankhyā B* **39,** 362-371.

JOSHI, P. C. (1979a). A note on the moments of order statistics from doubly truncated exponential distribution, *Ann. Inst. Statist. Math.* **31,** 321-324.

JOSHI, P. C. (1979b). On the moments of gamma order statistics, *Naval Res. Logist. Quart.* **26,** 675-679.

JOSHI, P. C. (1982). A note on the mixed moments of order statistics from exponential and truncated exponential distributions, *J. Statist. Plann. Inf.* **6,** 13-16.

JOSHI, P. C., and BALAKRISHNAN, N. (1981). An identity for the moments of normal order statistics with applications, *Scand. Actuarial J.* 1981, 203-213.

JOSHI, P. C., and BALAKRISHNAN, N. (1982). Recurrence relations and identities for the product moments of order statistics, *Sankhyā B* **44,** 39-49.

JUNG, J. (1955). On linear estimates defined by a continuous weight function, *Ark. Mat.* **3,** 199-209.

JUNG, J. (1962). Approximation of least-squares estimates of location and scale parameters, *in* Contributions to Order Statistics (A. E. Sarhan and B. G. Greenberg, eds.), pp. 28-33, John Wiley & Sons, New York.

KABIR, A. B. M. L., and AHSANULLAH, M. (1974). Estimation of the location and scale parameters of a power-function distribution by linear functions of order statistics, *Commun. Statist.* **3,** 463-467.

KABIR, A. B. M. L., and AHSANULLAH, M. (1979). Estimation of the location and scale parameters of a power function distribution based on a selected number of order statistics, *Commun. Statist.* **A8,** 139-149.

KAIGH, W. D., and LACHENBRUCH, P. A. (1982). A generalized quantile estimator, *Commun. Statist.—Theor. Meth.* **11,** 2217-2238.

KAMINSKY, K. S. (1972). Confidence intervals for the exponential scale parameter using optimally selected order statistics, *Technometrics* **14,** 371-383.

KAMINSKY, K. S. (1973). Comparison of approximate confidence intervals for the exponential scale parameter from sample quantiles, *Technometrics* **15,** 483-487.

KAMINSKY, K. S. (1974). Confidence intervals and tests for two exponential scale parameters based on order statistics in compressed samples, *Technometrics* **16,** 251-254.

KAPPENMAN, R. F. (1987). Improved distribution quantile estimation, *Commun. Statist.—Simul. Comput.* **16,** 307-320.

KARLIN, S., and TAYLOR, H. M. (1975). A First Course in Stochastic Processes, second edition, Academic Press, New York.

KEATING, J. P. (1983). Estimators of percentiles based on absolute loss, *Commun. Statist.—Theor. Meth.* **12,** 441-447.

KENDALL, M. G. (1948). The Advanced Theory of Statistics, Vol. 1, fourth edition, Charles Griffin and Co., London.

KENDALL, M. G., and STUART, A. (1973). The Advanced Theory of Statistics, Vol. 2, Charles Griffin and Co., London; Hafner, New York.

KIMBALL, B. F. (1947). Assignment of frequencies to a completely ordered set of sample data (discussion), *Trans. Amer. Geophysicists Union* **28,** 952.

KOUTROUVELIS, I. A. (1981). Large-sample quantile estimation in Pareto laws, *Commun. Statist.—Theor. Meth.* **10**(2), 189-201.

KUBAT, P., and EPSTEIN, B. (1980). Estimation of quantiles of location-scale distributions based on two or three order statistics, *Technometrics* **22,** 575-581.

KULLDORFF, G. (1961). Estimation from Grouped and Partially Grouped Samples, John Wiley & Sons, New York.

KULLDORFF, G. (1963a). Estimation of one or two parameters of the exponential distribution on the basis of suitably chosen order statistics, *Ann. Math. Statist.* **34,** 1419-1431.

KULLDORFF, G. (1963b). On the optimum spacing of sample quantiles from a normal distribution, Part I, *Skand. Aktuarietidskr.* **46,** 143-156.

KULLDORFF, G. (1964). On the optimum spacing of sample quantiles from a normal distribution, Part II, *Skand. Aktuarietidskr.* **47,** 71-87.

KULLDORFF, G. (1973). A note on the optimum spacing of sample quantiles from the six extreme value distributions, *Ann. Statist.* **1,** 562-567.

KULLDORFF, G., and VÄNNMAN, K. (1973). Estimating location and scale parameters of a Pareto distribution by linear functions of order statistics, *J. Amer. Statist. Assoc.* **68,** 218-227.

LAMBERT, J. A. (1964). Estimation of parameters in the three-parameter lognormal distribution, *Austral. J. Statist.* **6,** 29-32.

LAWLESS, J. F. (1982). Statistical Models & Methods For Lifetime Data, John Wiley & Sons, New York.

LEE, K. R., KAPADIA, C. H., and DWIGHT, B. B. (1980). On estimating the scale parameter of the Rayleigh distribution from doubly censored samples, *Statist. Hefte* **21,** 14-29.

LEONE, F. C., and HAMOUDA, E. M. (1973). Relative efficiencies of O-BLUE estimators in simple linear regression, *J. Amer. Statist. Assoc.* **68,** 953-959.

LIEBLEIN, J. (1954). A new method of analyzing extreme-value data, Nat. Advisory Comm. Aeronaut. Tech. Note 3053.

LIEBLEIN, J. (1955). On moments of order statistics from the Weibull distribution, *Ann. Math. Statist.* **26,** 330–333.

LIEBLEIN, J., and SALZER, H. E. (1957). Table of the first moment of ranked extremes, *J. Res., Nat. Bur. Stand.* **59,** 203–206.

LIEBLEIN, J., and ZELEN, M. (1956). Statistical investigation of the fatigue life of deep-grove ball bearings, *J. Res., Nat. Bur. Stand.* **57,** 273–316.

LIKEŠ, J. (1967). Distributions of some statistics in samples from exponential and power-function populations, *J. Amer. Statist. Assoc.* **62,** 259–271.

LIKEŠ, J. (1985). Estimation of the quantiles of Pareto's distribution, *Math. Operationsfors. und Statist., Ser. Statist.* **16,** 541–547.

LLOYD, E. H. (1952). Least-squares estimation of location and scale parameters using order statistics, *Biometrika* **39,** 88–95.

LWIN, T. (1976). Optimal linear estimators of location and scale parameters using order statistics and related empirical Bayes estimation, *Scand. Actuar. J.* 1976, 79–91.

MCCOOL, J. I. (1965). The construction of good linear unbiased estimates from the best linear estimates for a smaller sample size, *Technometrics* **7,** 543–552.

MCCOOL, J. I. (1974). Inferential techniques for Weibull populations, Aerospace Research Laboratories Report ARL TR 74-0180, Wright-Patterson Air Force Base, Ohio.

MCKAY, A. T. (1935). The distribution of the difference between the extreme observation and the sample mean in samples of n from a normal universe, *Biometrika* **27,** 466–471.

MAGUIRE, B. A., PEARSON, E. S., and WYNN, A. H. A. (1952). The time intervals between industrial accidents, *Biometrika* **39,** 168–180.

MALIK, H. J. (1970). Estimation of the parameters of the Pareto distribution, *Metrika* **15,** 126–132.

MALIK, H. J., BALAKRISHNAN, N., and AHMED, S. E. (1988). Recurrence relations and identities for moments of order statistics, I: Arbitrary continuous distribution, *Commun. Statist.—Theor. Meth.* (*Statist. Reviews*) **17**(8), 2623–2655.

MALMQUIST, S. (1950). On a property of order statistics from a rectangular distribution, *Skand. Aktuarietidskr.* **33,** 214–222.

MANN, N. R. (1967a). Tables for obtaining the best linear invariant estimates of parameters of the Weibull distribution, *Technometrics* **9,** 629–645.

MANN, N. R. (1967b). Results on location and scale parameter estimation with application to the extreme-value distribution, Aerospace Research Laboratories 67-0023, Wright-Patterson Air Force Base, Ohio.

MANN, N. R. (1968). Point and interval estimation procedures for the two-parameter Weibull and extreme-value distributions, *Technometrics* **10,** 231–256.

MANN, N. R. (1969a). A test for the hypothesis that two extreme-value scale parameters are equal, *Naval Res. Logist. Quart.* **16,** 207–216.

MANN, N. R. (1969b). Exact three-order-statistic confidence bounds on reliable life for a Weibull model with progressive censoring, *J. Amer. Statist. Assoc.* **64,** 306–315.

MANN, N. R. (1969c). Optimum estimators for linear functions of location and scale parameters, *Ann. Math. Statist.* **40,** 2149–2155.

MANN, N. R. (1970). Estimators and exact confidence bounds for Weibull parameters based on a few ordered observations, *Technometrics* **12,** 345–361.

MANN, N. R. (1971). Best linear invariant estimation for Weibull parameters under progressive censoring, *Technometrics* **13,** 521–534.

MANN, N. R., and FERTIG, K. W. (1973). Tables for obtaining Weibull confidence bounds and tolerance bounds based on best linear invariant estimates of parameters of the extreme-value distribution, *Technometrics* **15,** 87–102.

MANN, N. R., and FERTIG, K. W. (1975a). A goodness-of-fit test for the two-parameter vs. three-parameter Weibull; confidence bounds for threshold, *Technometrics* **17,** 237–245.

MANN, N. R., and FERTIG, K. W. (1975b). Simplified efficient point and interval estimators for Weibull parameters, *Technometrics* **17,** 361–368.

MANN, N. R., and FERTIG, K. W. (1977). Efficient unbiased quantile estimators for moderate-size complete samples from extreme-value and Weibull distributions: confidence bounds and tolerance and prediction intervals, *Technometrics* **19,** 87–94.

MANN, N. R., FERTIG, K. W., and SCHEUER, E. M. (1971). Confidence and tolerance bounds and a new goodness of fit test for the two-parameter Weibull or extreme value distribution with tables for censored samples of size 3(1)25, Aerospace Research Laboratories 71-0077, Wright–Patterson Air Force Base, Ohio.

MANN, N. R., SCHAFER, R. E., and SINGPURWALLA, N. D. (1974). Methods for Statistical Analysis of Reliability and Life Data, John Wiley & Sons, New York.

MARGOLIN, B. H., and WINOKUR, H. S., JR. (1967). Exact moments of the order statistics of the geometric distribution and their relation to inverse sampling and reliability of redundant systems, *J. Amer. Statist. Assoc.* **62,** 915–925.

MAXWELL, E. A. (1973). Application of the structural model to censored data, *J. Amer. Statist. Assoc.* **68,** 440–444.

MEHROTRA, K. G., and NANDA, P. (1974). Unbiased estimation of parameters by order statistics in case of censored samples, *Biometrika* **61,** 601–606.

MELNICK, E. L. (1964). Moments of Ranked Poisson Variates, M.S. Thesis, Department of Statistics, Virginia Polytechnic Institute, Blacksburg, Virginia, U.S.A.

MENDENHALL, W. (1983). Introduction to Probability and Statistics, PWS publication, Boston.

MENON, M. V. (1963). Estimation of the shape and scale parameters of the Weibull distribution, *Technometrics* **5,** 175–182.

MIKÉ, V. (1971). Efficiency-robust systematic linear estimates of location, *J. Amer. Statist. Assoc.* **66,** 594–601.

MILLER, R. G., JR. (1966). Simultaneous Statistical Inference, McGraw–Hill, New York.

MILLER, R. G., JR. (1977). Developments in multiple comparisons 1966–1976, *J. Amer. Statist. Assoc.* **72,** 779–788.

MIYAMOTO, Y. (1972). Three-parameter estimation for a generalized gamma distribution by sample quantiles in large samples, *J. Japan Statist. Soc.* **3,** 9–18.

MOOD, A. M., GRAYBILL, F. A., and BOES, D. C. (1974). Introduction to the Theory of Statistics, Third edition, McGraw-Hill, New York.

MOSTELLER, F. (1946). On some useful "inefficient" statistics, *Ann. Math. Statist.* **17,** 377–408.

MOSTELLER, F., and TUKEY, J. W. (1977). Data Analysis and Regression, Addison-Wesley, Reading, Massachusetts.

MOUSSA, E. A. (1972). Estimation and robustness of efficiency of linear models for small complete and censored samples, Ph.D. Thesis, University of Iowa, Iowa City, U.S.A.

MUNRO, A. H., and WIXLEY, R. A. J. (1970). Estimators based on order statistics of small samples from a three-parameter lognormal distribution, *J. Amer. Statist. Assoc.* **65,** 212–215.

MURPHY, R. B. (1951). On Tests For Outlying Observations, Ph.D. Thesis, Princeton University, Princeton, New Jersey, U.S.A.

MURTHY, V. K., and SWARTZ, G. B. (1975). Estimation of Weibull parameters from two-order statistics, *J. Roy. Statist. Soc. B* **37,** 96–102.

NELSON, W. (1972). Theory and application of hazard plotting for censored failure data, *Technometrics* **14,** 945–966.

NELSON, W., and SCHMEE, J. (1979). Inference for (log) normal life distributions from small single censored samples and BLUEs, *Technometrics* **21,** 43–54.

OGAWA, J. (1951). Contributions to the theory of systematic statistics, I, *Osaka Math. J.* **3,** 175–213.

OGAWA, J. (1952). Contributions to the theory of systematic statistics, II, *Osaka Math. J.* **4,** 41–61.

OGAWA, J. (1962). Optimum spacing and grouping for the exponential distribution, *in* Contributions to Order Statistics (A. E. Sarhan and B. G. Greenberg, eds.) pp. 371–380, John Wiley & Sons, New York.

OGAWA, J. (1976). A note on the optimal spacing of the systematic statistics-normal distribution, *in* Essays in Probability and Statistics (Ikeda *et al.*, eds.), pp. 467–474, Shinko Tsusho, Tokyo.

OPPENLANDER, J. E. (1986). Statistical Modeling and Simple Robust Estimators for Lower Percentiles in the Context of Life Data Analysis, Ph.D. Thesis, Union College, Schenectady, New York.

OPPENLANDER, J. E., SCHMEE, J., and HAHN, G. J. (1988). Some simple robust estimators of normal distribution tail percentiles and their properties, *Commun. Statist.—Theor. Meth.* **17**(7), 2279–2301.

PADGETT, W. J., and WEI, L. J. (1979). Estimation for the three-parameter inverse Gaussian distribution, *Commun. Statist.—Theor. Meth.* **8,** 129–137.

PARZEN, E. (1978). A density-quantile function perspective on robust estimation, *in* Robustness in Statistics (R. L. Laudner and G. N. Wilkinson, eds.), pp. 237–258, Academic Press, New York.

PARZEN, E. (1979). Nonparametric statistical data modelling, *J. Amer. Statist. Assoc.* **74,** 105–121.

PEARSON, E. S., and HARTLEY, H. O. (1970). Biometrika Tables for Statisticians, Vol. I, third edition (with additions), Cambridge University Press, Cambridge, England.

PEARSON, E. S., and HARTLEY, H. O. (1972). Biometrika Tables for Statisticians, Vol. II, Cambridge University Press, Cambridge, England.

PEARSON, K. (1934). Tables of the Incomplete *B*-Function, Cambridge University Press, Cambridge, England.

PEARSON, K., and PEARSON, M. V. (1931). On the mean character and variance of a ranked individual, and on the mean and variance of the intervals between ranked individuals. I (1931): Symmetrical distributions (normal and rectangular), *Biometrika* **23,** 364–397. II (1932): Case of certain skew curves, *Biometrika* **24,** 203–279.

PLACKETT, R. L. (1958). Linear estimation from censored data, *Ann. Math. Statist.* **29,** 131–142.

POLOVKO, A. M. (1968). Fundamentals of Reliability Theory, Academic Press, New York.

PRESCOTT, P. (1970). Estimation of the standard deviation of a normal population from doubly censored samples using normal scores, *Biometrika* **57,** 409–419.

PRESCOTT, P. (1974). Variances and covariances of order statistics from the gamma distribution, *Biometrika* **61,** 607–613.

PROSCHAN, F. (1963). Theoretical explanation of observed decreasing failure rate, *Technometrics* **5,** 375–383.

QUAYLE, R. J. (1963). Estimation of the scale parameter of the Weibull probability density function by use of one order statistic, M.S. Thesis, Air Force Institute of Technology, Wright-Patterson AFB, Dayton, Ohio.

RAGAB, A., and GREEN, J. (1984). On order statistics from the log-logistic distribution and their properties, *Commun. Statist.—Theor. Meth.* **13**(21), 2713–2724.

RAGAB, A., and GREEN, J. (1987). Estimation of the parameters of the loglogistic distribution based on order statistics, *Amer. J. Math. Manag. Sci.* **7,** 307–324.

RAGHUNANDANAN, K., and SRINIVASAN, R. (1970). Simplified estimation of parameters in a logistic distribution, *Biometrika* **57,** 677–678.

RAGHUNANDANAN, K., and SRINIVASAN, R. (1971). Simplified estimation of parameters in a double exponential distribution, *Technometrics* **13,** 689–691.

RAO, C. R. (1973). Linear Statistical Inference and Its Applications, second edition, John Wiley & Sons, New York.

RAO, C. R. (1975). Linear Statistical Inference and Its Applications, third edition, John Wiley & Sons, New York.

REISS, R.-D. (1980). Estimation of quantiles in certain non-parametric models, *Ann. Statist.* **8,** 87–105.

RÉNYI, A. (1953). On the theory of order statistics, *Acta Math. Acad. Sci. Hung.* **4,** 191–231.

RESEK, R. W. (1976). Estimation of the parameters of a general Student's t distribution, *Commun. Statist.* **A5,** 635–645.

ROBERTSON, C. A. (1977). Estimation of quantiles of exponential distributions with minimum error in predicted distribution functions, *J. Amer. Statist. Assoc.* **72,** 162–164.

RUBEN, H. (1954). On the moments of order statistics in samples from normal populations, *Biometrika* **41,** 200–227.

RUBEN, H. (1956). On the moments of the range and product moments of extreme order statistics in normal samples, *Biometrika* **43,** 458–460.

RUKHIN, A. L., and STRAWDERMAN, W. E. (1982). Estimating a quantile of an exponential distribution, *J. Amer. Statist. Assoc.* **77,** 159–162.

SALEH, A. K. MD. E. (1966). Estimation of the parameters of the exponential distribution based on order statistics in censored samples, *Ann. Math. Statist.* **37,** 1717–1735.

SALEH, A. K. MD. E. (1967). Determination of the exact optimum order statistics for estimating the parameters of the exponential distribution from censored samples, *Technometrics* **9,** 279–292.

SALEH, A. K. MD. E. (1981). Estimating quantiles of exponential distribution, *in* Statistics and Related Topics (Csorgo *et al.*, eds.), pp. 279–283, North-Holland, New York.

SALEH, A. K. MD. E., and ALI, M. M. (1966). Asymptotic optimum quantiles for the estimation of the parameters of the negative exponential distribution, *Ann. Math. Statist.* **37,** 143–151.

SALEH, A. K. MD. E., and HASSANEIN, K. M. (1986). Testing equality of location parameters and quantiles of s (≥ 2) location-scale distributions, *Bull. Inst. Math., Acad. Sinica* **14,** 39–49.

SALEH, A. K. MD. E., and SEN, P. K. (1985). Asymptotic relative efficiency of some joint-tests for location and scale parameters based on selected order statistics, *Commun. Statist.—Theor. Meth.* **14,** 621–633.

SALEH, A. K. MD. E., SCOTT, C., and JUNKINS, D. B. (1975). Exact first and second order moments of order statistics from the truncated exponential distribution, *Naval Res. Logist. Quart.* **22,** 65–77.

SALEH, A. K. MD. E., ALI, M. M., and UMBACH, D. (1982). Large sample estimation of Pareto quantiles using selected order statistics, *Metrika* **32,** 49–56.

SALEH, A. K. MD. E., ALI, M. M., and UMBACH, D. (1983). Estimating the quantile function of location-scale family of distributions based on few selected order statistics, *J. Statist. Plann. Inf.* **8,** 75–86.

SALEH, A. K. MD. E., ALI, M. M., and UMBACH, D. (1984). Tests of significance using selected sample quantiles, *Statist. Prob. Lett.* **2,** 295–297.

SALEH, A. K. MD. E., HASSANEIN, K. M., and BROWN, E. F. (1985). Optimum spacings for the joint estimation and tests of hypothesis of location and scale parameters of the Cauchy distribution, *Commun. Statist.—Theor. Math.* **14,** 247–254.

SALVOSA, L. R. (1930). Tables of Pearson's Type III function, *Ann. Math. Statist.* **1,** 191-198, Appendix, 1-125.

SARHAN, A. E., and GREENBERG, B. G. (1956). Estimation of location and scale parameters by order statistics from singly and doubly censored samples. Part I. The normal distribution up to samples of size 10, *Ann. Math. Statist.* **27,** 427-451. Correction **40,** 325.

SARHAN, A. E., and GREENBERG, B. G. (1957). Tables for best linear estimates by order statistics of the parameters of single exponential distributions from singly and doubly censored samples, *J. Amer. Statist. Assoc.* **52,** 58-87.

SARHAN, A. E., and GREENBERG, B. G. (1958a). Estimation of location and scale parameters by order statistics from singly and doubly censored samples. Part II. Tables for the normal distribution for samples of size $11 \leq N \leq 15$, *Ann. Math. Statist.* **29,** 79-105.

SARHAN, A. E., and GREENBERG, B. G. (1958b). Estimation problems in exponential distribution using order statistics, *Proc. of the Statistical Techniques in Missile Evaluation Symposium*, Blacksburg, Virginia, pp. 123-173.

SARHAN, A. E., and GREENBERG, B. G. (eds.) (1962). Contributions to Order Statistics, John Wiley & Sons, New York.

SÄRNDAL, C. E. (1962). Information from Censored Samples, Almqvist and Wiksell, Stockholm.

SÄRNDAL, C. E. (1964). Estimation of the parameters of the gamma distribution by sample quantiles, *Technometrics* **6,** 405-414.

SAW, J. G. (1959). Estimation of the normal population parameters given a singly censored sample, *Biometrika* **46,** 150-159.

SAW, J. G. (1960). A note on the error after a number of terms of the David-Johnson series for the expected values of normal order statistics, *Biometrika* **47,** 79-86.

SCHAEFFER, L. R., VAN VLECK, L. D., and VELASCO, J. A. (1970). The use of order statistics with selected records, *Biometrics* **26,** 854-859.

SCHNEIDER, B. E. (1978). Trigamma function, Algorithm AS121, *Appl. Statist.* **27,** 97-99.

SCHROEDINGER, E. (1915). Zur theorie der fall- und steigversuche an teilchen mit brownscher bewegung, *Physikalische Zeitschrift* **16,** 289-295.

SCHWEDER, T. (1976). Some "optimal" methods to detect structural shift or outliers in regression, *J. Amer. Statist. Assoc.* **71,** 491-501.

SEAL, K. C. (1956). On minimum variance among certain linear functions of order statistics, *Ann. Math. Statist.* **27,** 854-855.

SHAH, B. K. (1966). On the bivariate moments of order statistics from a logistic distribution, *Ann. Math. Statist.* **37,** 1002-1010.

SHAH, B. K. (1970). Note on moments of a logistic order statistics, *Ann. Math. Statist.* **41,** 2151-2152.

SHELNUTT, J. W., III (1966). Conditional Linear Estimation of the Scale Parameters of the Extreme Value Distribution by the Use of Selected Order Statistics, unpublished M.S. Thesis, Air Force Institute of Technology.

SHELNUTT, J. W., III, MOORE, A. H., and HARTER, H. L. (1973). Linear estimation of the scale parameter of the first asymptotic distribution of extreme values, *IEEE Trans. on Reliab.* **R-22,** 259–264.

SIDDIQUI, M. M., and RAGHUNANDANAN, K. (1967). Asymptotically robust estimators of location, *J. Amer. Statist. Assoc.* **62,** 950–953.

SINGH, M., and GUPTA, S. C. (1989). A non-iterative robust method of estimation in linear regression, Technical Report, Virginia Polytechnic Institute and State University, Blacksburg, Virginia, U.S.A.

SIOTANI, M. (1957). Order statistics for discrete case with a numerical application to the binomial distribution, *Ann. Inst. Statist. Math.* **8,** 95–104.

SMOLUCHOWSKY, M. W. (1915). Notiz uber die berechnung der browschen molekularbewegung bei der ehrenhaft-millikanschen versuchsanordnung, *Physikalische Zeitschrift* **16,** 318–321.

SRIKANTAN, K. S. (1962). Recurrence relations between the PDF's of order statistics, and some applications, *Ann. Math. Statist.* **33,** 169–177.

STEPHENS, M. A. (1975). Asymptotic properties for covariance matrices of order statistics, *Biometrika* **62,** 23–28.

STIGLER, S. M. (1974). Linear functions of order statistics with smooth weight functions, *Ann. Statist.* **2,** 676–693. Correction **7,** 466.

SUKHATME, P. V. (1937). Tests of significance for samples of the chi-square population with 2 degrees of freedom, *Ann. Eugen.* **8,** 52–56.

SWAMY, P. S. (1962). On the amount of information supplied by censored samples of grouped observations in the estimation of statistical parameters, *Biometrika* **49,** 245–249.

TADIKAMALLA, P. R. (1977). An approximation to the moments and the percentiles of gamma order statistics, *Sankhyā B* **39,** 372–381.

TAN, W. Y., and TABATABAI, M. A. (1988). A modified Winsorized regression procedure for linear models, *J. Statist. Comput. Simul.* **30,** 299–313.

TARTER, M. E. (1966). Exact moments and product moments of the order statistics from the truncated logistic distribution, *J. Amer. Statist. Assoc.* **61,** 514–525.

TARTER, M. E., and CLARK, V. A. (1965). Properties of the median and other order statistics of logistic variates, *Ann. Math. Statist.* **36,** 1779–1786. Correction *Ann. Statist.* **8,** 935.

TEICHROEW, D. (1956). Tables of expected values of order statistics and products of order statistics for samples of size twenty and less from the normal distribution, *Ann. Math. Statist.* **27,** 410–426.

THEIL, H. (1971). Principles of Econometrics, John Wiley & Sons, New York.

TIAO, G. C., and LUND, D. R. (1970). The use of OLUMV estimators in inference robustness studies of the location parameter of a class of symmetric distributions, *J. Amer. Statist. Assoc.* **65,** 370–386.

TIETJEN, G. L., KAHANER, D. K., and BECKMAN, R. J. (1977). Variances and covariances of the normal order statistics for sample sizes 2 to 50, *Selected Tables in Mathematical Statistics* **5,** 1–73.

TIKU, M. L. (1967). Estimating the mean and standard deviation from a censored normal sample, *Biometrika* **54,** 155–165.

TIKU, M. L. (1968). Estimating the parameters of normal and logistic distributions from censored samples, *Austral. J. Statist.* **10,** 64–74.

TIKU, M. L. (1970). Monte Carlo study of some simple estimators in censored normal samples, *Biometrika* **57,** 207–211.

TIKU, M. L. (1973). Testing group effects from Type II censored normal samples in experimental design, *Biometrics* **29,** 25–33.

TIKU, M. L. (1980). Robustness of MML estimators based on censored samples and robust test statistics, *J. Statist. Plann. Inf.* **4,** 123–143.

TIKU, M. L. (1981). Robust two-sample test and robust regression and analysis-of-variance via MML estimators, *in* Statistics and Related Topics (M. Csörgö, D. A. Dawson, J. N. K. Rao, and A. K. Md. E. Saleh, eds.), pp. 297–314, North-Holland, Amsterdam.

TIKU, M. L. (1982). Testing linear contrasts of means in experimental design without assuming normality and homogeneity of variances, *Biom. J.* **24,** 613–627.

TIKU, M. L., and MALIK, H. J. (1972). On the distribution of order statistics, *Austral. J. Statist.* **14,** 103–108.

TIKU, M. L., and STEWART, D. E. (1977). Estimating and testing group effects from Type I censored normal samples in experimental design, *Commun. Statist.* **A6**(15), 1485–1501.

TIKU, M. L., TAN, W. Y., and BALAKRISHNAN, N. (1986). Robust Inference, Marcel Dekker, New York.

TIPPETT, L. H. C. (1925). On the extreme individuals and the range of samples taken from a normal population, *Biometrika* **17,** 364–387.

TISCHENDORF, J. A. (1955). Linear Estimation Techniques Using Order Statistics, Ph.D. Thesis, Purdue University.

TUKEY, J. W. (1962). The future of data analysis, *Ann. Math. Statist.* **33,** 1–67.

TUKEY, J. W. (1970). Exploratory Data Analysis (Limited Preliminary Edition), Addison-Wesley, Reading, Massachusetts.

TUKEY, J. W., and MCLAUGHLIN, D. H. (1963). Less vulnerable confidence and significance procedures for location based on a single sample: trimming/Winsorization 1, *Sankhyā* **25,** 331–352.

TWEEDIE, M. C. K. (1956). Some statistical properties of the inverse Gaussian distribution, *Virg. J. of Sci.* **7,** 160–165.

TWEEDIE, M. C. K. (1957a). Statistical properties of the inverse Gaussian distribution. I, *Ann. Math. Statist.* **28,** 362–377.

TWEEDIE, M. C. K. (1957b). Statistical properties of the inverse Gaussian distribution. II, *Ann. Math. Statist.* **28,** 695–705.

UKITA, Y. (1955). On the efficiency of order statistics, *J. Hokkaido College Sci. and Art.* **6,** 54–65.

UMBACH, D., ALI, M. M., and HASSANEIN, K. M. (1981a). Small sample estimation of exponential quantiles with two order statistics, *Aligarth J. Statist.* **1,** 113–120.

UMBACH, D., ALI, M. M., and HASSANEIN, K. M. (1981b). Estimating Pareto quantiles using two order statistics, *Commun. Statist.—Theor. Meth.* **10**(19), 1933-1941.

UMBACH, D., ALI, M. M., and SALEH, A. K. MD. E. (1984). Hypothesis testing for the double exponential distribution based on optimal spacing, *Soochow J. Math.* **10,** 133-143.

VÄNNMAN, K. (1976). Estimators based on order statistics from a Pareto distribution, *J. Amer. Statist. Assoc.* **71,** 704-708.

VARADAN, J. (1989). Half logistic distribution: Type I censoring and estimation, M.Sc. project, McMaster University, Hamilton, Ontario, Canada.

WALD, A. (1944). On cumulative sums of random variables, *Ann. Math. Statist.* **15,** 283-296.

WALSH, J. E. (1957). Nonparametric mean estimation of percentage points and density function values, *Ann. Inst. Statist. Math.* **8,** 167-181.

WASAN, M. T. (1968). On an inverse Gaussian process, *Skand. Aktuarietidskr.* **51,** 69-96.

WASAN, M. T., and ROY, L. K. (1969). Tables of inverse Gaussian percentage points, *Technometrics* **11,** 590-603.

WATANABE, Y., *et al.* (1957). Some contributions to order statistics, *J. Gakugei, Tokushima Univ.* **8,** 41-90.

WATANABE, Y., *et al.* (1958). Some contributions to order statistics (continued), *J. Gakugei, Tokushima Univ.* **9,** 31-86.

WEISS, L. (1963). On the asymptotic distribution of an estimate of a scale parameter, *Naval Res. Logist. Quart.* **10,** 1-11.

WEISSMAN, I. (1978). Estimation of parameters and large quantiles based on the *K* largest observations, *J. Amer. Statist. Assoc.* **73,** 812-815.

WEISSMAN, I. (1980). Estimation of tail parameters under Type I censoring, *Commun. Statist.—Theor. Meth.* **9,** 1165-1175.

WHITE, J. S. (1964). Least squares unbiased censored linear estimation for the log Weibull (extreme value) distribution, *Indust. Math.* **14,** 21-60.

WHITE, J. S. (1969). The moments of log-Weibull order statistics, *Technometrics* **11,** 373-386.

WHITTEN, B. J., COHEN, A. C., and SUNDARAIYER, V. (1988). A pseudo-complete sample technique for estimation from censored samples, *Commun. Statist.—Theor. Meth.* **17**(7), 2239-2258.

WILCOXON, F. (1945). Individual comparisons by ranking methods, *Biometrics* **1,** 80-83.

WILCOXON, F. (1947). Probability tables for individual comparisons by ranking methods, *Biometrics* **3,** 119-122.

WILK, M. B., GNANADESIKAN, R., and HUYETT, M. J. (1962). Estimation of parameters of the gamma distribution using order statistics, *Biometrika* **49,** 525-545.

WILK, M. B., GNANADESIKAN, R., and HUYETT, M. J. (1963). Separate maximum likelihood estimation of scale or shape parameters of the gamma distribution using order statistics, *Biometrika* **50,** 217–221.

WILK, M. B., GNANADESIKAN, R., and LAUH, E. (1966). Scale parameter estimation from the order statistics of unequal gamma components, *Ann. Math. Statist.* **37,** 152–176.

WILKS, S. S. (1962). Mathematical Statistics, John Wiley & Sons, New York.

WILSON, E. B., and WORCESTER, J. (1945). The normal logarithmic transform, *Rev. Econ. Statist.* **27,** 17–22.

WINER, P. (1963). The estimation of the parameters of the iterated exponential distribution from singly censored samples, *Biometrics* **19,** 460–464.

WONG, K. H. T. (1988). Half logistic distribution: Order statistics and estimation, M.Sc. Project, McMaster University, Hamilton, Ontario, Canada.

YALE, C., and FORSYTHE, A. B. (1976). Winsorized regression, *Technometrics* **18,** 291–300.

YAMANOUCHI, Z. (1949). Estimates of the mean and standard deviation of a normal distribution from linear combinations of some chosen order statistics, *Bull. Math. Statist.* **3,** 52–57.

YAMAUTI, Z. (ed.) (1972). Statistical Tables and Formulas with Computer Applications, JSA-1972, Japanese Standard Association, Tokyo, Japan.

YOUNG, D. H. (1971). Moment relations for order statistics of the standardized gamma distribution and the inverse multinomial distribution, *Biometrika* **58,** 637–640.

YUAN, P. T. (1933). On the logarithmic frequency distributions and the semi-logarithmic correlation surface, *Ann. Math. Statist.* **4,** 30–74.

Author Index

A

Abdel-Aty, S. H., 17, 25, 341
Abe, S., 100, 341
Abramowitz, M., 39, 341
Adatia, A., 271, 341
Adichie, J. N., 316, 324, 341
Ahmed, S. E., 344, 357
Ahsanullah, M., 120, 271, 355
Aitchison, J., 278, 341
Ali, M. M., 94, 96, 267, 271, 341, 342, 361, 364, 365
Allen, D. M., 316, 320, 342
Ambagaspitiya, R. S., 299, 315, 316, 320, 322-324, 343, 344
Amin, N. A. K., 281, 347
Andrews, D. F., 316, 324, 342
Antle, C. E., 120, 290, 342, 345, 350
Arnold, B. C., 5, 10, 22, 23, 25, 29, 70, 109, 112, 120, 163, 259, 342

B

Bain, L. J., 120, 186, 187, 228, 342, 345, 351
Balakrishnan, N., 4, 5, 10, 22, 23, 25-30, 36, 39, 40, 43, 49, 60-63, 68, 70, 78, 80, 83-87, 92, 109, 112, 116, 120, 161-163, 165-168, 171, 176, 177, 181, 182, 185, 187, 191, 195, 197, 198, 201, 202, 206, 208, 255, 259, 299, 315, 316, 320, 322-324, 342-344, 355, 357, 364
Balmer, D. W., 271, 344
Barnett, V., 62, 120, 320, 344
Basu, D., 62, 344
Beckman, R. J., 363
Behnken, D. W., 320, 344
Belsley, D. A., 320, 344
Bennett, C. A., 215, 216, 233, 345
Benson, F., 239, 345
Bernardo, J. M., 39, 345
Beyer, J. N., 177, 271, 345
Billman, B. R., 120, 345
Birnbaum, A., 39, 271, 345
Bloch, D., 271, 345
Blom, G., 73, 100, 345
Boes, D. C., 359
Bofinger, E., 271, 345
Bondesson, L., 83, 345
Borenius, G., 57, 62, 345
Bose, R. C., 51, 345
Boulton, M., 344
Breakwell, J. V., 167, 345
Breiman, L., 271, 345
Breiter, M. C., 45, 345
Brown, E. F., 354, 361
Brown, J. A. C., 278, 341
Burr, I. W., 17, 345
Burrows, P. M., 61, 62, 346

C

Calitz, F., 127, 128, 346
Cane, G. J., 271, 346

Chan, L. K., 94, 96, 177, 187, 229, 234, 236–239, 263, 271, 341, 346, 347
Chan, M., 131, 135, 273, 281, 283, 284, 332, 347
Chan, N. N., 263, 271, 346, 347
Chan, P. S., 75, 76, 120, 161, 208, 210, 213, 344, 347
Cheng, R. C. H., 281, 347
Cheng, S. W., 177, 234, 236–239, 271, 346, 347
Chernoff, H., 94, 216, 222, 224, 271, 347
Chew, V., 117, 347
Chhikara, R. S., 281, 347, 348, 351
Chu, J. T., 119, 348
Clark, C. E., 70, 348
Clark, V. A., 39, 363
Cohen, A., 271, 348
Cohen, A. C., 3, 120, 123, 125–130, 133–135, 137, 138, 140–144, 147–149, 151, 153, 155, 157, 158, 273, 276–278, 280–285, 287, 288, 290, 292, 295, 296, 332–335, 347–349, 365
Cole, R. H., 25, 349
Craig, A. T., 236, 349
Crow, E. L., 278, 349
Cšorgo, M., 264, 349

D

D'agostino, R. B., 84, 85, 120, 187, 228, 349
Dalenius, T., 271, 349
Daly, J. F., 59, 349
Daniel, C., 316, 324–327, 349
Danziger, L., 120, 349
David, F. N., 21, 32, 60, 68–72, 349
David, H. A., 4, 7, 10, 119, 163, 259, 349
David, H. T., 51, 350
Davis, C. E., 271, 353
Davis, C. S., 60, 350
Davis, D. J., 117, 185, 350
Davis, H. T., 350
Ding, Y., 349
Dixon, W. J., 99, 114, 167, 215, 216, 249, 251, 252, 256, 261, 262, 350
Downton, F., 73, 83, 109, 112–114, 120, 350
Draper, N. R., 320, 344
Dubey, S. D., 163, 350
Dudman, J., 39, 345
Dumonceaux, R., 290, 350
Dwight, B. B., 356
Dyer, D. D., 75, 120, 162, 163, 165, 271, 350

E

Eastman, J., 342
Eisenberger, I., 271, 350
Eisenstat, S., 120, 352
Elashoff, J. D., 322, 350
Engelhardt, M., 120, 187, 228, 342, 350, 351
Epstein, B., 78, 92, 270, 271, 351, 356
Erto, P., 271, 351
Eubank, R. L., 271, 351

F

Fama, E. F., 271, 351
Fertig, K. W., 120, 187, 195, 271, 358
Fisher, R. A., 84, 98, 105, 107, 351
Folks, J. L., 281, 347, 348, 351
Forsythe, A. B., 316, 320, 322, 324, 326, 366
Fraser, D. A. S., 119, 351

G

Gajjar, A. V., 120, 351
Galambos, J., 4, 36, 351
Gastwirth, J. L., 347
Gibbons, D. E., 120, 351
Gins, J. D., 345
Gnanadesikan, M., 177, 271, 352
Gnanadesikan, R., 365, 366
Godwin, H. J., 51, 56, 236, 351
Govindarajulu, Z., 24, 25, 28, 48, 51, 58, 83, 120, 254, 352
Gravel, R., 120, 352
Graybill, F. A., 359
Green, J., 28, 120, 360
Greenberg, B. G., 3–5, 57, 61, 78, 83, 84, 87, 92, 94, 97, 99, 115, 120, 159, 167, 171, 174–176, 184, 224, 240, 352, 362
Grundy, P. M., 119, 352
Guida, M., 271, 351
Gumbel, E. J., 4, 120, 352
Gupta, A. K., 73, 83, 94, 97, 98, 100, 105, 158, 159, 167, 174, 175, 352
Gupta, S. C., 324, 363
Gupta, S. S., 4, 26, 28, 39, 43, 45, 47, 51, 87, 120, 177, 181, 182, 271, 345, 352

H

Hahn, G. J., 359
Hall, I. J., 120, 352

Halperin, M., 167, 352
Hammersley, J. M., 119, 271, 353
Hamouda, E. M., 299, 300, 304, 310, 353, 356
Harley, B. I., 114, 353
Harrell, F. E., 271, 353
Harter, H. L., 4, 5, 45, 48, 57, 120, 127, 129, 162, 167, 171, 174, 177, 181, 182, 184, 187, 195, 234, 262, 271, 274, 280, 281, 345, 353, 354, 363
Hartley, H. O., 5, 360
Hassanein, K. M., 177, 271, 341, 342, 354, 361, 364, 365
Hawkins, D. M., 62, 354
Helm, R., 290, 348, 349
Herd, R. G., 78, 354
Heyde, C. C., 280, 354
Higuchi, I., 271, 354
Hill, B. M., 127, 129, 280, 354
Hinich, M. J., 324, 354
Hoaglin, D. C., 320, 354
Hoeffding, W., 23, 354
Hosking, J. R. M., 271, 355
Huyett, M. J., 365

I

Isida, M., 167, 355

J

Johns, M. V., Jr., 228, 271, 347, 355
Johnson, N. L., 21, 32, 60, 68-72, 136, 139, 278, 283, 286, 349, 355
Jones, H. L., 51, 355
Joshi, M., 48, 51, 120, 352
Joshi, P. C., 28-30, 36, 43, 47, 60-62, 343, 355
Jung, J., 215, 216, 228-230, 233, 271, 355
Junkins, D. B., 361

K

Kabir, A. B. M. L., 120, 187, 271, 346, 355
Kahaner, D. K., 363
Kaigh, W. D., 271, 355
Kaminsky, K. S., 271, 355, 356
Kapadia, C. H., 356
Kappenman, R. F., 271, 356
Karlin, S., 34, 356
Keating, J. P., 271, 356
Kendall, M. G., 143, 164, 170, 194, 322, 356
Khatri, C. G., 120, 351
Kimball, B. F., 26, 356
Kocherlakota, S., 43, 49, 120, 343
Kotz, S., 36, 136, 139, 278, 283, 286, 351, 355
Koutrouvelis, I. A., 271, 356
Krishnaiah, P. R., 45, 345
Kubat, P., 270, 271, 356
Kuh, E., 344
Kulldorff, G., 120, 236, 237, 271, 356

L

Lachenbruch, P. A., 271, 355
Lambert, J. A., 128, 356
Laska, E. M., 271, 345
Lauh, E., 366
Lawless, J. F., 75, 79, 80, 85, 92, 196, 197, 356
Lee, A. F. S., 120, 228, 349
Lee, K. R., 163, 356
Leone, F. C., 299, 300, 353, 356
Leung, M. Y., 40, 120, 197, 198, 202, 206, 343
Lewis, T., 62, 320, 344
Lieberman, G. J., 94, 228, 271, 347, 355
Lieblein, J., 26, 48, 49, 120, 164, 187, 190, 356, 357
Likeš, J., 120, 271, 357
Lloyd, E. H., 80, 83, 216, 357
Lo, S.-H., 348
Lund, D. R., 120, 363
Lwin, T., 119, 357

M

Maguire, B. A., 249, 357
Malik, H. J., 23, 26-28, 40, 47, 116, 120, 182, 343, 344, 357, 364
Malmquist, S., 35, 357
Mann, N. R., 119, 120, 162, 187, 195, 197, 271, 357, 358
Margolin, B. H., 26, 358
Maxwell, E. A., 119, 358
McCool, J. I., 119, 290, 357
McDonald, G. C., 120, 351
McKay, A. T., 59, 357
McLaughlin, D. H., 99, 364
Mead, E. R., 187, 271, 346, 347
Meeden, G., 25, 342
Mehrotra, K. G., 120, 358
Meisner, M., 345

Melnick, E. L., 25, 358
Mendenhall, W., 316, 327, 328, 358
Menon, M. V., 295, 358
Miké, V., 238, 271, 345, 358
Miller, R. G., Jr., 4, 358
Mishriky, R. S., 119, 349
Miyamoto, Y., 271, 359
Mood, A. M., 164, 359
Moore, A. H., 120, 127, 129, 162, 167, 171, 177, 181, 184, 187, 195, 345, 353, 354, 363
Morton, K. W., 119, 271, 353
Mosteller, F., 3, 234, 238, 267, 359
Moussa, E. A., 299, 300, 302, 304-306, 359
Munro, A. H., 120, 359
Murphy, R. B., 62, 359
Murthy, V. K., 120, 359

N

Nanda, P., 120, 358
Nelson, W., 3, 120, 359
Norgaard, N. J., 273, 348

O

Ogawa, J., 215, 233, 238, 249, 271, 359
Oppenlander, J. E., 271, 359

P

Padgett, W. J., 133, 281, 359
Panchapakesan, S., 4, 352
Panjer, H. H., 347
Parzen, E., 265, 271, 359, 360
Pearson, E. S., 5, 114, 353, 357, 360
Pearson, K., 14, 50, 70, 283, 360
Pearson, M. V., 70, 360
Plackett, R. L., 70, 167, 360
Polovko, A. M., 162, 360
Posner, E. C., 271, 350
Prescott, P., 47, 99, 360
Proschan, F., 78, 360
Puthenpura, S., 28, 85, 86, 120, 208, 344

Q

Quayle, R. J., 163, 360
Qureishi, A. S., 352

R

Ragab, A., 28, 120, 360
Raghunandanan, K., 120, 177, 215, 249, 253-257, 271, 360, 363
Rao, C. R., 33, 34, 322, 360
Reiss, R.-D., 271, 360
Rényi, A., 34, 360
Resek, R. W., 120, 361
Revesz, P., 264, 349
Robertson, C. A., 271, 361
Roll, R., 271, 351
Roy, L. K., 281, 365
Ruben, H., 51, 54, 57, 361
Rukhin, A. L., 271, 361

S

Sack, R. A., 344
Saleh, A. K. Md. E., 26, 28, 271, 341, 342, 354, 361, 365
Salvosa, L. R., 287, 362
Salzer, H. E., 187, 190, 357
Sarhan, A. E., 3-5, 57, 61, 78, 83, 84, 87, 92, 94, 97, 99, 115, 120, 159, 167, 171, 174-176, 184, 224, 240, 352, 362
Särndal, C. E., 271, 362
Saw, J. G., 70, 167, 362
Schaeffer, L. R., 61, 362
Schafer, R. E., 358
Scheuer, E. M., 358
Schmee, J., 120, 359
Schneider, B. E., 39, 362
Schroedinger, E., 281, 362
Schweder, T., 324, 362
Scott, C., 361
Seal, K. C., 58, 362
Sen, P. K., 271, 361
Shah, B. K., 28, 39, 40, 352, 362
Shelnutt, J. W., III, 271, 362, 363
Shimizu, K., 278, 349
Shu, V. S., 10, 349
Siddiqui, M. M., 271, 363
Singh, K., 348
Singh, M., 324, 363
Singpurwalla, N. D., 358
Siotani, M., 17, 363
Smoluchowsky, M. W., 281, 363
Sobel, M., 78, 92, 351
Srikantan, K. S., 25, 26, 363

Srinivasan, R., 120, 177, 215, 249, 253-257, 360
Stegun, I. A., 39, 341
Stephens, M. A., 60, 84, 85, 99, 349, 350, 363
Stewart, D. E., 319, 364
Stigler, S. M., 96, 363
Stone, C. J., 345
Strawderman, W. E., 271, 361
Stuart, A., 164, 170, 194, 322, 356
Sugg, M., 349
Sukhatme, P. V., 34, 363
Sundaraiyer, V., 365
Swamy, P. S., 119, 363
Swartz, G. B., 120, 359

T

Tabatabai, M. A., 316, 320, 322, 324, 326-328, 363
Tadikamalla, P. R., 47, 363
Tagami, S., 167, 355
Talwar, P. P., 324, 354
Tan, W. Y., 316, 320, 322, 324, 326-328, 363, 364
Tarter, M. E., 39, 43, 363
Taylor, H. M., 34, 356
Teichroew, D., 28, 57, 113, 363
Theil, H., 320, 363
Tiao, G. C., 120, 363
Tietjen, G. L., 23, 57, 174, 262, 363
Tiku, M. L., 47, 84, 98, 167, 171, 176, 177, 181, 299, 310-312, 315, 317, 319, 364
Tippett, L. H. C., 57, 364
Tischendorf, J. A., 271, 364
Tukey, J. W., 3, 99, 224, 350, 359, 364
Tweedie, M. C. K., 281, 364

U

Ukita, Y., 271, 364
Umbach, D., 267, 271, 341, 342, 361, 364, 365

V

van Eeden, C., 120, 352
Vännman, K., 120, 236, 271, 356, 365
van Vleck, L. D., 362
Varadan, J., 161, 187, 191, 195, 208, 344, 365
Velasco, J. A., 362

W

Wald, A., 281, 365
Wallis, J. R., 271, 355
Walsh, J. E., 271, 365
Wasan, M. T., 281, 365
Watanabe, Y., 51, 365
Wei, L. J., 133, 281, 359
Weiss, L., 271, 365
Weissman, I., 271, 365
Welsch, R. E., 320, 344, 354
Whisenand, C. W., 75, 120, 162, 163, 165, 271, 350
White, J. S., 120, 187, 190, 365
Whitten, B. J., 3, 4, 123, 125-127, 129, 130, 133-135, 137, 138, 140-144, 273, 278, 280, 281, 283-285, 287, 288, 292, 295, 296, 329, 332-339, 347-349, 365
Wilcoxon, F., 4, 365
Wilk, M. B., 120, 365, 366
Wilks, S. S., 4, 366
Williams, G. T., 70, 348
Wilson, E. B., 128, 366
Winer, P., 120, 366
Winokur, H. S., Jr., 26, 358
Wixley, R. A. J., 120, 359
Wong, K. H. T., 68, 85, 86, 120, 161, 208, 271, 344, 366
Wood, F. S., 316, 324-327, 349
Worcester, J., 128, 366
Wynn, A. H. A., 357

Y

Ya'coub, K., 119, 348
Yale, C., 316, 320, 322, 324, 326, 366
Yamanouchi, Z., 238, 271, 366
Yamauti, Z., 57, 366
Young, D. H., 47, 366
Yuan, P. T., 279, 366

Z

Zelen, M., 120, 187, 357

Subject Index

A

AMLE (approximate maximum likelihood estimator)
  comparison with BLUEs, table of, 206
  definition, 161-162

B

Balakrishnan's AMLE, 255-264
  comparisons of AMLE and BLUEs, table of, 263
  likelihood function based on two symmetric order statistic, 257
    likelihood equations, 258
  unbiasing factors, 261
Bennett's optimal asymptotic estimators, 216-228
  optimal asymptotic weights for order statistic, 216
    Cauchy population, 225-228
    logistic population, 224-225
    normal population, 224-225
    symmetric cases, 221
Best linear estimators, *see* Linear estimators
Blom's unbiased linear estimators, *see* Linear estimators
BLUEs (best linear unbiased estimators)
  of location and scale parameters, 80-83
    examples, 83-87
  of scale parameter, 74-80
    examples, 78-79
    for one-parameter exponential distribution, 76-78
  of two-parameter exponential distribution, 87-92
    examples, 92-93
  variances and covariances of, 290

C

Cauchy-Schwartz inequality, 319
Censored sample
  completion technique, 329-340
  from normal distribution, 330-331
  from skewed distributions, 331-332
    gamma, 333-334
    inverse Gaussian, 332-333
    lognormal, 335
    Weibull, 334-335
  examples, 335-340
  random sample from inverse Gaussian distribution, 338
    successive interations of estimates, 339
    summary of estimates, table of, 338
  random sample from normal distribution, 336
    censored sample estimates, table of, 337
    successive iterations of sample estimates, table of, 337
  type II, 161

Cohen-Whitten estimators, 273-295
  Estimating equations, 274
Continuous weight functions, 215

D

David and Johnson's approximation, 68-72
Distribution of range, derivation of, 17
Dixon's simplified linear estimators, 216, 249-255
  mean of double exponential function, table of, 256
  mean of logistic population, 254
  $\hat{\sigma}$ of double exponential population, table of, 257
  standard deviation of logistic population, table of, 255
  standard deviation of normal distribution, table of, 252, 253
  variance and efficiency relative to sample mean, table of, 251
Downton's linear estimators with polynomial coefficients, 109-117
  example, 117-118

E

Exponential distribution, 144-146, 289-290
  BLUEs, 290
  examples, 290-295
  likelihood function, 144
  maximum likelihood estimators, 144
  modified MLE, 146
  singly right censored samples, 145
Extreme value distribution, 186-197
  best linear invariant estimators, 187
  bias, variances, and covariance of estimates
    $n = 10$, table of, 192
    $n = 20$, table of, 193
  comparison of bias, various estimators of parameters, table of, 194
  doubly censored sample, 187
    conditional bias of μ, 190
    likelihood equations, 188
    likelihood function, 187
    variances and covariances of estimates, 191
  examples, 195-197
  pdf and cdf, 186

F

Failure rate, linearly increasing, 162
First approximations of completed sample, 330
Fubini's theorem, 17

G

Gamma distribution, 286-289
  graphs of estimating function, 289
  modified moment estimators, 287-289
  parameters, 135-139
    asymptotic variances and covariances, 137-139
      table of, 138
    estimating equations, 136
    estimators, 136
      for two-parameter distribution, 137
    likelihood function, 136
  pdf and cdf, 286
  pdf of standardized gamma distribution, 287
  properties, 286
Gupta's simplified linear estimators, 94-97
  examples, 97-99

H

Half logistic distribution, 208-213
  approximate likelihood equation, 209
  example, 212-213
  likelihood equation, 208
  pdf and cdf, 208
  unbiasing factors, tables of, 210-212
Half normal distribution, 61

I

Inverse Gaussian distribution, 281-285
  cdf, 284
graphs of estimating function, 285
  mean, variance, skewness, and kurtosis, 283
  mode, 194
  modified moment estimators, 184
  pdf, 283
  pdf of standardized inverse Gaussian distribution, 183

Inverse Gaussian parameters,
asymptotic variances and covariances, 133-135
factors, table, 134
estimating equations, 132
estimators, 132
likelihood function, 131

J

Jung's optimal asymptotic estimators, 228-233
properties of, 228-231
Student's t distribution, 231-233

L

Linear estimators, 73-119
Blom's unbiased nearly best, 100-105
examples, 105-108
examples, 75-76
with multiple measurements, 300-301
Logistic distribution, 177-186
bias, variances, and covariance of estimates
simulated values, table of, 183
table of, 183
conditional bias of $\hat{\sigma}$, 180
examples, 184-186
likelihood equations, 178-179
modified MLEs of $\hat{\mu}$ and $\hat{\sigma}$ in censored samples, 177
pdf and cdf of type II censored sample, 177
Lognormal distribution, 278-281
estimators, 280
graphs of estimating function, 282
pdf and cdf, 278
pdf and cdf of standardized lognormal, 279
properties of, 279
regularity problems, 280
Lognormal parameters, 131
asymptotic variances and covariances, 128-131
factors, table of, 130
estimating equations, 127
estimators, 128
likelihood function, 127

M

Markov chain, 7
MLE (maximum likelihood estimator), 121-159, *see also* AMLE introduction, 121-122
Modified maximum likelihood estimation
with multiple measurements, 310-315
estimators, 312
example, 312-313
with single measurements, 315-321
asymptotic variance-covariance matrix, 321
bias and variance of MLEs and MMLEs, tables of, 323-324
examples, 324-329
likelihood equations, 317
residuals, 320
Monte Carlo simulations, 171
Moussa, 299, 300, 304, 305, 306
BLUES from data, table of, 306
data, table of, 305

N

n order statistics, joint distribution, 7
Normal distribution, 51-61, 146-159, 167-176
bias, variance, and covariance of estimators, table of, 172
simulated values, table of, 173
conditional bias of $\hat{\mu}$, 170
examples, 174-176
expected values of first order statistic, table of, 281
Saw's estimators, 167
singly truncated, 147-151
estimation curve, 149
estimation function, table to, 150-151
likelihood function, 147
maximum likelihood estimating equations, 147
maximum likelihood estimators, 148
pdf, 147
symmetric censoring, 171
Monte Carlo simulations, 171
Tiku's estimators, 167
approximate likelihood equations, 169

O

Ogawa's optimal estimators based on selected order statistic, 233-249
application of Gauss-Markov theorem, 238

Ogawa's optimal estimators based on selected order statistic—*contd.*
a review, 234–238
optimal spacings, 240
coefficients and relative efficiencies, table of, 241–247
example, 249
one-parameter exponential population, 240
Order statistics, applications, 2–3
exponential distribution results, 34–38
gamma distribution results, 43–47
half logistic results, 63–72
David and Johnson's approximation, 68-72
identities, 24
logistic distribution results, 38–43
means, variances, and covariances of, 30
normal distribution results, 51–61
recurrence relations, 23
single, distribution, 11
uniform distribution results, 30–33
Weibull distribution results, 47–51

P

Population quantiles, estimation of, 264–270
approximate BLUEs, 267
BLUEs based on k selected order statistic, 265–267
details of related work, 270–271
empirical distribution function, 264

R

Rayleigh distribution, 139–144, 162–166
bias, variance, and approximate variance of $\hat{\sigma}$, table of, 166
censored sample, 162–163
example, 165
failure rate, 162
likelihood function of random sample, 140
maximum likelihood estimating equations, 140
moment estimators, 141
moments and other characteristics, 140
pdf and cdf, 139
pdf of first order statistic, 140
Weibull, special case of, 139, 144
with origin at zero, 141
asymptotic variances of estimates, 143-144
reliability of estimates, 143
singly right censoring, 142
truncation, 141
Robust inference, 98, 167, 176

S

Saw's estimators, *see* Normal distribution, Saw's estimators
Simple linear regression mode., 302–305
example, 305
BLUEs from Moussa's data, table of, 306
Moussa's data, table, 305
K-sample with equal variance, 305–308
K-sample with unequal variance, 308-310
variances of parameter estimates, 304
Singly censored samples
type I, 152
estimation function, table of, 154–155
likelihood function, 152
maximum likelihood estimation, 152–153
type II, 153

T

Taylor series, 163
Threshold parameter, 3
Tiku's estimators, *see* Normal distribution, Tiku's estimators
Two order statistics, joint distribution, 8–11
Type I generalized logistic distribution, 197-108
pdf and cdf, 197
type II doubly censored sample, 198–208
approximate conditional bias of $\hat{\sigma}$, 201
approximate likelihood equations, 200
bias, variances, and covariance of estimates
table of $n = 15$ and $b = 1{\cdot}0$, 203
table of $n = 15$ and $b = 2{\cdot}0$, 204
table of $n = 15$ and $b = 3{\cdot}0$, 205
comparison of AMLEs with BLUES, table of, 206
efficiencies, 202
estimates and standard errors for censored samples, table of, 207
exact conditional bias of $\hat{\mu}$, 201
example, 206

Type I generalized logistic distribution,
  type II doubly censored sample—*contd.*
    likelihood equations, 198
    mean square errors of estimates, 202
    variances and covariances of estimates,
      201

W

Weibull distribution, 274-278
  graphs of estimating functions, 276-277
  modified moment estimates, 275-278
  pdf and cdf, 274
  pdf of first order statistic, 275
  properties of, 275
Weibull parameters, 122-126
  errors of estimates, 124
    variance-covariance factors
      table of, 125
      estimates, formulas, 126
  estimating equations, 123
  estimators, 123
  likelihood function, 122
Windsorized estimation, 316, 326-327
  mean, 224

STATISTICAL MODELING AND DECISION SCIENCE

Gerald J. Lieberman and Ingram Olkin, editors

Samuel Eilon, *The Art of Reckoning: Analysis of Performance Criteria*
Enrique Castillo, *Extreme Value Theory in Engineering*
Joseph L. Gastwirth, *Statistical Reasoning in Law and Public Policy: Volume 1, Statistical Concepts and Issues of Fairness; Volume 2, Tort Law, Evidence, and Health*
Takeaki Kariya and Bimal K. Sinha, *Robustness of Statistical Tests*
Marvin Gruber, *Regression Estimators: A Comparative Study*
Alexander von Eye, editor, *Statistical Methods in Longitudinal Research: Volume I, Principles and Structuring Change; Volume II, Time Series and Categorical Longitudinal Data*
M. M. Desu and D. Raghavarao, *Sample Size Methodology*
N. Balakrishnan and A. Clifford Cohen, *Order Statistics and Inference: Estimation Methods*